Index

Author´s notes

This guide, text book, or as you prefer to name it has been developed after having used the similar materials prepared one and two years ago (Mathematics 1st, 2nd, 3rd and 4th ESO).

In the same line than in previous materials, it contains enough theoretical development and proposed exercises with solutions, in order to promote our students' autonomy. Moreover, for almost each introduced concept, some exercises are solved step by step.

As imagination is not my best feature, I have been oriented by materials I already had got and used in previous years with my students (in Spanish), as well as:

- Spanish text books:
    - Matemáticas 3º ESO (Autores: J. Cólera y otros; Ed.: Anaya).
    - Notes and work sheets (in English and Spanish): Alfonso Gonzalez.
    - English Dictionary (Oxford Edition).

I have tried not to make mistakes, but there may be some of them. I would be very pleased to read your suggestions or comments. (javiersanchezpi@gmail.com).

Finishing, I would like to thank my family, the best and only I have had, have or will have.

I really hope this material to be useful. *Enjoy Maths!!!*

Javier Sánchez Pina.
Murcia. August, 2015.

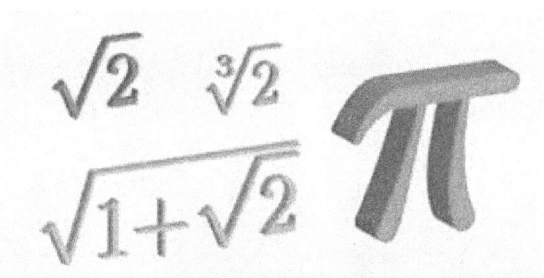

Unit 1.- Real numbers

1. Sets of numbers

Quantities are classified into different sets or type of numbers.

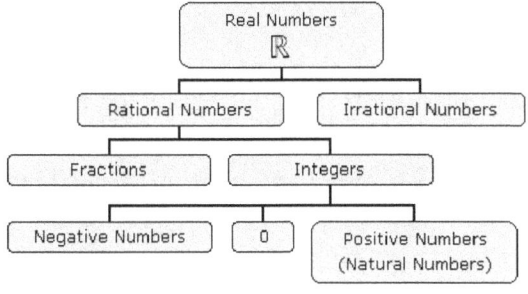

Rational numbers (Q): They are those numbers that can be written as a fraction. A fraction is the indicated quotient of two integer numbers. There are several types of rational numbers:

- Integer numbers (Z): $Z = \{...., -3, -2, -1, 0, 1, 2, 3, ...\}$.

The set of the integer number can be classified, as well, into two sets, called ***natural numbers*** **(N)** = {0 , 1 , 2 , 3 , ...} and their opposites **(Z⁻)** = {-1 , -2 , -3 ,...} , called as ***negative integers***.

All these numbers are rational because we can write them as a fraction.

- **Non integer rational numbers:** They are all those fractions whose quotient is NOT a whole number. For example, $\frac{1}{2}$, $\frac{4}{3}$ or $\frac{7}{10}$.

When calculating their decimal expressions, we always obtain a decimal number (an exact or a recurring decimal number):

– **Exact or terminating** decimal: It finishes, so you can write down all its digits.

 For example: $\frac{7}{4} = 7 \div 4 = 1.75$

– **Recurring or repeating** decimal: It does not finish, it goes on forever, but <u>some of the digits are repeated over and over again</u>. For example:

$5.6666....... = 5.\overline{6}$ $\qquad$ $14.353535......... = 14.\overline{35}$ $\qquad$ $87.42222....... = 87.4\overline{2}$

 We can distinguish:

- **Pure recurring decimals**: Recurring decimals in which period starts just after the decimal point. For example: $5.6666....... = 5.\overline{6}$ $\qquad$ $14.353535........ = 14.\overline{35}$

- **Mixed recurring decimals**: Recurring decimals in which period does not start just after the decimal point. $\quad$ For example: $87.42222....... = 87.4\overline{2}$

− **Irrational numbers (I):** They are those numbers that CANNOT be written as a fraction. Irrationals are those decimal numbers having an infinite quantity decimal digits, non being periodically repeated. Some examples of irrational numbers:

3,1010010001....;   0,3737737773…..;   0,123456789101112……

Some other more famous irrational numbers are:

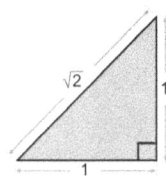

Л = 3,141592654...... is obtained when dividing the length of a circumference by its diameter.

$\sqrt{2}$ = 1,414213562....... is obtained when calculating the diagonal of a square whose side is 1 and $\sqrt[3]{2}$ = 1,25992105...... is obtained when calculating the edge of a cube whose volume is 2.

e = 2.718281828…….. is the base of the natural logarithms.

They are also irrational numbers those roots whose result is not an integer or a recurring decimal, as well as the results of adding, subtracting, multiplying or dividing a rational and an irrational number.

1.1. Rational expression of decimal numbers

The decimal expression of a rational number can be converted into a fraction:

All decimal numbers can be expressed as a fraction.

In the following example we show the process to find out these fractions.

- **Exact decimal:** Express as a fraction $x = 2.34$

Solution: We must build a fraction whose
  - Numerator: Decimal number without decimal point.
  - Denominator: A "1" followed by as many "0" as decimal digits decimal number has.

In our case: $2.34 = \dfrac{234}{100}$ If possible, we must simplify this fraction to give it in its simplest

form: So, $2.34 = \dfrac{224}{100} = \dfrac{112}{50} = \boxed{\dfrac{56}{25}}$

- **Recurring decimal number:** Express as a fraction $x = 14.\overline{35} = 14.3535355.........$

Solution: We must apply the following process, consisting on multiplying it by 1, 10, 100, 1000, ..., till finding two decimals having *identical decimal parts*, so that they can be subtracted, to cancel their decimal parts and converting them into integers.

$$100 \cdot x = 1435.353535....$$
$$10 \cdot x = 143.535353.....$$
$$x = 14.353535........$$

$$100 \cdot x = 1435.353535....$$
$$\cancel{10 \cdot x = 143.535353......}$$
$$x = 14.353535........$$

As you can see, x and 100·x have the same decimal part. So, we subtract them:

$$99 \cdot x = 1421 \implies x = \dfrac{1421}{99}$$

Example: Express as a fraction: a) $8.\overline{5}$ b) $25.4\overline{52}$

Solution:

a)

$$10 \cdot x = 85.555.......$$
$$x = 8.5555......$$
_____

As you can see, both decimals have the same decimal part. So, we subtract them:

$$10 \cdot x = 85.555.......$$
$$x = 8.5555......$$
_____
$$9 \cdot x = 77 \quad \Longrightarrow \quad x = \frac{77}{9}$$

b)

$$1000 \cdot x = 25452.5252......$$
$$100 \cdot x = 2545.25252......$$
$$10 \cdot x = 254.525252....$$
$$x = 25.4525252....$$
_____

As you can see, $1000 \cdot x$ and $10 \cdot x$ have the same decimal part. So, we subtract them:

$$1000 \cdot x = 25452.5252......$$
$$\cancel{100 \cdot x = 2545.25252......}$$
$$\cancel{10 \cdot x = 254.525252....}$$
$$x = 25.4525252....$$
_____
$$990 \cdot x = 25198 \quad \Longrightarrow \quad x = \frac{25198}{990}$$

<table>
<tr><td rowspan="2">Exercises</td><td>

1. Classify the following numbers into rational or irrational numbers, indicating the reason for your answer:

$$\frac{4}{5}; \quad \frac{10}{5}; \quad -2,333...; \quad \sqrt{7}; \quad \sqrt{36}; \quad \frac{\pi}{2}; \quad -5; \quad 7,\dot{4}$$

2. Classify the following numbers into rational or irrational numbers, indicating the reason for your answer:

$$\frac{1}{8} \quad \frac{\pi}{3} \quad \sqrt{5} \quad 2,6 \quad 0 \quad -3 \quad -\frac{25}{3} \quad \sqrt{13} \quad 0,1 \quad 6,\dot{4} \quad 534 \quad 1,41421356\,2...$$

</td></tr>
</table>

Exercises

3. Locate the following numbers into the diagram:

$3,42; \quad \dfrac{5}{6}; \quad -\dfrac{3}{4}; \quad \sqrt{81}; \quad \sqrt{5}; \quad -1; \quad \dfrac{\pi}{4}; \quad 1,4555...$

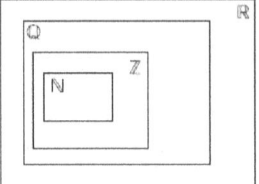

4. Work out, directly and writing first as fractions, and check you obtain the same results:

a) $0.\overline{3} + 0.\overline{6}$ b) $0.\overline{3} - 0.\overline{15}$ c) $3.\overline{41} + 2.37\overline{8}$ d) $0.\overline{4} \cdot 0.1$ e) $3.\overline{1} + 2.0\overline{3}$

f) $0.\overline{3} + 0.1\overline{6}$ g) $4 \cdot 2.\overline{5}$ h) $4.\overline{89} - 3.\overline{78}$ i) $8 - 2.\overline{7}$ j) $4.5 \cdot 0.0\overline{2} + 0.\overline{4}$

1.2. Representation of real numbers on the Number Line

Real numbers are represented on the Number Line by mean of different methods, depending on the type of number. We are seeing it with some examples:

a) Exact decimals:

Example: Represent the numbers a) 2 and b) 3.47:

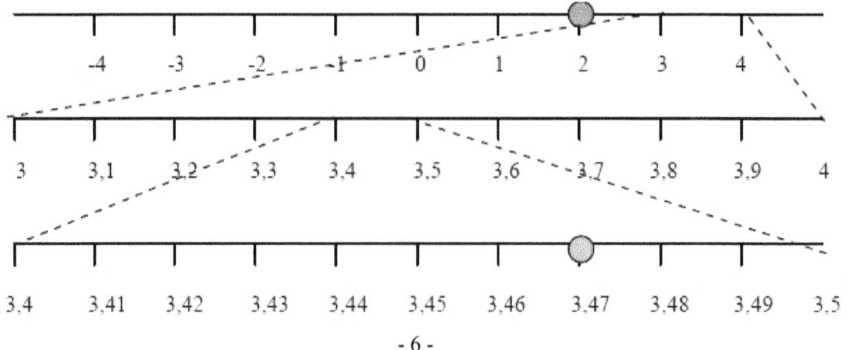

b) Recurring decimals: We must write it as a fraction, divide the unit into the quantity of parts indicated by the denominator, and taking the parts indicated by the numerator.

Example: Represent the number 0.833333333....:

Solution: $0.8333333.... = \dfrac{5}{6}$

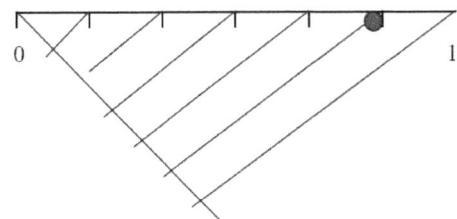

c) Irrational (squared roots): We will use the Pythagoras´ Theorem to draw a right triangle whose hypotenuse is this number.

Example: Represent $\sqrt{10}$: Notice $10 = 3^2 + 1^2 \quad \rightarrow \quad \sqrt{10} = \sqrt{3^2 + 1^2}$

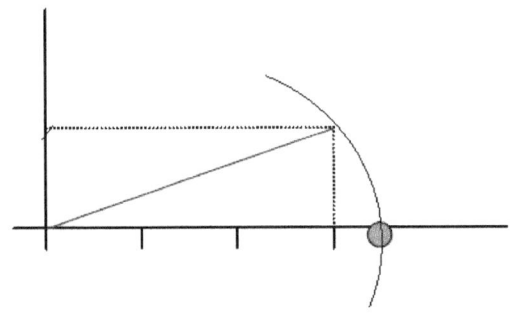

Sometimes, we must repeat this process more than once:

Example: Represent $\sqrt{102}$: Notice $\quad 101 = 10^2 + 1^2 \qquad \rightarrow \qquad \sqrt{101} = \sqrt{10^2 + 1^2} \qquad$ (step 1)

$$102 = \sqrt{101}^2 + 1^2 \qquad \rightarrow \qquad \sqrt{102} = \sqrt{\sqrt{101}^2 + 1^2} \qquad \text{(step 2)}$$

5. Represent the following rational numbers on the same Number Line:

$$\dfrac{3}{2} \qquad -3 \qquad 0,\overline{6} \qquad \dfrac{5}{6} \qquad -\dfrac{3}{4} \qquad \dfrac{11}{5} \qquad 2,25 \qquad \dfrac{19}{6} \qquad 3,\overline{9}$$

6. Locate $\sqrt{2},\sqrt{3},\sqrt{5},\sqrt{6},\sqrt{7},\sqrt{8}$ *and* $\sqrt{10}$ on the Number Line, by using the Pythagoras' Theorem and you are allowed to draw more than one Number Line. Use your rulers and compas.

7. Represent the following rational numbers on the Number Line: 2.3; $\dfrac{7}{4}$; (-3).

8. Represent on the Number Line, by using the Pythagoras' Theorem:

a) $\sqrt{50}$ b) $\sqrt{82}$ c) $\sqrt{18}$ d) $\sqrt{46}$

9. Represent the following rational numbers on the Number Line: 4.7 and 4.777777....

2. Intervals and semi-lines

Intervals are segments and semi-lines into the Real Line.

There are different types of intervals:

The open interval **(a,b)** includes the real numbers x between a and b, but a and b are NOT included. It is expressed as a $<$ x $<$ b.

The closed interval **[a,b]** includes the real numbers x between a and b, being a and b included. It is expressed as a $\le$ x $\le$ b.

In the same way, the interval **[a,b)** is expressed as a $\leq$ x < b, being a included but not b;

while the interval **(a,b]** is expressed as a < x $\leq$ b, being b included but not a.

The open semi-line **(a, ∞)** includes all the real numbers being higher than a, where a is not included. It is expressed as a < x;

while the closed semi-line **[a, ∞)** includes all the real numbers being higher than a, being a included. It is expressed as a $\leq$ x.

In the same way, semi-line **(-∞, b)** includes all the real numbers being lower than b, where b is not included. It is expressed as x < b;

while the semi-line **[-∞, b]** includes all the real numbers being lower than b, being b included. It is expressed as x $\leq$ b.

10. Complete the following chart:

	GRAPHIC REPRESENTATION	INTERVAL	DEFINITION
1		[-1,3]	
2	(graph: open at 0, open at 2)		
3	(graph: closed at -2, open at 4)		
4		[-2,1)	
5			$\{x \in IR / 1 < x \leq 5\}$
6	(graph: closed at -1, to ∞)		
7			$\{x \in IR / x < 2\}$
8		$(0,\infty)$	
9	(graph: from -∞ to closed 3)		
10		(-1,5)	
11			$\{x \in R / x \leq 0\}$
12		$[2/3,\infty)$	

11. Write in all their forms, the following intervals and semi-lines:

a) $\{x \ / \ -2 \leq x < 3\}$ b) $(-\infty, -2]$ c) Numbers bigger than (-1)

d)

12. Write as intervals the values of x that verify the following relationships:

a) $|x + 2| \geq 3$ b) $|x - 4| < 2$

13. Complete the following chart:

Exercises

	GRAPHIC REPRES.	INTERVAL	DEFINITION
13			$\{x\in IR/ -2<x\le2\}$
14			$\{x\in IR/ \|x\|<3\}$
15			$\{x\in IR/ \|x\|\ge3\}$
16	 2 ∞		
17		$[-1,1]$	
18			$\{x\in IR/ x<-1\}$
19	 -4 4		
20		$(-\infty,-2)\cup(2,\infty)$	
21		$(-\infty,2)\cup(2,\infty)$	
22			$\{x\in IR/ \|x\|\le5\}$
23		$[-2,2]$	

3. Operations with rational numbers

As you have widely studied operations with fractions, in the next chart you are offered a brief summary of them, as well as some examples.

3.1.- Addition and subtraction of fractions

- If they have the same denominator, we will leave the denominator and add or subtract the numerators.

$$\frac{a}{b}+\frac{c}{b}=\frac{a+c}{b}$$

Example: $\dfrac{2}{5}+\dfrac{1}{5}=\boxed{\dfrac{3}{5}}$

- If they have different denominators, first of all, we will transform them onto new fractions having all of them the same denominator, which will be the LCM of the original denominators.

Example: $\dfrac{1}{4}-\dfrac{2}{3}+\dfrac{7}{5} \Rightarrow LCM(4,3,5)=60 \Rightarrow \dfrac{15}{60}-\dfrac{40}{60}+\dfrac{84}{60}=\boxed{\dfrac{59}{60}}$

3.2.- Product of fractions

We must multiply, separately, numerator by numerator and denominator by denominator.

$$\frac{a}{b}\cdot\frac{c}{d}=\frac{a\cdot c}{b\cdot d}$$

Example: Calculate $\dfrac{6}{8}\cdot\dfrac{10}{15}$

Solution: Simplifying at the end) $\dfrac{6}{8}\cdot\dfrac{10}{15}=\dfrac{6\cdot10}{8\cdot15}=\dfrac{60}{120}=\dfrac{6}{12}=\dfrac{6\div2}{12\div2}=\dfrac{3\div3}{6\div3}=\boxed{\dfrac{1}{2}}$

     Simplifying before) $\dfrac{6}{8}\cdot\dfrac{10}{15}=\dfrac{6\cdot10}{8\cdot15}=\dfrac{(2\cdot3)\cdot(2\cdot5)}{(2\cdot2\cdot2)\cdot(3\cdot5)}=\dfrac{2\cdot3\cdot2\cdot5}{2\cdot2\cdot2\cdot3\cdot5}=\boxed{\dfrac{1}{2}}$

3.3.- Inverse of a fraction

We obtain the inverse of a fraction, changing the places of numerator and denominator.

Inverse of $\dfrac{a}{b}$ is $\boxed{\dfrac{b}{a}}$

3.4.- Division of fractions

We multiply the first fraction by the inverse of the second one.

$$\boxed{\frac{a}{b}:\frac{c}{d}=\frac{a}{b}\cdot\frac{d}{c}=\frac{a\cdot d}{b\cdot c}}$$

Example: Calculate $\dfrac{6}{8}:\dfrac{15}{10}$

Solution:

Simplifying at the end) $\dfrac{6}{8}:\dfrac{15}{10}=\dfrac{6}{8}\cdot\dfrac{10}{15}=\dfrac{6\cdot10}{8\cdot15}=\dfrac{60}{120}=\dfrac{6}{12}=\dfrac{6\div2}{12\div2}=\dfrac{3\div3}{6\div3}=\boxed{\dfrac{1}{2}}$

Simplifying before) $\dfrac{6}{8}:\dfrac{15}{10}=\dfrac{6}{8}\cdot\dfrac{10}{15}=\dfrac{6\cdot10}{8\cdot15}=\dfrac{(2\cdot3)\cdot(2\cdot5)}{(2\cdot2\cdot2)\cdot(3\cdot5)}=\dfrac{2\cdot3\cdot2\cdot5}{2\cdot2\cdot2\cdot3\cdot5}=\boxed{\dfrac{1}{2}}$

3.5.- Mixed operations with fractions

They will be done following this order:

- 1st) Products and divisions (from left to right).

- 2nd) Additions and subtractions.

- If there are parentheses or brackets, we will solve first into them.

Exercises

14. Arrange the following numbers, from lower to higher:

a) $\dfrac{1}{2}$ $\dfrac{3}{4}$ $\dfrac{5}{6}$

b) $\dfrac{1}{2}$ $\dfrac{3}{5}$ $\dfrac{7}{15}$

c) $\dfrac{1}{5}$ $\dfrac{3}{4}$ $\dfrac{2}{7}$ $\dfrac{9}{8}$ $\dfrac{6}{5}$ $\dfrac{5}{6}$

15. Work out:

a) $\dfrac{4}{5}+\dfrac{1}{6}-\dfrac{2}{15}$

b) $2+\dfrac{5}{6}+\dfrac{1}{3}$

c) $\dfrac{-3}{10}+\dfrac{5}{6}+\dfrac{4}{3}$

d) $\dfrac{-5}{12}+\dfrac{2}{9}+\dfrac{-7}{15}$

e) $1-\dfrac{-2}{3}+3$

f) $\left(\dfrac{3}{4}-\dfrac{1}{5}\right)-\left(\dfrac{-7}{2}+\dfrac{5}{2}\right)$

16. Work out:

a) $\dfrac{2}{7}\cdot\dfrac{3}{5}$

b) $\dfrac{2}{3}\cdot\dfrac{5}{7}\cdot\dfrac{6}{9}$

c) $\dfrac{2}{7}:\dfrac{3}{5}$

d) $\left(\dfrac{2}{7}:\dfrac{4}{5}\right)\dfrac{4}{6}$

e) $\dfrac{15}{2}:(-5)$

f) $\left(\dfrac{3}{4}\cdot\dfrac{1}{5}\right)+\left(\dfrac{-7}{2}:\dfrac{5}{2}\right)$

17. A student says: «During a day, I spend 1/3 of time sleeping and 3/8 at school. Classes take the 2/3 of the time I spend at school». What fraction of my day do classes take?

18. Work out by two methods, directly and extracting a common factor:

a) $\dfrac{2}{3}\cdot\dfrac{1}{5}+\dfrac{2}{3}\cdot\dfrac{5}{6}$

b) $\dfrac{3}{4}\cdot\dfrac{5}{6}+\dfrac{5}{6}\cdot\dfrac{1}{9}$

c) $\dfrac{-3}{11}\cdot\dfrac{1}{4}-\dfrac{6}{5}\cdot\dfrac{-3}{11}$

19. Work out the following expressions:

a) $3-4\left[\dfrac{1}{3}-\dfrac{1}{2}\left(\dfrac{1}{4}-\dfrac{1}{5}\right)+3:\left(\dfrac{1}{3}:\dfrac{1}{2}\right)\right]$

b) $(3-4)\left[\left(\dfrac{1}{3}-\dfrac{1}{2}\right)\left(\dfrac{1}{4}-\dfrac{1}{5}\right)+\left(3:\dfrac{1}{3}\right):\dfrac{1}{2}\right]$

c) $\left(\dfrac{1}{3}+\dfrac{1}{2}\right)\left(\dfrac{1}{2}-\dfrac{1}{4}\right)+5-3\left(4:\dfrac{3}{5}+1\right)$

d) $\left[\dfrac{1}{3}+\dfrac{1}{2}\left(\dfrac{1}{2}-\dfrac{1}{4}\right)+5\right]-3\left[4:\left(\dfrac{3}{5}+1\right)\right]$

e) $\left[\dfrac{3}{8}\left(\dfrac{5}{3}-\dfrac{1}{2}\right)-\dfrac{4}{11}\left(\dfrac{3}{4}-\dfrac{1}{5}\right)\right]:\left[\dfrac{5}{9}-\left(\dfrac{-3}{4}+\dfrac{1}{2}\right)+\dfrac{10}{3}\left(\dfrac{1}{2}-\dfrac{3}{5}\right)\right]$

f) $\left(\dfrac{3}{5}:\dfrac{2}{3}-\dfrac{4}{5}\cdot\dfrac{4}{3}+\dfrac{1}{3}-\dfrac{3}{4}:\dfrac{3}{7}\right)\left(\dfrac{2}{3}+\dfrac{-7}{2}-\dfrac{5}{6}+\dfrac{1}{4}\right)\left(\dfrac{-4}{3}+\dfrac{2}{3}-\dfrac{1}{6}\right)$

4. Scientific notation

When we observe Nature out of quotidian life, for example going into microscopic world or going out, to other planets and galaxies, we are finding quantities that are not comfortable to use. Sometimes because they are very little (size of a cell = 0.0000000018 cm) and sometimes because they are too large (light speed = 300 000 000 m/s).

These quantities, containing long chains of zeros, to the right of to the left of significant digits, make them uncomfortable to use.

In order to make them more comfortable to handle, it is used ***scientific notation***, which consists on transforming numbers with long chains, into shorter expressions. In the following examples, we are showing how scientific notation works.

Parts of a number in scientific notation, $N = a.bcc... \times 10^k$:

> ➤ An integer part, with an only digit (Not zero).
> ➤ A decimal part.
> ➤ A power with base 10 and integer exponent. (k is an integer, positive or negative).

Example: Express 322400000000000 in scientific notation.

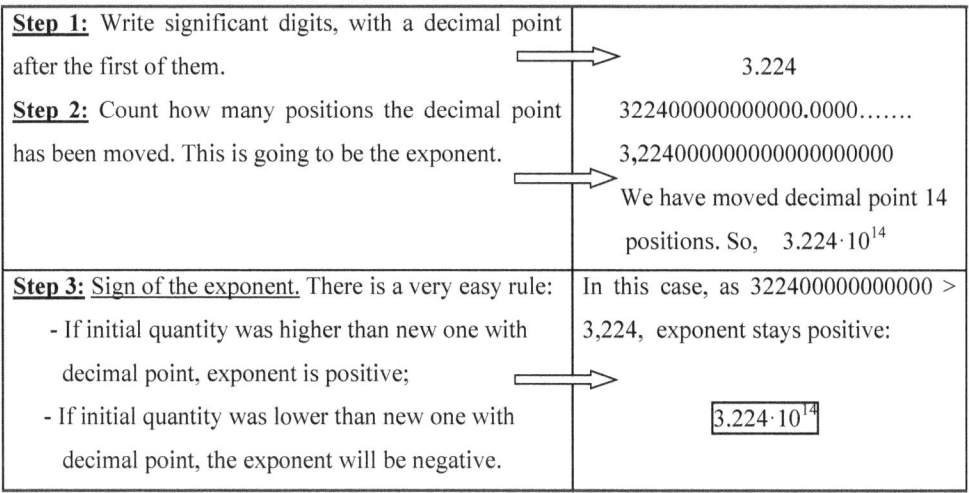

Step 1: Write significant digits, with a decimal point after the first of them.	3.224
Step 2: Count how many positions the decimal point has been moved. This is going to be the exponent.	322400000000000.0000....... 3,22400000000000000000000 We have moved decimal point 14 positions. So, $3.224 \cdot 10^{14}$
Step 3: Sign of the exponent. There is a very easy rule: - If initial quantity was higher than new one with decimal point, exponent is positive; - If initial quantity was lower than new one with decimal point, the exponent will be negative.	In this case, as 322400000000000 > 3,224, exponent stays positive: $3.224 \cdot 10^{14}$

This transformation can also be done by using the definition of power:

Example: Express in scientific notation these quantities: a) 529000   b) 0.00025

Solution:

a) $529000 = 5.29 \cdot 100000 = \boxed{5.29 \cdot 10^5}$ b) $0.00025 = \dfrac{2.5}{100000} = 2.5 \cdot \dfrac{1}{100000} = \boxed{2.5 \cdot 10^{-5}}$

Exercises	**20.** Write in scientific notation:				
	a) 657	b) 0.00058	c) 1258000	d) 0.0021	e) 321000
	f) 0.000012	g) 0.0012	h) 7800	i) 9757000	j) 0.00023

Product and quotient in scientific notation

To ***multiply*** or ***divide*** two quantities given in scientific notation, we multiply or divide, separately, numeric parts and powers of 10. You will have to use the laws of powers.

Example: Calculate: a) $2.7 \cdot 10^{-5} \cdot 3.2 \cdot 10^8$ b) $\dfrac{4.62 \cdot 10^7}{2 \cdot 10^{-4}}$ c) $8.31 \cdot 10^{-5} \cdot 5.2 \cdot 10^9$

Solution:

a) $2.7 \cdot 10^{-5} \cdot 3.2 \cdot 10^8 = \begin{vmatrix} 2.7 \cdot 3.2 = 8.64 \\ 10^{-5} \cdot 10^8 = 10^{-5+8} = 10^3 \end{vmatrix} = \boxed{8.64 \cdot 10^3}$.

b) $\dfrac{4.62 \cdot 10^7}{2 \cdot 10^{-4}} = \begin{vmatrix} \dfrac{4.62}{2} = 2.31 \\ 10^7 : 10^{-4} = 10^{7-(-4)} = 10^{11} \end{vmatrix} = \boxed{2.31 \cdot 10^{11}}$.

c) $8.31 \cdot 10^{-5} \cdot 5.2 \cdot 10^9 = \begin{vmatrix} 8.31 \cdot 5.2 = 43.212 \\ 10^{-5} \cdot 10^9 = 10^{-5+9} = 10^4 \end{vmatrix} = 43.212 \cdot 10^4$, but in scientific notation, we

can write only one significant digit before decimal point, so, we will write $4.3212 \cdot 10 \cdot 10^4 =$

$= \boxed{4.3212 \cdot 10^5}$.

Exercis	**21.** Work out:
	a) $4,1 \cdot 10^3 \times 2 \cdot 10^5$ b) $2,8 \cdot 10^{-4} : 2 \cdot 10^3$ c) $3,2 \cdot 10^5 : 2 \cdot 10^{-2}$

Addition and subtraction of numbers in scientific notation

When ***adding*** or ***subtracting*** numbers in scientific notation, we need all the powers of 10 to have the ***same exponents***. In this case, we will add or subtract the numerical part, leaving the same power of 10.

* If they do not have the same exponent, previously, we must reduce to *common exponent*.

Example: Calculate: a) $3.75 \cdot 10^8 + 2.11 \cdot 10^8$ b) $2.72 \cdot 10^5 - 3.1 \cdot 10^3$

Solution:

a) $3.75 \cdot 10^8 + 2.11 \cdot 10^8 = (3.75 + 2.11) \cdot 10^8 = \boxed{5.86 \cdot 10^8}$.

b) $2.72 \cdot 10^5 - 3.1 \cdot 10^3 =$ Notice they do not have the same exponents. So, we must reduce to common exponent. And we can choose:

$2.72 \cdot 10^5 - 3.1 \cdot 10^3 = \left| 2.72 \cdot 10^5 = 2.72 \cdot 10^2 \cdot 10^3 = 272 \cdot 10^3 \right| = 272 \cdot 10^3 - 3.1 \cdot 10^3 =$

$= (272 - 3.1) \cdot 10^3 = 268.9 \cdot 10^3 = \boxed{2.689 \cdot 10^5}$.

$2.72 \cdot 10^5 - 3.1 \cdot 10^3 = \left| 3.1 \cdot 10^3 = 3.1 \cdot 10^{-2} \cdot 10^5 = 0.031 \cdot 10^5 \right| = 2.72 \cdot 10^5 - 0.031 \cdot 10^5 =$

$= (2.72 - 0.031) \cdot 10^5 = \boxed{2.689 \cdot 10^5}$.

Exercises	**22.** Express as scientific notation:
	a) $32 \cdot 10^5$ b) $75 \cdot 10^{-4}$ c) $843 \cdot 10^7$ d) $458 \cdot 10^{-7}$ e) $0,03 \cdot 10^6$ f) $0,0025 \cdot 10^{-5}$
	23. Calculate, without using your paper and pencil:
	a) $(1,5 \cdot 10^7) \cdot (2 \cdot 10^5)$ b) $(3 \cdot 10^6):(2 \cdot 10^{11})$ c) $(4 \cdot 10^{-7}):(2 \cdot 10^{-12})$ d) $\sqrt{4 \cdot 10^8}$

24. Calculate, expressing the result as scientific notation, and check your answer by using your calculator:

a) $(3,5 \cdot 10^7) \cdot (4 \cdot 10^8)$ b) $(5 \cdot 10^{-8}) \cdot (2,5 \cdot 10^5)$ c) $(1,2 \cdot 10^7):(5 \cdot 10^{-6})$ d) $(6 \cdot 10^{-7})^2$

25. Calculate, expressing the result as scientific notation, and check your answer by using your calculator:

a) $5,3 \cdot 10^{12} - 3 \cdot 10^{11}$ b) $3 \cdot 10^{-5} + 8,2 \cdot 10^{-6}$ c) $6 \cdot 10^{-9} - 5 \cdot 10^{-8}$

d) $7,2 \cdot 10^8 + 1,5 \cdot 10^{10}$

5. Powers and roots

In this part of the unit, we are going to work with the function "root", which is reciprocal to a power.

For example:

POWERS	ROOTS
A square has a side of 5 cm. What is its area? $$A = s^2 = 5^2 = 25 \text{ cm}^2.$$	The area of a square is 49 cm^2. What is the length of its side? $s = 7$ cm, because $7^2 = 49$. This is a root, a square root: $$\sqrt{49} = 7 \iff 7^2 = 49$$
A cube has a side of 4 m. What is its volume? $$V = s^3 = 4^3 = 64 \text{ m}^3.$$	The volume of a cube is cube 8 cm^3. What is the length of its side? $s = 2$ cm, because $2^3 = 8$. This is a root, a cubic root: $$\sqrt[3]{8} = 2 \iff 2^3 = 8$$

Root of a real number *a*, written $\sqrt[n]{a}$ or $a^{1/n}$, where *n* is a natural number, is another real number, *b*, so that $\qquad \sqrt[n]{a} = b \ \ or \ \ a^{1/n} \Leftrightarrow b^n = a$

In this expression, *n* is named **index** and *a* is named **radicand**.

26. Calculate the value of the following roots WITHOUT USING A CALCULATOR:

a) $\sqrt{9}$ e) $\sqrt{25}$ i) $\sqrt{49}$ m) $\sqrt{100}$ o) $\sqrt{1}$

b) $\sqrt{0}$ f) $\sqrt{\dfrac{1}{4}}$ j) $\sqrt{\dfrac{1}{9}}$ n) $\sqrt{\dfrac{4}{25}}$ p) $\sqrt{\dfrac{16}{100}}$

c) $\sqrt{0,25}$ g) $\sqrt{0,09}$ k) $\sqrt{0,0081}$ ñ) $\sqrt{0,49}$ q) $\sqrt{7^6}$

d) $\sqrt{5^{24}}$ h) $\sqrt{2^{10}}$ l) $\sqrt{9^{-10}}$

27. Calculate the value of the following roots WITHOUT USING A CALCULATOR:

a) $\sqrt[3]{8}$ e) $\sqrt[3]{27}$ i) $\sqrt[3]{64}$ m) $\sqrt[3]{1000}$

b) $\sqrt[3]{-1}$ f) $\sqrt[3]{-8}$ j) $\sqrt[3]{-27}$ n) $\sqrt[3]{-1000}$

c) $\sqrt[3]{\dfrac{1}{8}}$ g) $\sqrt[3]{\dfrac{1}{125}}$ k) $\sqrt[3]{-\dfrac{64}{125}}$ ñ) $\sqrt[3]{\dfrac{64}{1000}}$

d) $\sqrt[3]{0,125}$ h) $\sqrt[3]{0,027}$ l) $\sqrt[3]{0,001}$ o) $\sqrt[3]{-0,216}$

28. Calculate the value of the following roots BY USING THE DEFINITION OF ROOT:

a) $\sqrt[3]{-8}$ **b)** $\sqrt{-8}$ **c)** $\sqrt[6]{-1}$ **d)** $\sqrt[5]{-32}$

e) $\sqrt[4]{81}$ **f)** $\sqrt{5^2}$ **g)** $\sqrt[6]{2^6}$ **h)** $\sqrt{\dfrac{625}{81}}$

I) $\sqrt[3]{\dfrac{27}{64}}$ **j)** $\sqrt[4]{\dfrac{81}{16}}$ **k)** $\sqrt[5]{3^{15}}$ **l)** $\sqrt[3]{0,064}$

m) $\sqrt{0,1}$ **n)** $\sqrt{2,25}$ **o)** $\sqrt{2,7}$

29. Calculate the value of k:

a) $\sqrt[3]{k} = 2$ **b)** $\sqrt[5]{-243} = -3$ **c)** $\sqrt[5]{k} = \dfrac{2}{3}$ **d)** $\sqrt[3]{1,331} = 1,1$

30. Write as a root:

a) $25^{\frac{1}{3}}$ b) $12^{\frac{1}{4}}$ c) $a^{\frac{3}{5}}$ d) $\left(\dfrac{1}{2}\right)^{\frac{-2}{3}}$ e) $b^{\frac{2}{7}}$ f) $2^{\frac{3}{2}}$ g) $2^{\frac{1}{2}}$ h) $\left(a^2 + b^2\right)^{\frac{1}{2}}$

i) $(a-b)^{\frac{-1}{2}}$ j) $2x^{\frac{3}{5}}$ k) $\left(2x^3 y\right)^{\frac{2}{5}}$ l) $(8x)^{\frac{-2}{3}}$ m) $(-8)^{\frac{1}{5}}$ n) $-3^{\frac{1}{2}}$

31. Write as a power:

a) $\left(\sqrt[3]{3}\right)^6$ b) $\left(\sqrt[3]{4}\right)^8$ c) $\left(\sqrt{3}\right)^2$ d) $\left(\sqrt[5]{31}\right)^5$ e) $\left(\sqrt{n^2}\right)^7$

f) $\left(\sqrt[3]{m^3}\right)^5$ g) $\left(\sqrt[5]{m^2}\right)^3$ h) $\left(\sqrt[4]{5}\right)^2$ i) $\sqrt[3]{\sqrt{128}}$

32. Use your calculator to find out the value of the following roots, rounding to the thousandths:

   a) $\sqrt[4]{8}$ **b)** $\sqrt[5]{9}$ **c)** $\sqrt[6]{25}$ **d)** $\sqrt[3]{10}$

   e) $\sqrt[5]{-15}$ **f)** $\sqrt[6]{-40}$ **g)** $\sqrt[4]{2^3}$ **h)** $\sqrt[5]{3^2}$

   i) $\sqrt[6]{5^2}$ **j)** $\sqrt[8]{256}$ **k)** $\sqrt[3]{64}$

33. Write the following powers as roots and calculate their values WITHOUT USING YOUR CALCULATOR:

   a) $4^{1/2}$ **b)** $125^{1/3}$ **c)** $625^{1/4}$

   d) $8^{2/3}$ **e)** $64^{5/6}$ **f)** $81^{3/4}$

   g) $8^{-2/3}$ **h)** $27^{-1/3}$

Exercises

Equivalent roots. Simplification of roots

> ➤ In roots having a power as radicand, *index* and *exponent* can be *multiplied* or *divided* by a *same number*, in the same way you are used to do it with fractions.
> ➤ When we divide index and exponent by a same number, we are *simplifying* the root.

As we have seen above, a root can be written as a power, being its exponent a fraction. So, the same operations we use with fractions, can be used with roots (simplifying, ordering, ...).

Example: Simplify: a) $\sqrt[6]{8}$ b) $\sqrt[12]{625}$ c) $\sqrt[18]{a^{12}}$

Solution (as root):

a) $\sqrt[6]{8} = \sqrt[6]{2^3} = \sqrt[2]{2^1} = \sqrt{2}$, where we have divided by 3.

b) $\sqrt[12]{625} = \sqrt[12]{5^4} = \sqrt[3]{5}$, where we have divided by 4.

c) $\sqrt[18]{a^{12}} = \sqrt[3]{a^2}$, where we have divided by HCF(18, 12) = 6.

Solution (as power):

a) $\sqrt[6]{8} = \sqrt[6]{2^3} = 2^{\frac{3}{6}} = 2^{\frac{1}{2}} = \sqrt{2}$, where numerator and denominator have been divided by 3.

b) $\sqrt[12]{625} = \sqrt[12]{5^4} = 5^{\frac{4}{12}} = 5^{\frac{1}{3}} = \sqrt[3]{5}$, where numerator and denominator have been divided by 4.

c) $\sqrt[18]{a^{12}} = a^{\frac{12}{18}} = a^{\frac{2}{3}} = \sqrt[3]{a^2}$, where numerator and denominator have been divided by

HCF(18, 12) = 6.

Exercises

34. Simplify the following roots:

a) $\sqrt[4]{3^2}$ b) $\sqrt[8]{5^4}$ c) $\sqrt[9]{27}$ d) $\sqrt[5]{1024}$

e) $\sqrt[6]{8}$ f) $\sqrt[9]{64}$ g) $\sqrt[8]{81}$ h) $\sqrt[12]{x^9}$

i) $\sqrt[12]{x^8}$ j) $\sqrt[5]{x^{10}}$ k) $\sqrt[6]{a^2b^4}$ l) $\sqrt[10]{a^4b^6}$

m) $\sqrt[6]{5^3}$ n) $\sqrt[15]{2^{12}}$ o) $\sqrt[10]{a^8}$ p) $\sqrt[12]{x^4y^8z^4}$

35. Decide if the following roots are equivalent, and check with your calculator:

a) $\sqrt{5}$, $\sqrt[4]{25}$, $\sqrt[6]{125}$, $\sqrt[8]{625}$ b) $\sqrt{9}$, $\sqrt[3]{27}$, $\sqrt[4]{81}$, $\sqrt[5]{243}$ c) $\sqrt{2}$, $\sqrt[4]{4}$, $\sqrt[6]{8}$, $\sqrt[8]{16}$

36. Rewrite the following roots so that they have the same index and arrange them from lower to higher:

a) $\sqrt{5}$, $\sqrt[5]{2^3}$, $\sqrt[15]{7^2}$ b) $\sqrt[3]{5}$, $\sqrt[5]{7^3}$, $\sqrt[15]{3^2}$ c) $\sqrt[4]{3}$, $\sqrt[6]{16}$, $\sqrt[15]{9}$

d) $\sqrt{2}$, $\sqrt[3]{32}$, $\sqrt[5]{27}$ e) $\sqrt{2}$, $\sqrt[3]{3}$, $\sqrt[4]{4}$, $\sqrt[5]{5}$, $\sqrt[6]{6}$ f) $\sqrt[3]{16}$, $\sqrt[4]{125}$, $\sqrt[6]{243}$

Operations with roots. Laws of roots

Law	Example	Description
$\sqrt[n]{a}\cdot\sqrt[n]{b}=\sqrt[n]{a\cdot b}$	$\sqrt{2}\cdot\sqrt{3}=\sqrt{6}$	**Product** of two roots having the same index is a root having the common index and whose radicand is the product of radicands.
$\sqrt[n]{a}:\sqrt[n]{b}=\sqrt[n]{a:b}$	$\sqrt[5]{12}:\sqrt[5]{4}=\sqrt[5]{3}$	**Quotient** of two roots having the same index is a root having the common index and whose radicand is the quotient of radicands.
$\left(\sqrt[n]{a}\right)^{m}=\sqrt[n]{a^{m}}$	$\left(\sqrt[5]{4}\right)^{2}=\sqrt[5]{16}$	**Power** of a root is another root having the same index and whose radicand is the power of the original radicand.
$\sqrt[m]{\sqrt[n]{a}}=\sqrt[n\cdot m]{a}$	$\sqrt{\sqrt[3]{8}}=\sqrt[6]{8}$	**Root** of a root is another root having the same radicand and whose index is the product of the original index.

These laws can be easily demonstrated by applying the power rules:

a) $\sqrt[n]{a}\cdot\sqrt[n]{b}=a^{\frac{1}{n}}\cdot b^{\frac{1}{n}}=(a\cdot b)^{\frac{1}{n}}=\sqrt[n]{a\cdot b}$

b) $\sqrt[n]{a}:\sqrt[n]{b}=a^{\frac{1}{n}}:b^{\frac{1}{n}}=(a:b)^{\frac{1}{n}}=\sqrt[n]{a:b}$

c) $\left(\sqrt[n]{a}\right)^{m}=\left(a^{\frac{1}{n}}\right)^{m}=a^{\frac{m}{n}}=\left(a^{m}\right)^{\frac{1}{n}}=\sqrt[n]{a^{m}}$

d) $\sqrt[m]{\sqrt[n]{a}}=\left(a^{\frac{1}{n}}\right)^{\frac{1}{m}}=a^{\frac{1}{n\cdot m}}=\sqrt[n\cdot m]{a}$

Exercises

37. Calculate:

a) $\sqrt{3}\cdot\sqrt{5}\cdot\sqrt{7}$ b) $\sqrt[3]{4}\cdot\sqrt[3]{5}\cdot\sqrt[3]{2}$ c) $\sqrt[5]{12}\div\sqrt[5]{6}$ d) $\sqrt{4}\div\sqrt{2}$

e) $\sqrt[3]{\sqrt{5}}$ f) $\sqrt[6]{\sqrt[3]{6}}$ g) $\sqrt{5}\cdot\sqrt{6}$ h) $\sqrt{2a}\cdot\sqrt{a}$

i) $\sqrt{18}\div\sqrt{50}$ j) $\sqrt{12}\cdot\sqrt{\dfrac{3}{4}}\cdot\sqrt{\dfrac{12}{5}}\cdot\sqrt{\dfrac{15}{4}}$

38. Work out the following products of roots having the same index, and simplify as far as possible:

a) $\sqrt{2}\,\sqrt{32}$

b) $\sqrt{2}\,\sqrt{15}$

c) $\sqrt[3]{3}\,\sqrt[3]{9}$

d) $\sqrt{2}\,\sqrt{8}$

e) $\sqrt{3}\,\sqrt{4}$

f) $\sqrt[3]{2}\,\sqrt[3]{5}$

g) $\sqrt{12}\,\sqrt{6}\,\sqrt{50}$

h) $\sqrt{21}\,\sqrt{7}$

i) $4\sqrt{3}\cdot 2\sqrt{27}$

39. Work out the following products of roots having different index, and simplify as far as possible:

a) $\sqrt{2}\,\sqrt[3]{32}$

b) $\sqrt[3]{2}\,\sqrt[4]{8}$

c) $\sqrt[3]{2}\,\sqrt[5]{2}$

d) $\sqrt[3]{9}\,\sqrt[6]{3}$

e) $\sqrt[3]{2^2}\,\sqrt[4]{2}$

f) $\sqrt[4]{a^3}\,\sqrt[6]{a^5}$

g) $\sqrt[3]{2}\,\sqrt{3}\,\sqrt[4]{8}$

40. Simplify, applying, first of all, the laws of roots:

a) $\dfrac{\sqrt{32}}{\sqrt{2}}$

b) $\dfrac{\sqrt{8}}{\sqrt{2}}$

c) $\dfrac{\sqrt[3]{81}}{\sqrt[3]{9}}$

d) $\dfrac{\sqrt{15}}{\sqrt{3}}$

e) $\dfrac{\sqrt{3}}{\sqrt{4}}$

f) $\dfrac{\sqrt[3]{16}}{\sqrt[3]{2}}$

g) $\sqrt{\dfrac{256}{729}}$

h) $\dfrac{\sqrt{21}}{2\sqrt{7}}$

i) $\dfrac{\sqrt{33}}{\sqrt{3}}$

j) $\sqrt[3]{\dfrac{125}{512}}$

k) $\sqrt[4]{\dfrac{16}{625}}$

l) $\dfrac{\sqrt{2}\,\sqrt{8}}{\sqrt{32}}$

m) $\sqrt{\dfrac{154}{9}+23}-\sqrt{4\dfrac{144}{9}}$

n) $\sqrt{\left(-\dfrac{3}{2}\right)^2+\left(\dfrac{3\sqrt{3}}{2}\right)^2}$

41. Work out the following quotients of roots having different index, transforming them, first of all, into common index. Simplify as far as possible:

a) $\dfrac{\sqrt{8}}{\sqrt[4]{2}}$

b) $\dfrac{\sqrt[3]{9}}{\sqrt[6]{3}}$

c) $\dfrac{\sqrt{2}}{\sqrt[3]{32}}$

d) $\dfrac{\sqrt[4]{4}}{\sqrt[6]{8}}$

e) $\dfrac{\sqrt[3]{7^2}}{\sqrt{7}}$

f) $\dfrac{\sqrt{9}}{\sqrt[3]{3}}$

g) $\dfrac{\sqrt[5]{16}}{\sqrt{2}}$

h) $\dfrac{\sqrt{ab}}{\sqrt[3]{ab}}$

i) $\dfrac{\sqrt[4]{a^3b^5c}}{\sqrt{ab^3c^3}}$

j) $\dfrac{\sqrt[6]{a^3}}{\sqrt[3]{a^2}}$

k) $\dfrac{\sqrt[3]{-2000}}{3\sqrt{2}}$

l) $\dfrac{\sqrt[3]{4}\,\sqrt{3}}{\sqrt[6]{12}}$

m) $\dfrac{\sqrt[8]{8}}{\sqrt[4]{4}\,\sqrt{2}}$

n) $\dfrac{\sqrt[3]{5}\cdot\sqrt{125}}{\sqrt[4]{25}}$

o) $\dfrac{\sqrt[3]{2}\cdot\sqrt{3}\cdot\sqrt[12]{2}}{\sqrt[12]{18}}$

p) $\dfrac{\sqrt[3]{4}\cdot\sqrt{3}\cdot\sqrt[12]{2}}{\sqrt[4]{2}}$

q) $\dfrac{\sqrt[6]{54}\cdot\sqrt[12]{27}}{\sqrt[12]{4}\cdot\sqrt[4]{12}}$

r) $\dfrac{\sqrt[4]{abc^2}\cdot\sqrt[12]{a^3b^5c^2}}{\sqrt[6]{a^2b^2c}}$

42. Simplify:

a) $\left(\sqrt[3]{a^2}\right)^6$ b) $\left(\sqrt[6]{ab^2}\right)^2$ c) $\left(\sqrt{x}\right)^3 \cdot \sqrt[3]{x}$ d) $\dfrac{\left(\sqrt[3]{2}\right)^4}{\left(\sqrt[4]{2}\right)^2}$

e) $\dfrac{\sqrt{2}\left(\sqrt[3]{2}\right)^4}{\left(\sqrt[4]{2}\right)^3}$ f) $\sqrt{2}\left(\sqrt[4]{2}\right)^3\left(\sqrt[3]{2}\right)^2$ g) $\dfrac{\left(\sqrt[4]{3}\right)^5}{\left(\sqrt{3}\right)^2\left(\sqrt[3]{3}\right)^4}$ h) $\sqrt{2}\left(\sqrt[4]{2\sqrt[3]{4}}\right)^3$

i) $\sqrt{\sqrt{2^6}}$ j) $\sqrt{\sqrt{12}}$ k) $\left(\sqrt{\sqrt{\sqrt{2}}}\right)^8$ l) $\sqrt[3]{\sqrt[4]{x^5 x^7}}$

m) $\sqrt[3]{\sqrt[4]{x^{15}}}$ n) $\left(\sqrt[3]{\sqrt[7]{\sqrt{8x^3}}}\right)^7$ o) $\left(\sqrt[3]{\sqrt{5}}\right)^5\left(\sqrt[4]{5}\right)^3$

p) $\dfrac{\left(\sqrt{x}\right)^3}{\left(\sqrt[3]{\sqrt[4]{x}}\right)^6}$ q) $\dfrac{\left(\sqrt[3]{2}\right)^4\cdot\left(\sqrt[4]{8}\right)^3}{\sqrt{\left(\sqrt[3]{4}\right)^2}}$ r) $\dfrac{\sqrt[3]{\sqrt{a^2}}\cdot\left(\sqrt{a^3}\right)^3}{\left(\sqrt{a}\right)^3\cdot\sqrt[3]{a^4}}$ s) $\dfrac{\left(\sqrt{27}\right)^3\cdot\sqrt{\sqrt[3]{9}}}{\sqrt[3]{81}\cdot\left(\sqrt{3}\right)^3}$

Introducing and extracting factors from roots

These laws allow us to introduce or extract factors from roots.

Example: Introduce factors into the following roots:

Solution: a) $2\sqrt{3}$ b) $2\sqrt[3]{5}$

a) $2\sqrt{3} = \sqrt{2^2}\cdot\sqrt{3} = \sqrt{4}\cdot\sqrt{3} = \boxed{\sqrt{12}}$

b) $2\sqrt[3]{5} = \sqrt[3]{2^3}\cdot\sqrt[3]{5} = \sqrt[3]{8}\cdot\sqrt[3]{5} = \boxed{\sqrt[3]{40}}$

Example: Extract from the following roots: a) $\sqrt{200}$ b) $\sqrt[3]{250}$

Solution:

a) $\sqrt{200} = \sqrt{2^3\cdot 5^2} = \sqrt{2\cdot 2^2\cdot 5^2} = 2\cdot 5\cdot\sqrt{2} = \boxed{10\sqrt{2}}$

b) $\sqrt[3]{250} = \sqrt[3]{5^3\cdot 2} = \sqrt[3]{5^3}\cdot\sqrt[3]{2} = \boxed{5\sqrt[3]{2}}$

43. Introduce in the roots:

a) $5\sqrt{3}$ b) $2\sqrt[3]{4}$ c) $3\sqrt[4]{2}$ d) $2\sqrt[3]{2}$ e) $2\sqrt{5}$ f) $7\sqrt{a}$

g) $2a\sqrt{3a}$ h) $x\sqrt{\dfrac{1}{x}}$ i) $x^3 y\sqrt{xy}$ j) $\dfrac{1}{3}\sqrt[4]{\dfrac{27}{2}}$ k) $\dfrac{3}{2}\sqrt{\dfrac{2}{3}}$ l) $\dfrac{2}{a}\sqrt{\dfrac{ax}{2}}$

m) $\dfrac{3}{2xy}\sqrt{\dfrac{2xz}{y}}$

44. Extract all possible factors from the following roots:

a) $\sqrt{3^5}$ b) $\sqrt[4]{5^{10}}$ c) $\sqrt[3]{81}$ d) $\sqrt{300}$ e) $\sqrt{18}$ f) $\sqrt{32}$

g) $\sqrt{8}$ h) $\sqrt{75}$ i) $\sqrt{200}$ j) $\sqrt[3]{625}$ k) $\sqrt[4]{32}$ l) $\sqrt[4]{243}$

45. Introduce in the roots:

a) $5\sqrt{3}$ b) $2\sqrt[3]{4}$ c) $3\sqrt[4]{2}$ d) $2\sqrt[3]{2}$ e) $2\sqrt{5}$ f) $7\sqrt{a}$ g) $2a\sqrt{3a}$

h) $x\sqrt{\dfrac{1}{x}}$ i) $x^3 y\sqrt{xy}$ j) $\dfrac{1}{3}\sqrt[4]{\dfrac{27}{2}}$ k) $\dfrac{3}{2}\sqrt{\dfrac{2}{3}}$ l) $\dfrac{2}{a}\sqrt{\dfrac{ax}{2}}$ m) $\dfrac{3}{2xy}\sqrt{\dfrac{2xz}{y}}$

46. Extract all possible factors from the following roots:

a) $\sqrt{\dfrac{27}{4}}$ b) $\sqrt[5]{\dfrac{5x^{10}}{y^8}}$ c) $\sqrt[3]{\dfrac{8x^4 y^3 z}{n^6}}$ d) $3\sqrt{8a^3}$ e) $2x^2 y\sqrt{x^4 y^3}$

f) $\dfrac{xy^2}{3}\sqrt{27xy^3}$ g) $\sqrt{8}$ h) $\sqrt{12}$ i) $\sqrt[3]{16}$ j) $\sqrt[3]{54}$

k) $\sqrt[5]{64}$ l) $\sqrt{12x^3 y^5 z^2}$ m) $\sqrt[3]{\dfrac{8x^4}{81y^6}}$

47. Work out:

a) $\sqrt{\sqrt[3]{b^2}}$ b) $\sqrt{\sqrt{\sqrt{x}}}$ c) $\sqrt[3]{\sqrt{\sqrt[4]{a^3}}}$ d) $\sqrt{x\sqrt{x\sqrt{x}}}$ e) $\sqrt{x\sqrt{x\sqrt{x^2}}}$ f) $\sqrt[3]{4\sqrt{4\sqrt[3]{4}}}$

g) $\sqrt[3]{\sqrt[4]{\sqrt{a^{24}b^{12}c^6}}}$ h) $\sqrt[3]{2\sqrt{(1-a)\sqrt{(1-a)}}}$ i) $\sqrt[4]{\sqrt[3]{\sqrt{\sqrt[3]{2^{144}}}}}$

Exercises

Like roots

Two or more radicals are said to be *like radicals* if they have the *same index* and the *same radicand*.

For example, these radicals are like radicals: $\sqrt[5]{14}$, $8\sqrt[5]{14}$ and $(-2)\sqrt[5]{14}$.

To *add* or *subtract* radicals, they must be like radicals.

Example: Calculate: a) $\sqrt[5]{14}$ - $8\sqrt[5]{14}$ + $10\sqrt[5]{14}$

Solution:

a) $\sqrt[5]{14}$ - $8\sqrt[5]{14}$ + $10\sqrt[5]{14}$ = $(1 - 8 + 10)\sqrt[5]{14}$ = $\boxed{3\sqrt[5]{14}}$.

b) $6\sqrt[3]{5} - \sqrt[3]{40} = 6\sqrt[3]{5} - \sqrt[3]{2^3 \cdot 5} = 6\sqrt[3]{5} - 2\sqrt[3]{5} = \boxed{4\sqrt[3]{5}}$

c) $\sqrt{2} + \sqrt{8} + \sqrt{18} - \sqrt{32} = \sqrt{2} + \sqrt{2^3} + \sqrt{2 \cdot 3^2} - \sqrt{2^5} = \sqrt{2} + 2\sqrt{2} + 3\sqrt{2} - 4\sqrt{2} = (1 + 2 + 3 - 4)\sqrt{2} =$
$= \boxed{2\sqrt{2}}$

Exercises

48. Add the following radicals, extracting previously all possible factors and making them like radicals:

a) $\sqrt{7} + \sqrt{28} - \sqrt{63}$ b) $\sqrt{121} + \sqrt{169} - \sqrt{225}$ c) $\sqrt{2} + \sqrt{8} + \sqrt{18} - \sqrt{32}$

d) $\sqrt{5} + \sqrt{45} + \sqrt{180} - \sqrt{80}$ e) $\sqrt{24} - 5\sqrt{6} + \sqrt{486}$ f) $\sqrt[3]{54} - 2\sqrt[3]{16}$

g) $\sqrt{2} + 3\sqrt{18} - \sqrt{32}$ h) $\sqrt{5} + \sqrt{180} - \sqrt{80}$ i) $\sqrt{2} + \sqrt{32} + 5\sqrt{8}$

j) $3\sqrt{20} + 2\sqrt{5}$ k) $2\sqrt{27} - 4\sqrt{12}$ l) $3\sqrt{28} - 2\sqrt{27}$

m) $\sqrt[3]{16} - \sqrt[3]{54}$

49. Transform into like radicals and simplify:

a) $2a\sqrt{2} - \sqrt{8} + 3\sqrt{2}$ b) $2a\sqrt{3} - \sqrt{27a^2} + a\sqrt{12}$ c) $(3+a)\sqrt{5} - \sqrt{125} + \sqrt{5a^2}$

d) $4\sqrt{12} - \dfrac{3}{2}\sqrt{48} + \dfrac{2}{3}\sqrt{27} + \dfrac{3}{5}\sqrt{75}$ e) $7\sqrt{54} - 3\sqrt{18} + \sqrt{24} - \dfrac{3}{5}\sqrt{50} - \sqrt{6}$

f) $\sqrt[4]{144} + 3\sqrt{27} - \sqrt{48}$ g) $5\sqrt[6]{256} - 3\sqrt[3]{16} - \sqrt[3]{128}$ h) $\sqrt{3} - \sqrt{108} + \sqrt{648} - \sqrt{1875}$

i) $\dfrac{3\sqrt{3}}{2} - \dfrac{7\sqrt{108}}{4} + \sqrt{648} - \dfrac{2}{3}\sqrt{1875}$

Rationalization

Sometimes, when operating with radicals, there are fractions with radicals at the denominator. *Rationalizing* consists on finding another equivalent fraction, without radicals at the denominator.

To rationalize a fraction, you must multiply **numerator** and **denominator** by a same expression that makes radicals at denominator disappear.

➤ If denominator is $\sqrt{a}$, we will multiply numerator and denominator by $\sqrt{a}$.

➤ If denominator is $a + \sqrt{b}$, $a - \sqrt{b}$, $\sqrt{a} + \sqrt{b}$ or $\sqrt{a} - \sqrt{b}$ we will multiply numerator and denominator by their **conjugated** (the same expression, changing the central sign).

Example: Rationalize: a) $\dfrac{5}{\sqrt{3}}$ b) $\dfrac{3}{2\sqrt{2}}$ c) $\dfrac{3}{\sqrt{2} - 1}$

Solution:

a) $\dfrac{5}{\sqrt{3}}$ $\rightarrow$ Multiply numerator and denominator by $\sqrt{3}$ $\rightarrow$ $\dfrac{5\cdot\sqrt{3}}{\sqrt{3}\cdot\sqrt{3}} = \dfrac{5\cdot\sqrt{3}}{(\sqrt{3})^2} = \boxed{\dfrac{5\cdot\sqrt{3}}{3}}$

b) $\dfrac{3}{2\sqrt{2}}$ $\rightarrow$ Multiply numerator and denominator by $\sqrt{2}$ $\rightarrow$ $\dfrac{3\cdot\sqrt{2}}{2\cdot\sqrt{2}\sqrt{2}} = \dfrac{3\cdot\sqrt{2}}{2\cdot(\sqrt{2})^2} = \boxed{\dfrac{3\cdot\sqrt{2}}{4}}$

c) $\dfrac{3}{\sqrt{2} - 1}$ $\rightarrow$ Multiply numerator and denominator by $(\sqrt{2} + 1)$ $\rightarrow$

$\rightarrow \dfrac{3\cdot(\sqrt{2}+1)}{(\sqrt{2}-1)\cdot(\sqrt{2}+1)} = \dfrac{3\cdot(\sqrt{2}+1)}{(\sqrt{2})^2 - 1} = \dfrac{3\cdot(\sqrt{2}+1)}{2-1} = \boxed{3\sqrt{2} + 3}$

50. Rationalize the following expressions:

a) $\dfrac{2}{\sqrt{3}}$ b) $\dfrac{1}{3\sqrt{2}}$ c) $\dfrac{5}{\sqrt{5}}$ d) $\dfrac{2\sqrt{6}}{\sqrt{2}}$ e) $\dfrac{\sqrt{27}}{\sqrt{8}}$ f) $\sqrt{\dfrac{2}{3}}$ g) $\sqrt{\dfrac{5}{2}}$ h) $\dfrac{3}{\sqrt{5}}$

51. Multiply by its conjugated each of the following expressions:

a) $2-\sqrt{x}$ b) $\sqrt{x}-2$ c) $\sqrt{x}+\sqrt{y}$

52. Rationalize the following expressions:

a) $\dfrac{2}{3-\sqrt{5}}$ b) $\dfrac{8\sqrt{5}}{\sqrt{7}-\sqrt{3}}$ c) $\dfrac{9}{\sqrt{x}-\sqrt{y}}$ d) $\dfrac{1+\sqrt{x}}{1-\sqrt{x}}$

53. Rationalize the following expressions:

a) $\dfrac{2}{1-\sqrt{3}}$ b) $\dfrac{5}{\sqrt{7}-\sqrt{2}}$ c) $\dfrac{\sqrt{2}}{5-\sqrt{2}}$ d) $\dfrac{\sqrt{3}}{\sqrt{12}+\sqrt{2}}$

54. Rationalize and simplify:

a) $\dfrac{2}{1+\sqrt{2}}$ b) $\dfrac{4}{3-\sqrt{2}}$ c) $\dfrac{23}{5-\sqrt{2}}$ d) $\dfrac{1}{1-\sqrt{3}}$ e) $\dfrac{1}{\sqrt{5}+3}$

f) $\dfrac{1}{\sqrt{3}-\sqrt{2}}$ g) $\dfrac{10}{\sqrt{3}+\sqrt{2}}$ h) $\dfrac{\sqrt{2}}{\sqrt{2}+3}$ i) $\dfrac{1+\sqrt{3}}{1-\sqrt{3}}$

6. Logarithms

Let a and N two real and positive numbers. ***Logarithm*** of N at the base a, denoted as $\log_a N$, is the number at which a must be raised to obtain N.

$$\log_a N = x \longleftrightarrow a^x = N$$

Example:

a) $\log_{10} 100 = 2 \leftrightarrow 10^2 = 100$ b) $\log_5 125 = 3 \leftrightarrow 5^3 = 125$

c) $\log_2 32 = 5 \leftrightarrow 2^5 = 32$ d) $\log_2 \frac{1}{64} = -6 \leftrightarrow 2^{-6} = \frac{1}{64}$

Exercises

55. Find out the value of x:

a) $\log_2 x = 3$ b) $\log_5 x = 0$ c) $\log_3 x = 2$ d) $\log_{\frac{1}{2}} x = -1$

e) $\log_{0,3} x = -2$ f) $\log_2 x = -\dfrac{1}{2}$ g) $\log_x 27 = 3$ h) $\log_x 16 = -4$

i) $\log_x \dfrac{1}{4} = 2$ j) $\log_x \dfrac{1}{3} = \dfrac{1}{2}$ k) $\log_2 32 = x$ l) $\log_3 \dfrac{1}{81} = x$

m) $\log_{\frac{1}{2}} 16 = x$ n) $\log_{\frac{1}{125}} 625 = -x$ ñ) $\log_{0,01} 0,1 = x$ o) $\log_{\frac{1}{4}} \dfrac{1}{128} = x$

.

2 important logarithms:

➢ **Decimal logarithm:** Its base is 10. Sometimes, the base is not indicated: $\log_{10} N = \log N$

➢ **Natural logarithm:** Its base is the irrational number, e. $LnN = \log_e N$

Rules for logarithms

➢ **Logarithm of a product:** Let M and N positive real numbers, then:

$$\log_a M.N = \log_a M + \log_a N$$

➢ **Logarithm of a quotient:** Let M and N positive real numbers, then:

$$\log_a M:N = \log_a M - \log_a N$$

➢ **Logarithm of a power:** If N is a positive real number and a a real number, then:

$$\log_a N^\alpha = \alpha.\log_a N$$

➢ **Logarithm of a root:** If N is a positive real number and n a natural number bigger than 1, then:

$$\log_a \sqrt[n]{N} = \frac{1}{n}\log_a N$$

Example: Knowing that $\log_{10}2 = 0.301$ and that $\log_{10}3 = 0.477$, calculate:

a) $\log_{10}6$ b) $\log_{10}9$ c) $\log_{10}64$ d) $\log_{10}\sqrt[3]{36}$

Solution:

a) $\log(6) = \log(2\cdot3) = \log(2) + \log(3) = 0.301 + 0.477 = \boxed{0.778}$.

b) $\log(9) = \log(3^2) = 2\cdot \log(3) = 2\cdot 0.477 = \boxed{0.954}$.

c) $\log(64) = \log(2^6) = 6\cdot \log(2) = 6\cdot 0.301 = \boxed{1.806}$.

d) $\log(\sqrt[3]{36}) = \log(\sqrt[3]{2^2\cdot3^2}) = \log\left((2^2\cdot3^2)^{\frac{1}{3}}\right) = \log\left((2\cdot3)^{\frac{2}{3}}\right) = \frac{2}{3}\cdot \log(2\cdot3) =$

$= \frac{2}{3}\cdot\{\log(2) + \log(3)\} = \frac{2}{3}\cdot\{0.301 + 0.477\} = \boxed{0.519}$.

---

56. Knowing that $\log2 = 0.3010$, $\log3 = 0.4771$ and $\log7 = 0.85$, calculate:

a) $\sqrt[5]{21}$ b) $\sqrt{441}$ c) $\sqrt[3]{6}$ d) $\sqrt[4]{\dfrac{1323}{627}}$

57. Develop the following logarithms by applying their rules:

a) $\log(2ab)$ b) $\log\dfrac{3a}{4}$ c) $\log\dfrac{2a^2}{3}$ d) $\log a^5 b^4$ e) $\log\dfrac{2}{ab}$

f) $\log\sqrt{ab}$ g) $\log\dfrac{\sqrt{x}}{2y}$ h) $\log 2a\sqrt{b}$ i) $\log\dfrac{3a\sqrt[3]{b}}{c}$ j) $\log\dfrac{5a^2 b^4\sqrt{c}}{2xy}$

k) $\log(abc)^3$ l) $\log(\dfrac{a\sqrt{c}}{2})^4$ m) $\log 7ab^3\sqrt{5c^2}$ n) $\log\sqrt{\dfrac{2ab}{x^2 y}}$ ñ) $\log(a^2 - b^2)$

o) $\log\dfrac{\sqrt[3]{a^2}}{\sqrt[5]{b^3}}$ p) $\log\dfrac{\sqrt{a}\cdot\sqrt[3]{b}}{\sqrt[4]{cd}}$ r) $\log(x^4 - y^4)$ s) $\log\dfrac{m-n}{2}$ t) $\log\sqrt{\dfrac{a(b-c)}{d^2 m}}$

58. Write as an only logarithm:

a) $\log a + \log b$ b) $\log x - \log y$ c) $\dfrac{1}{2}\log x + \dfrac{1}{2}\log y$

d) $\log a - \log x - \log y$ e) $\log p + \log q - \log r - \log s$ f) $\log 2 + \log 3 + \log 4$

g) $\dfrac{1}{3}\log a - \dfrac{1}{2}\log b - \dfrac{1}{2}\log c$

Exercises

59. Knowing that log 2 = 0.301, log 3 = 0.477 and log 7 = 0.85, calculate:

a) log 8   b) log 9   c) log 5   d) log 54   e) log 75   f) log 0,25

g) log (1/6)   h) log (1/98)   i) log (1/36)   j) log (2/3)   k) log 0.3   l) log 1.25

Review exercises

Sets of numbers

1. Classify the following numbers into rational or irrational numbers:

$$\frac{1}{8} \qquad \frac{\pi}{3} \qquad \sqrt{5} \qquad 2,6 \qquad 0 \qquad -3 \qquad -\frac{25}{3} \qquad \sqrt{13} \qquad 0,1 \qquad 6,\hat{4} \qquad 534 \qquad 1,414213\ldots$$

2. Classify the following numbers into rational or irrational numbers:

$$\frac{\pi}{2} \qquad \sqrt{3} \qquad \sqrt{4} \qquad 0,0015 \qquad -10 \qquad \frac{5}{6} \qquad 2,\overset{\frown}{3} \qquad 2,020020002\ldots$$

3. Point out which of the following numbers are irrational and why:

a) 3,629629629....	d) 0,123456789...	g) 0,130129128...
b) 0,128129130...	e) 7,129292929...	
c) 5,216968888...	f) 4,101001000...	

4. Work out, directly and writing first as fractions, and check you obtain the same results:

k) $0,\bar{6} : 0,0\bar{5} + 0,25$ l) $1,25 - 1,16 + 1,\hat{1}$ m) $2,\overset{\frown}{7} \cdot 1,8 + 2,\overset{\frown}{26} : 0,1\overset{\frown}{13}$

n) $1,9\bar{2} + 0,25 (0,2\bar{5} + 0,\bar{5})$ o) $\sqrt{2,7}$ p) $0,83 - 0,8 : 0,6$

q) $4,08\bar{3} \cdot 1,1\bar{1} - 0,15 : 0,3$ r) $0,\bar{6} + 1,3\bar{8} \cdot 0,72$ s) $0,\bar{5} - 0,15 + 1,2\bar{3}$

Intervals and semi-lines

5. Work out the and the of the following sets. First of all, draw them:

a) A=[-2,5)	c) E=(0,3]	e) I=[-5,-1)	g) M=(2,5)	i) Q=(-3,7)
B=(1,7)	F=(2,∞)	J=(2,7/2]	N=(5,9]	R=(2,4]
b) C=(-1,3]	d) G=(-∞,0]	f) K=(-∞,0)	h) O=[-3,-1)	j) S=[-3,2)
D=(1,6]	H=(-3,∞)	L=[0,∞)	P=(2,7]	T=(0,∞)
				U=[1,4]

6. Fill in the next chart:

GRAPHICAL REPRESENTATION	INTERVAL	Math. expression
 -1 3	[-1,3]	{x∈ IR/ -1≤x≤3}
0 2		
-2 4		
	[-2,1)	
		{x∈ IR/ 1<x≤5}
 -1 ∞		
		{x∈ IR/ x<2}
	(0,∞)	

7. Fill in the next chart:

GRAPHICAL REPRESENTATION	INTERVAL	Math. expression
	(-1,5)	
		{x∈ R/ x≤0}
	[2/3,∞)	
		{x∈ IR/ -2<x≤2}
		{x∈ IR/ |x|<3}
		{x∈ IR/ |x|≥3}
 2 ∞		
	[-1,1]	
		{x∈ IR/ x<-1}

8. Fill in the next chart:

GRAPHICAL REPRESENTATION	INTERVAL	Math. expression
 -4 4		
	$(-\infty,-2)\cup(2,\infty)$	
	$(-\infty,2)\cup(2,\infty)$	
		$\{x\in IR/\ |x|\le5\}$
	$[-2,2]$	
 -3 3		

9. Work out the and the of the following sets. First of all, draw them:

a) A=[-2,5)
    B=(1,7)

b) C=(-1,3]
    D=(1,6]

c) E=(0,3]
    F=(2,∞)

d) G=(-∞,0]
    H=(-3,∞)

e) I=[-5,-1)
    J=(2,7/2]

f) K=(-∞,0)
    L=[0,∞)

g) M=(2,5)
    N=(5,9]

h) O=[-3,-1)
    P=(2.7]

i) Q=(-3,7)
    R=(2,4]

j) S=[-3,2)
    T=(0,∞)
    U=[1,4]

Fractions

10. Work out:

a) $\dfrac{7}{4}-\left(\dfrac{5}{3}+\dfrac{2}{3}\cdot\dfrac{1}{5}\right)+2$

b) $1-\dfrac{3}{5}\left(\dfrac{2}{3}+\dfrac{1}{2}\right)$

c) $3-\dfrac{4}{5}:2+\dfrac{1}{2}\cdot\left(1-\dfrac{14}{3}\right)$

d) $\dfrac{5}{6}:\left(\dfrac{2}{3}+1\right)-\dfrac{3}{4}\left(\dfrac{2}{3}-1\right)$

e) $\dfrac{\dfrac{7}{5}-\dfrac{3}{4}\cdot\dfrac{2}{5}}{3-\dfrac{1}{4}}$

f) $\dfrac{\left(\dfrac{1}{4}-\dfrac{7}{8}\right):\dfrac{2}{3}+1}{\dfrac{5}{6}\cdot\left(\dfrac{2}{3}-\dfrac{3}{4}\right)}$

- 33 -

11. Work out:

a) $-\left(\dfrac{3}{4}-1\right)-\left(\dfrac{1}{2}-\dfrac{1}{4}+\dfrac{1}{5}-2\right)$ b) $3-\dfrac{1}{4}+\left(-2-\dfrac{1}{2}+\dfrac{3}{5}\right)$ c) $-\left(\dfrac{1}{2}-\dfrac{1}{4}\right)-\left(3-\dfrac{1}{2}+\dfrac{5}{3}-1\right)$

d) $\dfrac{1}{2}+3-\left(-\dfrac{1}{4}+\dfrac{5}{2}-8+1\right)$ e) $-\left(\dfrac{4}{5}-1\right)-\left(2-\dfrac{3}{2}+3-\dfrac{4}{5}\right)$ f) $-\left(-1+\dfrac{3}{7}-3+\dfrac{4}{3}\right)-\left(-3+\dfrac{1}{2}\right)$

g) $\dfrac{1-\dfrac{3}{4}+\dfrac{1}{2}}{\dfrac{2}{3}+3}$ h) $\dfrac{\left(1-\dfrac{4}{5}\right)-(-2+1)}{\dfrac{4}{5}-3+\dfrac{1}{2}}$ i) $\dfrac{29}{7}-\left(2-\dfrac{4}{5}\right):\left(\dfrac{3}{5}+\dfrac{1}{2}-\dfrac{3}{4}\right)$

Scientific notation

12. Calculate, expressing the result as scientific notation, and check your answer by using your calculator:

a) $(2,8\cdot10^{-5}):(6,2\cdot10^{-12})$ b) $(7,2\cdot10^{-6})^3:(5,3\cdot10^{-9})$

c) $7,86\cdot10^5 - 1,4\cdot10^6 + 5,2\cdot10^4$ d) $(3\cdot10^{-10} + 7\cdot10^{-9}):(7\cdot10^6 - 5\cdot10^5)$

13. Calculate, expressing the result as scientific notation, and check your answer by using your calculator:

a) $2,5\cdot10^7+3,6\cdot10^7$ b) $4,6\cdot10^{-8}+5,4\cdot10^{-8}$ c) $1,5\cdot10^6+2,4\cdot10^5$

d) $2,3\cdot10^9+3,25\cdot10^{12}$ e) $3,2\cdot10^8-1,1\cdot10^8$ f) $4,25\cdot10^7-2,14\cdot10^5$

g) $7,28\cdot10^{-3}-5,12\cdot10^{-3}$ h) $(2\cdot10^9)\cdot(3,5\cdot10^7)$ i) $(2\cdot10^5)^2$

j) $(1,4\cdot10^{15} + 2,13\cdot10^{18})\cdot2\cdot10^{-5}$ k) $2,23\cdot10^{-3} + 3\cdot10^{-4} - 5\cdot10^{-5}$

l) $(0,55\cdot10^{23} - 5\cdot10^{21})\cdot2\cdot10^{-13}$

14. Calculate the volume (m^3) of the Earth, if its average radius is 6378 km, giving the result with only two digits.

15. One type of human cell having a cylinder shape, has a diameter of about 7 millionths of a meter and about 2 millionths of a meter in height. Calculate its volume in scientific notation.

16. With a laboratory instrument, the weight of one hundred rice seeds has been determined and it is 0,0000277 kg. How many of these seeds are there in 1000 ton of rice?

17. The light from the Sun takes 8 minutes and 20 seconds to come to the Earth. Calculate the distance Sun-Earth.

Powers and roots

18. Simplify:

a) $\dfrac{\sqrt{12}}{\sqrt{3}}$
b) $\dfrac{\sqrt[3]{4}}{\sqrt{2}}$
c) $\sqrt[4]{\dfrac{5}{12}}:\sqrt[4]{\dfrac{20}{3}}$
d) $\dfrac{\sqrt[4]{a^2}}{\sqrt[4]{a}}$
e) $\sqrt{\dfrac{3}{2}}:\sqrt{\dfrac{2}{3}}$
f) $\dfrac{\sqrt[6]{20}}{\sqrt[4]{10}}$

g) $\sqrt[3]{2^2}\cdot\sqrt[4]{2}$
h) $\sqrt[4]{a^3}\cdot\sqrt[6]{a^5}$
i) $\dfrac{\sqrt[8]{8}}{\sqrt[4]{2}\cdot\sqrt{2}}$

19. Extract all possible factors from the following roots:

a) $\sqrt{900}$
b) $\sqrt[3]{64x^9y^4}$
c) $\sqrt[4]{8^2x^{10}y^6z^8}$

20. Introduce factors in these roots:

a) $3\sqrt[3]{5}$
b) $2^2\sqrt[4]{3}$
c) $\dfrac{3}{5}\sqrt[5]{\dfrac{5}{3}}$

21. Calculate:

a) $\sqrt[4]{x^2y^2}\cdot\sqrt{xz}\cdot\sqrt{yz}$
b) $\left(2+\sqrt5\right)\left(3-\sqrt5\right)$
c) $2a\sqrt{a}\cdot ab^2\sqrt[3]{b^2}$
d) $\dfrac{1}{3}\sqrt{\dfrac{a}{b}}\cdot 6\sqrt[3]{ab^2}\cdot\sqrt[4]{\dfrac{a^2}{b}}$

e) $4\sqrt{72}:\sqrt{8}$
f) $\sqrt[4]{xy^2z^3}:\sqrt[6]{x^2z^3}$
g) $\dfrac{\sqrt[3]{ab^2c^2}}{\sqrt[4]{a^2bc}}$
h) $\dfrac{\sqrt{2a}}{\sqrt[3]{a}}$

i) $28\sqrt{x^4y^3}:7\sqrt{x^3y}$
j) $\dfrac{\sqrt[4]{x^3y^2z^3w}}{\sqrt[3]{x^2y^2zv^2}}$

22. Calculate:

a) $\sqrt{63}-\dfrac{5}{2}\sqrt{28}+\sqrt{112}$
b) $\sqrt3+3\sqrt3-5\sqrt3$
c) $2\sqrt8+4\sqrt{72}-7\sqrt{18}$

d) $3\sqrt2+4\sqrt8-\sqrt{32}+\sqrt{50}$
e) $5\sqrt{12}+\sqrt{27}-8\sqrt{75}+\sqrt{48}$
f) $\sqrt2+\dfrac{3\sqrt2}{4}-\dfrac{5\sqrt2}{3}$

23. Calculate:

a) $\sqrt{320}+\sqrt{80}-\sqrt{500}$
b) $\sqrt{125}+\sqrt{54}-\sqrt{45}$
c) $\sqrt[3]{40}+\sqrt[3]{135}-\sqrt[3]{5}$

24. Calculate:

a) $\dfrac{2\sqrt{3}-5}{\sqrt{3}-2}$
b) $\dfrac{1+\sqrt{3}}{1-\sqrt{3}}$
c) $\dfrac{\sqrt{5}+2\sqrt{3}}{2\sqrt{5}-\sqrt{3}}$
d) $\dfrac{3\sqrt{2}-4}{3\sqrt{2}+4}$

e) $\dfrac{2\sqrt{8}-3\sqrt{2}}{2\sqrt{8}+3\sqrt{2}}$
f) $\dfrac{\sqrt{5}+\sqrt{3}}{\sqrt{5}-\sqrt{3}}$
g) $\dfrac{3\sqrt{5}-4}{\sqrt{5}-2}$
h) $\dfrac{24-13\sqrt{3}}{2\sqrt{3}-3}$

i) $\dfrac{2\sqrt{2}}{\sqrt{3}-\sqrt{2}}$
j) $\dfrac{4-\sqrt{6}}{\sqrt{6}-2}$
k) $\dfrac{2-\sqrt{8}}{2+\sqrt{2}}$
l) $\dfrac{-\sqrt{3}-1}{1-\sqrt{3}}$

m) $\dfrac{9+4\sqrt{3}}{3\left(4-\sqrt{3}\right)}$
n) $\dfrac{\sqrt{2}+4}{2-\sqrt{2}}$
o) $\sqrt{x}+\dfrac{x}{2\sqrt{x}}$

Logarithms

25. Use the definition of logarithm to calculate:

a) $\log_2(4)$
b) $\log_2(\frac{1}{2})$
c) $\log_2(\sqrt{2})$
d) $\log_3(\frac{1}{27})$
e) $\log_3(81)$

f) $\log_3(\sqrt{27})$
g) $\log_3(\sqrt{27})$
h) $\log_3(\sqrt{3})$
i) $\log(1)$
j) $\log(\frac{1}{1000})$

k) $\log(0,1)$
l) $\log(0,01)$
m) $\log(\sqrt{10})$

26. Write as an only logarithm:

a) $\dfrac{3}{2}\log a+\dfrac{5}{2}\log b$
b) $\log a+\dfrac{1}{2}\log b-2\log c$
c) $\log(a+b)+\log(a-b)$

d) $\dfrac{1}{2}\log x-\dfrac{1}{3}\log y+\dfrac{1}{4}\log z$
e) $\log(a-b)-\log 3$
f) $\dfrac{p}{n}\log a+\dfrac{q}{n}\log b$

g) $\log a-4\log b+\dfrac{1}{5}(\log c-2\log d)$

27. Knowing that $\log 2 = 0.3$; $\log 3 = 0.47$; $\log 5 = 0.69$ and $\log 7 = 0.84$, calculate:

a) $\log 4$
b) $\log 6$
c) $\log 27$
d) $\log 14$
e) $\log\sqrt{2}$
f) $\log\sqrt[3]{15}$

h) $\log\dfrac{2}{3}$
i) $\log 3,5$
j) $3\log\dfrac{2}{5}-4\log\dfrac{1}{7}$
k) $\log 18-\log 16$

Unit 2.- Progressions

This unit starts with a review of *arithmetic* and *geometric progressions* you have already studied in previous years (points 1 to 3). After that, we will study the *limits of the progressions*.

1. Progressions of real numbers. General term

In the following drawings, observe the process carrying to the creation of new black triangles. Following the process, how many black triangles will Figure 5 have? And, in general, Figure *n*?

Figure	Number of triangles
1	$1 = 3^0$
2	$3 = 3^1$
3	$9 = 3^2$
4	$27 = 3^3$

In figure n, the number of black triangles is 3^{n-1}. So, for example, using this expression, we can calculate the number of black triangles that figure 10 will have: $3^{(10-1)} = 3^9 = 19683$.

The ordinated number sequence we have obtained, 1, 3, 9, 27, 81, 243, ... is named a *progression of real numbers* and the expression 3^{n-1} is known as the **general term** of the progression.

> ➤ A *real numbers progression* $a_1, a_2, a_3, a_4, \ldots, a_n, \ldots$ is an ordinated sequence of infinite real numbers.
> ➤ Each number in the sequence is named **term**.
> ➤ The numbers 1, 2, 3, ... are named **index**, and indicate the position of a term in the sequence.
> ➤ The term a_n is named **n-th term** or **general rule** or **formula**, and it is the expression that allows to know the value of a term of the sequence by only knowing its position in the sequence.

Exercises

1. Calculate $a_2, a_5, a_{40}, a_{n+1}, a_{2n}$ of the following progressions defined by:

   a) $a_n = 1 - 2n$ b) $b_n = \dfrac{3n+1}{4n}$ c) $c_n = 1 - \dfrac{2n}{3}$

   d) $d_n = (-1)^{n+1} \cdot \dfrac{n-2}{n+1}$ e) $e_n = \sqrt{1+4n}$

2. Has the progression defined by $a_n = n^2 - 16$ any term whose value is 33?, 0 ?, (-12) ?, 8 ?, (-16)?

3. Write the first four terms of the progressions:

   a) $a_n = \dfrac{2n+3}{3 \cdot 2^n}$ b) $b_n = (-1)^n \cdot (n+1)^2$ c) $c_n = \left(1 + \dfrac{1}{n}\right)^n$

There are sequences that do not have a general term, and each term is calculated from the previous ones. They are named ***recurring sequences***.

Example: The sequence 1, 1, 2, 3, 5, 8, ... is defined by the recurrence rule $a_n = a_{n-1} + a_{n-2}$.

Exercises

4. Form a recurring sequence with the rule $j_1 = 2$; $j_2 = 3$; $j_n = j_{n-2} + j_{n-1}$

5. Build a recurring sequence with the rule $a_1 = 5$; $a_n = a_{n-1} + n$.

6. Add one new term to each sequence and find out the recurring rule for the following sequences:

a) 1, –4, 5, –9, 14, –23, ... (Difference)

b) 1, 2, 3, 6, 11, 20, ... (Relation of each term with the three previous)

c) 1; 2; 1,5; 1,75; ... (Semi-addition)   d) 1, 2, 2, 1, 1/2, 1/2, 1, ... (Quotient)

7. What is the seventh term of the following recurring progressions:

a) $a_1 = 1$, $a_2 = 1$, $a_n = a_{n-1} + a_{n-2}$ b) $a_n = 2.a_{n-1} - 3.a_{n-2} + 1$, $a_1 = 0$, $a_2 = 1$

8. Calculate the general term of the following progressions:

a) 1, 2, 4, 8, 16, 32, 64, ...
b) 3, 6, 9, 12, 15, ...
c) 5, 7/2, 3, 11/4, 13/5, 5/2, ...
d) 0, 3, 0, 3, 0, 3, ...
e) 0, 2, 6, 12, 20, 30, 42, 56, ...
f) 2, 1, 1/2, 1/4, 1/8, 1/16, ...
g) 12, 7, 2 , -3 , ...
h) 2, -4, 6, -8, 10, ...
i) 2, 7, 2, 7, 2, 7, ...
j) 1/3, 5/4, 7/9, 9/16, ...
k) 2,5,20,17,26,....
l) 20,10,5,2'5, 1'25,....
m) 1/2, 4/3, 9/4, 16/5, 25/6, ...
n) –9/8, 16/15, -25/24, 36/35, -49/48, 64/63, ...

ñ) 1, -3, 5, -7, 9, -11, 13, -15, ...
o) 1, 8, 27, 64, 125, ...
p) 1, 3, 6, 10, 15, 21, 28, 36, ...
q) 2/3, -4/9, 8/27, -16/81, 32/243, ...
r) 1, 1, 2, 3, 5, 8, 13, 21, 34, ...
s) 5, -5, 5, -5, 5, ...
t) 2, 9, 28, 65, 126, 217, ...
u) –3, 3, -3, 3, -3, 3, ...
v) 1/2, 2/3, 3/4, 4/5, ...
w) 2, 3'5, 5, 6'5, 8, ...
x) 1/2, -5/4, 9/8, -13/16, 17/32,....
y) –2/5, 5/7, -8/9, 1, -14/13, ...
z) ¼,1,9/12,1,25/28,1,......

2. Arithmetic sequences

Look at the following sequence: 3, 8, 13, 18, 23, 28, 33, 38, ...

As you can see, each term is obtained by adding 5 to the previous one.

An *arithmetic sequence* is made by adding a constant quantity to the previous term. This constant quantity is named *difference*.

So, an arithmetic sequence is perfectly defined by its *first term (a_1)* and its *difference (d)*.

Example: An arithmetic sequence is, for example, 5, 8, 11, 14, 17, 20, ...
Notice that $a_1 = 5$ and $d = 3$.

Now, we are going to find out the general term of this arithmetic sequence.

$$a_1 = 5$$
$$a_2 = 5 + 3 = a_1 + d$$
$$a_3 = 5 + 6 = 5 + 3 \cdot 2 = a_1 + d \cdot 2$$
$$a_4 = 5 + 9 = 5 + 3 \cdot 3 = a_1 + d \cdot 3$$

$$a_n = a_1 + d \cdot (n - 1)$$

The general term of an *arithmetic sequence* is $a_n = a_1 + d \cdot (n - 1)$

9. Find the general term and the a_{28} for the following arithmetic sequences:

a) 2, 5, 8, 11, …..;   b) 10, 5, 0, -5, -10, -15, …..

c) 120, 140, 160, 180, …   d) 9, 7, 5, 3, 1, -1, -3, …..

10. Calculate a_{31} for the previous sequences.

11. Write the five first terms and the general term of the following arithmetic sequences:

a) Natural numbers   b) Natural odd numbers   c) Natural even numbers.

12. Write the general term of the following sequences:

a) 3, 7, 11, 15, …   b) 6, 4, 2, 0, -2, …

13. Write the five first terms and the general term of the following arithmetic sequences:

a) $a_1 = 1.5$; $d = 2$ b) $a_1 = 32$; $d = -5$

c) $a_1 = 5$; $d = 0.5$ d) $a_1 = -3$; $d = -4$

14. Write the general term and calculate a_{50} for:

a) 25, 18, 11, 4, …   b) $-13, -11, -9, -7, …$

c) 1,4; 1,9; 2,4; 2,9; …   d) $-3, -8, -13, -18, …$

15. Find out the first term and the difference of the following arithmetic sequences:

a) $a_2 = 18$; $a_7 = -17$ b) $a_4 = 15$; $a_{12} = 39$

c) Even numbers   d) Multiples of 3.

16. What is the position of a term whose value is 56 in the arithmetic sequence given by $a_1 = 8$ and $d = 3$?

17. John is going to run a competition. He starts his training running for 3 km, and every day he is increasing his distance in 1.5 km. How many days must he continue his training to run 21 km?

18. If two terms of an arithmetic sequence are $s_1 = 6$ and $s_3 = 9$, find out the value of the difference and write the general term.

Sum of the n first terms of an arithmetic sequence

Imagine one day you start a training jogging for a popular race. This day, you may run for 5 minutes, and the following days your resistance will increase and increase......

Imagine each day you can run for 5 minutes more than previous. The minutes you run each day are given by an arithmetic sequence, 5, 10, 15, 20, 25,......., whose general term is:

$a_n = 5 + 5(n - 1)$.

But imagine now you want to know the total quantity of minutes you have run during your 7 days training. This is the concept of *sum of the n first terms*.

In the following lines we are going to see a very easy method to calculate it.

You only have to write twice the sequence of terms you want to add. First one, in its natural sense, and second one, in an inverse sense.

$$S_7 = 5 + 10 + 15 + 20 + 25 + 30 + 35$$
$$S_7 = 35 + 30 + 25 + 20 + 15 + 10 + 5$$

Notice that if you add each pair of terms, below and above, you have the same result, 40

$$
\begin{array}{ccccccccccccc}
S_7 = & 5 & + & 10 & + & 15 & + & 20 & + & 25 & + & 30 & + & 35 \\
S_7 = & 35 & + & 30 & + & 25 & + & 20 & + & 15 & + & 10 & + & 5 \\
\hline
2S_7 = & 40 & + & 40 & + & 40 & + & 40 & + & 40 & + & 40 & + & 40
\end{array}
$$

$$2 \cdot S_7 = 7 \cdot (5 + 35) \quad \rightarrow \quad S_7 = \frac{7 \cdot (5 + 35)}{2} \quad \rightarrow \quad S_7 = \frac{7 \cdot (a_1 + a_7)}{2}$$

So, in general:

$$\boxed{S_n = \frac{n \cdot (a_1 + a_n)}{2}}$$

The sum of the n first terms of an arithmetic sequence is $S_n = \dfrac{n \cdot (a_1 + a_n)}{2}$

Example: Calculate the addition of the 500 first even numbers.

<u>Solution:</u>

The sequence of even numbers is 2, 4, 6, 8, 10, ... and it is an arithmetic sequence where $a_1 = 2$ and $d = 2$.

The term $a_{500} = a_1 + (500 - 1)d = 2 + 499 \times 2 = 1.000$

Therefore, the sum of the 500 first values is:

$$S_{500} = \frac{500 \cdot (a_1 + a_{500})}{2} = \frac{500 \cdot (2 + 1000)}{2} = \frac{500 \cdot 1002}{2} = \boxed{250,500}.$$

Exercises

19. Calculate the sum of the first 100 terms of the arithmetic sequence whose $a_1 = 7$ and $d = 5$.

20. Calculate the sum of the first 16 terms of the arithmetic sequence given by
$a_n = 5 + 4 \cdot (n-1)$

21. The dose of a drug was 100 mg the first day, and 5 less each day, during 12 days. How many mg of the drug will the person have taken at the end of his treatment?

22. a) Calculate the sum of all the odd numbers less than 100.

b) If $a_1 = 5$ and $a_2 = 7$, calculate a_{40} and S_{40}.

c) If $b_1 = 5$ and $b_2 = 12$, calculate S_{32}.

23. In a sequence, $c_1 = 17$ and $c_5 = 9$, find out the sum S_{20}.

24. Sum $1 + 2 + 3 + 4 + ... + n$ of the n first naturals *is* 60 378. Calculate the value of n.

3. Geometric sequences

Look at the following sequences:

$$2, \; 6, \; 18, \; 54, \; 162, \ldots;$$

$$2, \; 1, \; 1/2, \; 1/4, \; 1/8, \ldots;$$

$$3, \; -3, \; 3, \; -3, \; 3, \; -3, \ldots$$

Notice that, in all of them, each term has been obtained by multiplying the previous one by a constant number. These sequences are named *geometric sequences*. The constant number is named ratio (r).

A *geometric sequence* is made by multiplying a constant quantity to the previous term. This constant quantity is named *ratio*.

So, a geometric sequence is perfectly defined by its *first term (a_1)* and its *ratio (r)*.

Example: A geometric sequence is, for example, 5, 10, 20, 40, 80, 160, ...
Notice that $a_1 = 5$ and $r = 2$.

Now, we are going to find out the general term of this arithmetic sequence.

$a_1 = 5$

$a_2 = 5 \cdot 2 = a_1 \cdot r$

$a_3 = 5 \cdot 2 \cdot 2 = 5 \cdot 2^2 = a_1 \cdot r^2$

$a_4 = 5 \cdot 2 \cdot 2 \cdot 2 = 5 \cdot 2^3 = a_1 \cdot r^3$

$$a_n = a_1 \cdot r^{n-1}$$

The general term of a *geometric sequence* is $\boxed{a_n = a_1 \cdot r^{n-1}}$

Exercises

25. The two first terms of a geometric sequence are 5 and 10. Write down the 3 following terms and the ratio.

26. In a geometric sequence, s_1=100 and s_3=400. Calculate the 5 first terms.

27. In a bacteria population, the number of bacteria is duplicated every day. If the first day there were 4 bacteria, find out:

a) the general term of the geometric sequence.

b) the number of bacteria existing in one month.

28. Find out the general term of the sequence: 3, 6, 12, 24, 48, ...

29. Write the 5 first terms of the following geometric sequences:

a) $a_1 = 0.3$; $r = 2$ b) $a_1 = -3$; $r = \frac{1}{2}$ c) $a_1 = 200$; $r = -0.1$ d) $a_1 = 1/81$; $r = 3$

30. Find the general term of the following sequences:

    a) 20; 8; 3.2; 1.28; ...   b) 40, 20, 10, 5, ...
    c) 6; -9; 13.5; -20.25; ...   d) 0.48; 4.8; 48; 480; ...

31. Find the general term of the following sequences:

    a) 20; 8; 3.2; 1.28; ...   b) 40, 20, 10, 5, ...

    c) 6; -9; 13.5; -20.25; ...   d) 0.48; 4.8; 48; 480; ...

32. Find the general term of the following sequences:

    a) $a_1 = 1/81$; $a_3 = 1/9$ b) $a_2 = 0.6$; $a_4 = 2.4$

    c) $a_3 = 3$; $r = 1/10$ d) $a_4 = 20.25$; $r = -1.5$

33. Find the general term of the following sequences:

    a) 20; 8; 3.2; 1.28; ...   b) 40, 20, 10, 5, ...

    c) 6; -9; 13.5; -20.25; ...   d) 0.48; 4.8; 48; 480; ...

34. Find the general term of the following sequences:

    a) $a_1 = 1/81$; $a_3 = 1/9$ b) $a_2 = 0.6$; $a_4 = 2.4$

    c) $a_3 = 3$; $r = 1/10$ d) $a_4 = 20.25$; $r = -1.5$

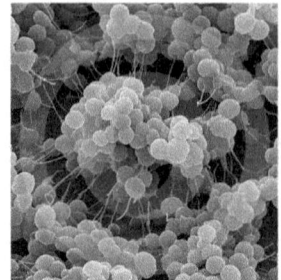

35. Certain kind of bacteria reproduces by bipartition each quarter of an hour. How many bacteria will be produced by each, after 6 hours?

36. Build a geometric sequence whose first term is 125 and ratio = 0.4.

37. In a geometric sequence, $a_1 = 0.625$ and $a_3 = 0.9$. Calculate r and the six first terms.

Sum of the n first terms of a geometric sequence

We are going to name $a_1, a_2, a_3, ..., a_n, ...$ a geometric sequence with ratio r and we are going to calculate the sum of its n first terms.

$$S_n = a_1 + a_2 + a_3 + \ldots\ldots + a_{n-2} + a_{n-1} + a_n$$

Now, we are going to multiply S_n by r:

$$r \cdot S_n = r \cdot a_1 + r \cdot a_2 + r \cdot a_3 + \ldots\ldots + r \cdot a_{n-2} + r \cdot a_{n-1} + r \cdot a_n =$$
$$= a_2 + a_3 + a_4 + \ldots\ldots + a_{n-1} + a_n + r \cdot a_n$$

And, now, we are going to calculate the difference $r \cdot S_n - S_n$:

$r\,S_n$	=				a_2	+	a_3	+	a_4	+	$\ldots$	+	a_{n-1}	+	a_n	+	$a_n r$
S_n	=		a_1	+	a_2	+	a_3	+	a_4	+	$\ldots$	+	a_{n-1}	+	a_n		
$r\,S_n - S_n$	=	$-$	a_1													+	$a_n r$

$$r \cdot S_n - S_n = r \cdot a_n - a_1 = r \cdot a_1 \cdot r^{n-1} - a_1 = a_1 \cdot r^n - a_1$$

$$(r - 1) \cdot S_n = a_1 \cdot r^n - a_1$$

$$S_n = \frac{a_1 \cdot r^n - a_1}{r - 1} = \frac{a_1 \cdot (r^n - 1)}{r - 1} = \boxed{a_1 \cdot \frac{r^n - 1}{r - 1}}$$

- 46 -

The sum of the n first terms of a geometric sequence is $a_1 \cdot \dfrac{r^n - 1}{r - 1}$

Example: Knowing that in a geometric sequence, the terms $a_1 = 3$ and $a_5 = 1875$, calculate the sum of the 5 first terms.

Solution:

First of all, we will calculate the ratio. As $a_5 = a_1 \cdot r^4$, $1875 = 3 \cdot r^4$, $r^4 = 625$; $r = 5$.

The 5 first terms are 3, 15, 75, 375, 1.875, and their sum is 2,343.

Using the expression, $S_n = \dfrac{a_1 \cdot r^n - a_1}{r - 1}$ $\rightarrow$ $S_5 = \dfrac{3 \cdot 5^5 - 3}{5 - 1} = \boxed{2,343}$.

Exercises	**38.** Calculate the sum of the ten first terms of the following geometric sequences: a) $a_1 = 5$; $r = 1.2$ $\quad\quad\quad$ b) $a_1 = 5$; $r = -2$ **39.** Calculate the sum of the 5 first terms of a geometric sequence in which $a_1 = 1\ 000$ and $a_4 = 8$. **40.** Using the given expression, calculate $3 + 6 + 12 + 24 + 48 + 96 + 192 + 384$. **41.** Calculate the sum of the ten first terms of a geometric sequence in which $a_1 = 8.192$ y $r = 2.5$. **42.** Using the given expression, calculate the sum $1 + 3 + 9 + 27 + \ldots + 3^7$.

Sum of the infinite terms of a geometric sequence with /r/<1

In gometric sequences with $/r/ < 1$, the absolute value of their terms is decreasing. And, moreover, when n becomes relatively large, their values are almost zero.

Example: Calculate the four first terms of the geometric sequence $a_n = 10 \cdot (0.5)^n$ and notice that the absolute value of its terms is decreasing.

<u>Solution:</u> a_n = 5; 2.5; 1.25; 0.625, ...

As its values are decreasing, till having values very closed to zero, these last terms are insignificant when calculating the sum. So, in these cases, it is useful to calculate the sum of the infinite values, and it is used the expression:

$$S_n = \frac{a_1}{1-r}$$

Example: In a geometric sequence with a_1 = 25 and r = 0.5, calculate the sum of its infinite

terms.   <u>Solution:</u> $S_n = \dfrac{a_1}{1-r} = \dfrac{25}{1-0.5} = \dfrac{25}{0.5} = \boxed{50}$.

Exercises

43. In a geometric sequence with a_1 = 8 and r = 0.75, calculate the sum of its infinite terms.

44. In a geometric sequence with a_1 = 30 and r = -0.2, calculate the sum of its infinite terms.

45. In a geometric sequence with a_4 = 10 and a_6 = 0.4, calculate:

a) The ratio, r   b) The first term, a_1 c) The 8th term, a_8.

d) The sum of its 8 first terms, S_8 e) The sum of its infinite terms.

---

4. Limit of a progression

A progression can be graphically represented by calculating its terms and representing them on a coordinated axes. These values will be represented as isolated points.

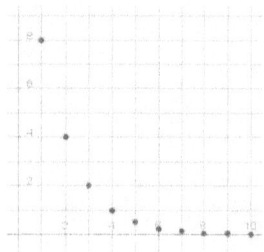

> The limit of a progression is the value to which terms are getting closer and closer, when "n" is taking increasing values.

- If the terms are getting closer to a number, *l*, we say ***lim a_n = l***. And it is read: ***"The limit of a_n is l".***

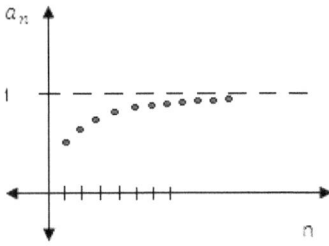

- If it increases and its values are greater than whatever number, we say ***lim a_n = +∞***. And it is read: ***"The limit of a_n is +∞ ".***

- If it decreases and its values are lower than whatever number, we say ***lim a_n = -∞***. And it is read: ***"The limit of a_n is -∞ ".***

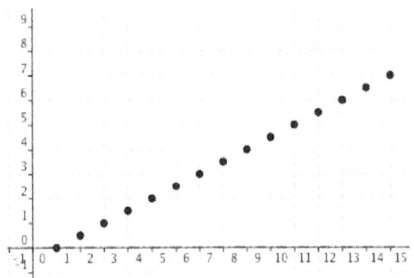
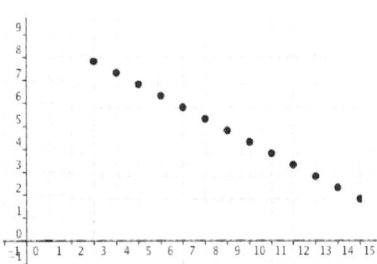

- There are some other progressions whose behaviour in not any of the previous. They are said *not to have limit* and they are named *oscillating progressions*.

46. Represent the 7 first terms of the following progressions and calculate its limit if it exists:

a) $a_n = n^3 - 9n^2 + 18n$

b) $a_n = \dfrac{n^2 - 5}{2}$

c) $a_n = -3n + 10$

d) $a_1 = -1, a_{n+1} = -2.a_n + 3$

e) $a_n = (-1)^n + 2$

f) $a_{n+1} = (-1)^n.a_n, a_3 = 5$

g) $a_n = n.(n-6)$

h) $1/2, 2/3, 3/4, 4/5, \dots$

i) $b_n = \dfrac{2^n + 1}{2^{n-1}}$

Convergent and divergent progressions

➤ When the limit of a_n is a number, *l*, this progression is said to be a ***convergent*** progression.

➤ When the limit of a_n is $+\infty$ *or* $-\infty$, this progression is said to be a ***divergent*** progression.

➤ When a progression has no limit, it is said to be an ***oscillating*** progression.

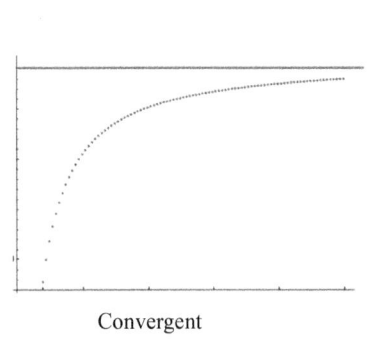

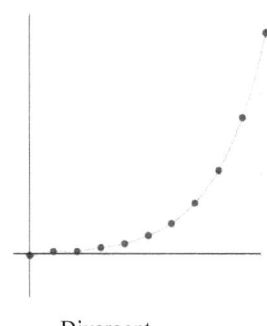

Convergent   Divergent

Example: Calculate the limits of the following progressions and indicate if they are convergent or divergent progressions: a) $a_n = 3n + 2$ b) $b_n = 4 + n^2$ c) $c_n = 5 - 2n^3$

Solution:

a) n is a very big number. When multiplied by 3, it becames bigger. So, we have:

$\lim a_n = \infty + 2 = \infty$ → a_n is a divergent progression.

b) n is a very big number. Its square, n^2, is bigger. So, we have:

$\lim b_n = 4 + \infty = \infty$ → b_n is a divergent progression.

c) n is a very big number. Its cube, n^3, is bigger. So, we have:

$\lim c_n = 5 - \infty = -\infty$ → b_n is a divergent progression.

5. Indeterminate form of a limit

Imagine you are calculating the limit of the progression $a_n = n^3 - n$.

You would say: *$\lim a_n = \infty - \infty$ What is the limit in this case?*

Or imagine you are calculating the limit of the progression $a_n = \dfrac{n^2 + 2}{n^5 + 1}$.

You would say: $lim\ a_n = \dfrac{\infty}{\infty}$ *What is the limit in this case?*

This type of limits are named ***indeterminate form of a limit***, but we can solve them by using the following rules.

5.1. Indeterminate form $\dfrac{\infty}{\infty}$

We name $\dfrac{P(n)}{Q(n)}$

> ➤ If degree of P(n) > degree of Q(n), the limit of a_n is ∞ *(+∞ or -∞,* depending on the signs).
> ➤ If degree of Q(n) > degree of P(n), the limit of a_n is 0.
> ➤ If degree of P(n) = degree of Q(n), the limit of a_n is the quotient of the coefficients of the monomials of higher degrees in numerator and denominator.

Example: Calculate the limits of the progressions $a_n = \dfrac{n^2 + 2}{n^5 + 1}$, $b_n = \dfrac{n^2 - 4}{n + 1}$ and $c_n = \dfrac{3n^2 + 5n - 2}{2n^2 - 4n + 1}$

Solution:

- As degree of numerator < degree of denominator, the limit of a_n is *0*.

- As degree of numerator > degree of denominator, the limit of b_n is $+\infty$.

- As degree of numerator = degree of denominator, the limit of c_n is $\dfrac{3}{2}$.

Exercises

47. Calculate the limit of the following progressions:

a) $a_n = \dfrac{1-2n}{n}$ b) $a_n = \dfrac{n+3}{n-1}$ c) $a_n = \dfrac{1+n}{1-2n}$ d) $a_n = \dfrac{n-2}{n+4}$

e) $a_n = \dfrac{3+8n}{2n+5}$ f) $a_n = \dfrac{3n+5}{12n+1}$ g) $a_n = \dfrac{6n-1}{3n+3}$ h) $a_n = \dfrac{3-2n}{n-2}$

48. Calculate the limit of the following progressions:

a) $\lim\limits_{n\to\infty} \dfrac{\sqrt{n+3}-\sqrt{n-3}}{2}$ b) $\lim\limits_{n\to\infty} \dfrac{n^2-2n+1}{n-3}$ c) $\lim\limits_{n\to\infty} \dfrac{3n+2}{n^2-1}$

d) $\lim\limits_{n\to\infty} \dfrac{n-4}{n+2}$ e) $\lim\limits_{n\to\infty} \dfrac{3n^2+5}{6n^2-2}$ f) $\lim\limits_{n\to\infty} \dfrac{(n-2)^2}{n^2+1}$

49. Calculate the limit of the following progressions:

a) $\lim\limits_{n\to\infty} \dfrac{(n^2+2)(n^2-2)}{(n+2)^2(2n-1)^2}$ b) $\lim\limits_{n\to\infty} \dfrac{3n^2-n+1}{n+5}$ c) $\lim\limits_{n\to\infty} \dfrac{-n^3+2n-1}{n^2+n+2}$

50. Calculate the limit of the following progressions:

a) $\lim\limits_{n\to\infty} \dfrac{n^2+2n-1}{\sqrt{n^2-n+4}}$ b) $\lim\limits_{n\to\infty} \dfrac{3n^5+2n^3-1}{\sqrt{n^6-n^4+4}}$ c) $\lim\limits_{n\to\infty} \dfrac{3n^2-5n}{\sqrt{4n^4-2n+3}}$

d) $\lim\limits_{n\to\infty} \sqrt{\dfrac{n^4-2n-3}{2n^5-1}}$ e) $\lim\limits_{n\to\infty} \dfrac{-n^4+3}{\sqrt{3n^4-2n}}$ f) $\lim\limits_{n\to\infty} \dfrac{5n^2-n+2}{\sqrt{7n^6+3n^3}}$

g) $\lim\limits_{n\to\infty} \dfrac{-n^3+3n}{\sqrt{9n^4+4}}$ h) $\lim\limits_{n\to\infty} \dfrac{7n^3-n+2}{\sqrt{7n^6+3n^3}}$ i) $\lim\limits_{n\to\infty} \dfrac{\sqrt{8n^8-4n^2}}{2n^3-4n^4}$

5.2. Indeterminate form $\infty - \infty$ We name $P(n)-Q(n)$

➢ If degree of $P(n)$ > degree of $Q(n)$, the limit of a_n is $+\infty$ *(The sign is determined by $P(n)$)*.

➢ If degree of $Q(n)$ > degree of $P(n)$, the limit of a_n is $-\infty$ *(Sign is determined by $Q(n)$)*.

➢ If degree of $P(n)$ = degree of $Q(n)$, we will operate in different ways, depending if there are radicals or not:

a) If there are not radicals, we will operate the difference of fractions.

b) If there are radicals, we must multiply and divide the expression by its *conjugated* expression. The conjugated expression of *(a + b)* is *(a – b)*.

Example: Calculate the limits of the following progressions:

a) $a_n = n^2 - \sqrt{n^2 + 1}$

   Solution: $\lim a_n = \infty - \infty$. Degree($n^2$) = 2 and degree($\sqrt{n^2 + 1}$) = 1. So, $\lim a_n = \boxed{+\infty}$.

b) $a_n = n^2 - \sqrt{n^7 - 2}$

   Solution: $\lim a_n = \infty - \infty$. Degree($n^2$) = 2 and degree($\sqrt{n^7 - 2}$) = $\frac{7}{2}$. So, $\lim a_n = \boxed{-\infty}$.

c) $a_n = \dfrac{n^2 + 1}{n} - \dfrac{n^3 - 2}{n^2 - 1}$

   Solution: $\lim a_n = \infty - \infty$. Degree($\dfrac{n^2 + 1}{n}$) = 1 and degree($\dfrac{n^3 - 2}{n^2 - 1}$) = 1.

   As there are not radicals, we must operate the difference of fractions:

   $$\lim \frac{n^2 + 1}{n} - \frac{n^3 - 2}{n^2 - 1} = \lim \frac{n^4 - 1 - n^4 + 2n}{n \cdot (n^2 - 1)} = \lim \frac{-1 + 2n}{n \cdot (n^2 - 1)} = \boxed{0}.$$

d) $a_n = n - \sqrt{n^2 + 1}$

   Solution: $\lim a_n = \infty - \infty$. Degree($n$) = 1 and degree($\sqrt{n^2 + 1}$) = 1. So, we must multiply

   and divide the expression by its conjugated expression, $n + \sqrt{n^2 + 1}$:

   $$\lim a_n = n - \sqrt{n^2 + 1} = \lim \frac{\left(n - \sqrt{n^2 + 1}\right)\left(n + \sqrt{n^2 + 1}\right)}{n + \sqrt{n^2 + 1}}$$

   Remember this identity: $\boxed{(a + b) \cdot (a - b) = a^2 - b^2}$

   $$\lim \frac{\left(n - \sqrt{n^2 + 1}\right)\left(n + \sqrt{n^2 + 1}\right)}{n + \sqrt{n^2 + 1}} = \lim \frac{n^2 - \left(\sqrt{n^2 + 1}\right)^2}{n + \sqrt{n^2 + 1}} = \lim \frac{n^2 - \left(n^2 + 1\right)}{n + \sqrt{n^2 + 1}} = \lim \frac{n^2 - n^2 - 1}{n + \sqrt{n^2 + 1}} =$$

   $$= \lim \frac{-1}{n + \sqrt{n^2 + 1}} = \frac{-1}{\infty} = \boxed{0}$$

51. Calculate the limit of the following progressions:

a) $\lim\limits_{n\to\infty}\left(\dfrac{6n-5}{3n^2-2}-\dfrac{5n-3}{4n^2+1}\right)$

b) $\lim\limits_{n\to\infty}\left(\dfrac{n^2}{2n-1}-\dfrac{n^2+1}{2n+1}\right)$

c) $\lim\limits_{n\to\infty}\left(\dfrac{n^2+1}{3n+1}-\dfrac{n^2}{3n-1}\right)$

d) $\lim\limits_{n\to\infty}\left(\dfrac{n^3+2n^2}{3n^2-n}-\dfrac{2n^3-1}{3n^2-3}\right)$

e) $\lim\limits_{n\to\infty}\left(3n+2-\dfrac{7+3n^2}{n+1}\right)$

f) $\lim\limits_{n\to\infty}\left(2n^2-1-\dfrac{2n^4+5n^3-n^2}{n^2+2}\right)$

52. Calculate the limit of the following progressions:

a) $\lim\limits_{n\to\infty}\sqrt{2n^2+3n-2}-\sqrt{2n^2+2}$

b) $\lim\limits_{n\to\infty}\sqrt{n+a}-\sqrt{n}$

c) $\lim\limits_{n\to\infty}\left(\sqrt{n^2+4n+1}-\sqrt{n^2+8n+1}\right)$

d) $\lim\limits_{n\to\infty}\left(\sqrt{n^2+2n}-\sqrt{n^2+n}\right)$

e) $\lim\limits_{n\to\infty}\left(n-\sqrt{n^2+10n}\right)$

f) $\lim\limits_{n\to\infty}\left(\sqrt{3+4n^2}-\sqrt{n+4n^2}\right)$

g) $\lim\limits_{n\to\infty}\left(\sqrt{4n^2-1}-\sqrt{4n^2+2n}\right)$

h) $\lim\limits_{n\to\infty}\left(\sqrt{2n^2-1}-\sqrt{2n^2+2n}\right)$

i) $\lim\limits_{n\to\infty}\left(\sqrt{n^2-3n+2}-n\right)$

j) $\lim\limits_{n\to\infty}\left(\sqrt{2n^4-1}-2n^2\right)$

k) $\lim\limits_{n\to\infty}\left[\sqrt{n^2-10n+8}-(n-3)\right]$

5.3. Indeterminate form 1^{∞} or "e" limit

When calculating the limit of progression $a_n=\left(\dfrac{n+4}{n}\right)^{n+2}$, we have: $lim\ a_n=1^{\infty}$. This kind of indeterminate form are solved by applying the following rule:

If $lim\ (a_n)^{b_n} = 1^{\infty}$, *its limit is* e^L , *where* $L = lim\ [a_n - 1] b_n$

Example: Calculate the limit of the progression $a_n = \left(\dfrac{n+4}{n}\right)^{n+2}$

Solution: $lim \left(\dfrac{n+4}{n}\right) = 1$ and $lim\ (n+2) = \infty$. So, this limit is 1^{∞} and it is an **"e" limit.**

We must apply the previous expression:

$$L = lim \left(\dfrac{n+4}{n} - 1\right) \bullet (n+2) = lim \left(\dfrac{n+4-n}{n}\right) \bullet (n+2) = lim \left(\dfrac{4}{n}\right) \bullet (n+2) = lim \left(\dfrac{4n+8}{n}\right) = 4$$

As these limits are e^L , the limit of our progression is $\boxed{e^4}$.

---

53. Calculate the limit of the following progressions:

a) $\lim\limits_{n \to \infty} \left(\dfrac{4n-2}{4n-3}\right)^{4n+3}$

b) $\lim\limits_{n \to \infty} \left(\dfrac{n^2+5}{n^2}\right)^{2n^2}$

c) $\lim\limits_{n \to \infty} \left(1 - \dfrac{1}{n}\right)^{n}$

d) $\lim\limits_{n \to \infty} \left(\dfrac{3n^2+1}{3n^2-1}\right)^{\frac{n^2}{n+1}}$

e) $\lim\limits_{n \to \infty} \left(\dfrac{n-1}{n-2}\right)^{n^2-3}$

f) $\lim\limits_{n \to \infty} \left(\dfrac{3n^2+5}{3n^2-n}\right)^{n^2-2}$

g) $\lim\limits_{n \to \infty} \left(\dfrac{4n-3}{4n}\right)^{\frac{2n^2-1}{2n}}$

h) $\lim\limits_{n \to \infty} \left(\dfrac{2n+7}{2n-3}\right)^{\frac{n^4}{n^3+1}}$

i) $\lim\limits_{n \to \infty} \left(\dfrac{3n-2}{3n+2}\right)^{3n}$

Exercises

6. Limits of geometrical progressions a^n (with a ≠ 1)

For example, when calculating the limit of the progression $a_n = \left(\dfrac{6n^2 - n + 5}{2n^2 + 2n + 1}\right)^{n-3}$, we have:

$lim \left(\dfrac{6n^2 - n + 5}{2n^2 + 2n + 1}\right) = 3$ and $lim\ (n-3) = \infty$. So, this limit is 3^{∞} and it is **NOT** an *"e" limit*.

Notice if you multiply $3 \cdot 3 \cdot 3 \cdot 3 \cdot \dots$.. a lot of times, this product has no limit, its limit is ∞.

Another example, when calculating the limit of the progression $a_n = \left(\dfrac{n+5}{2n+1}\right)^{n+1}$, we have:

$lim \left(\dfrac{n+5}{2n+1}\right) = \dfrac{1}{2}$ and $lim\ (n+1) = \infty$. So, this limit is $\left(\dfrac{1}{2}\right)^{\infty}$ and it is **NOT** an *"e" limit*.

Notice if you multiply $\dfrac{1}{2} \cdot \dfrac{1}{2} \cdot \dfrac{1}{2} \cdot \dfrac{1}{2} \cdot$ ..... a lot of times, this product decreases to 0, its limit is 0.

> ➤ If $lim\ a_n = a^{\infty}$, its limit is $\begin{cases} \infty & if\ a > 1 \\ 0 & if\ a < 1 \end{cases}$

> ➤ If $lim\ a_n = a^{-\infty}$, its limit is always 0. *(Remember $a^{-\infty} = 1/a^{\infty} = 1/\infty = 0$)*

Be careful, if $a = 1$, this is a *"e" limit (section 5.3)*.

Exercises

54. Calculate the limit of the following progressions:

a) $\displaystyle\lim_{n\to\infty} \left(\dfrac{2n^2 + 2}{n^2 + n + 1}\right)^{-n^2 - n + 1}$

b) $\displaystyle\lim_{n\to\infty} \left(\dfrac{9n^2 - n + 1}{3n^2 + 2n - 3}\right)^{-n^3 + 2n^2 - n}$

c) $\displaystyle\lim_{n\to\infty} \left(\dfrac{2n^2 - n + 2}{4n^2 + 2n - 3}\right)^{n^3 + 2n^2 - n}$

d) $\displaystyle\lim_{n\to\infty} \left(5 + 2n^3 - 3n\right)^{\frac{4n^2 - 4n^3}{1 + 2n^3}}$

e) $\displaystyle\lim_{n\to\infty} \left(\dfrac{2n^4 - 7n}{5n^4 - 11}\right)^{n - 2n^3 + 1}$

f) $\displaystyle\lim_{n\to\infty} \left(\dfrac{4n^2 - 2n}{3n^2 + 6}\right)^{\frac{-2n^2 + 3}{3n - 1}}$

The next limits you must calculate are mixed:

55. Calculate the limit of the following progressions:

a) $\lim \left[(-1)^n \cdot \dfrac{1}{n+1} \right]$

b) $\lim \left[(-1)^n \cdot (n+1) \right]$

c) $\lim \left[(-1)^n \cdot \dfrac{n}{n+1} \right]$

d) $\lim \left[\sqrt[3]{n^2} + \dfrac{1}{2n+3} \right]$

e) $\lim \dfrac{n^2}{1+\dfrac{1}{n}}$

f) $\lim \left[\dfrac{1}{2n} + \dfrac{1}{3n^2} + 1 \right]$

g) $\lim \left(\dfrac{1}{n} \right)^{-n^2}$

h) $\lim (n^2 - 7n^3 + 19n - 1)$

i) $\lim (n^{10} - n^8 - n^6)$

j) $\lim \dfrac{3n-1}{\dfrac{1}{2n}+1}$

k) $\lim \dfrac{5n^3 - n^2 - n}{2n^3 + 4n - 1}$

l) $\lim \dfrac{-3n^3 - n^2 - n - 1}{-n^2 - n - 1}$

m) $\lim \dfrac{n^2 + 7n + 5}{n^5}$

n) $\lim \dfrac{3n^3 - n^2 - n - 1}{-n^2 - n - 1}$

ñ) $\lim \left(\dfrac{4n^3 + 2n}{5n^3 - 2} \right)^{\frac{2n+1}{n^2}}$

o) $\lim \left(\dfrac{8n^3 - 1}{2n^3 + n + 1} \right)^{\frac{n}{2n-1}}$

p) $\lim \left(\dfrac{n+3}{2n^3 - 1} \right)^{2n}$

q) $\lim \left(\dfrac{2n^2 + 2}{n^2 + n + 1} \right)^{-n^2 - n - 1}$

r) $\lim \left(\dfrac{1}{2n^2 - 1} \right)^{\frac{-n^2}{n+1}}$

s) $\lim \dfrac{5n}{\sqrt{2n^4 - 1}}$

t) $\lim \dfrac{5n^3}{\sqrt{7n^2 - 3} + 1}$

u) $\lim \left[\sqrt{n^2 - 1} - \sqrt{n^2 + 1} \right]$

v) $\lim \left[n - \sqrt{n^2 + 10n} \right]$

w) $\lim \left[\sqrt{4n^2 + 3n + 1} - 2n - 1 \right]$

x) $\lim \dfrac{\sqrt{25n^2 + 1} - \sqrt{9n^2 + 1}}{\sqrt{4n^2 + 1} - 1}$

y) $\lim \left[\sqrt{n} \cdot \left(\sqrt{n+2} - \sqrt{n+1} \right) \right]$

z) $\lim \left(2 + \dfrac{1}{n} \right)^n$

56. Calculate the limit of the following progressions:

1) $\lim \left(1 - \dfrac{1}{n} \right)^{-n}$

2) $\lim \left(1 + \dfrac{2}{5n} \right)^{2n}$

3) $\lim \left(1 + \dfrac{1}{n+5} \right)^{n+1}$

4) $\lim \left(\dfrac{3n-1}{2n+4} - \dfrac{n+1}{n-1} \right)^n$

5) $\lim \left(\dfrac{3n^2 + 1}{3n^2 - 1} \right)^{\frac{n^2}{n+1}}$

6) $\lim \dfrac{\sqrt{n+1}}{n+2}$

7) $\lim \left(\dfrac{n^2}{2n-1} - \dfrac{n^2 + 1}{2n+1} \right)$

8) $\lim \left(\dfrac{2n^2 - n + 1}{2n^2 - 3n + 2} \right)^{-n+1}$

9) $\lim \left(\dfrac{n}{n+5} \right)^{n^2}$

10) $\lim_{n \to \infty} \sqrt{2n^2 + 3n - 2} - \sqrt{2n^2 + 2}$

11) $\lim_{n \to \infty} \dfrac{(n^2 + 2)(n^2 - 2)}{(n+2)^2 (2n-1)^2}$

12) $\lim_{n \to \infty} \left(\dfrac{6n-5}{3n^2 - 2} - \dfrac{5n-3}{4n^2 + 1} \right)$

13) $\lim_{n \to \infty} \left(\dfrac{4n-2}{4n-3} \right)^{4n+3}$

14) $\lim_{n \to \infty} \left(\sqrt{n^2 + 4n + 1} - \sqrt{n^2 + 8n + 1} \right)$

15) $\lim_{n \to \infty} \left(\dfrac{n^2 + 5}{n^2} \right)^{2n^2}$

(Continuation of 56)

16) $\lim_{n\to\infty}\left(\dfrac{3n^2+1}{3n^2-1}\right)^{\frac{n^3}{n+1}}$

17) $\lim_{n\to\infty}\dfrac{3n^2-n+1}{n+5}$

18) $\lim_{n\to\infty}\left(\sqrt{n^2+2n}-\sqrt{n^2+n}\right)$

19) $\lim_{n\to\infty}\left(\dfrac{n^2}{2n-1}-\dfrac{n^2+1}{2n+1}\right)$

20) $\lim_{n\to\infty}\left(\dfrac{2n^2+2}{n^2+n+1}\right)^{-n^2-n+1}$

21) $\lim_{n\to\infty}\left(n-\sqrt{n^2+10n}\right)$

22) $\lim_{n\to\infty}\left(\dfrac{2n^2-2}{2n^2+1}\right)^{\frac{n^3}{n-1}}$

23) $\lim_{n\to\infty}\left(\sqrt{3+4n^2}-\sqrt{n+4n^2}\right)$

24) $\lim_{n\to\infty}\left(\dfrac{n-1}{n-2}\right)^{n^2-3}$

25) $\lim_{n\to\infty}\left(\dfrac{9n^2-n+1}{3n^2+2n-3}\right)^{-n^3+2n^2-n}$

26) $\lim_{n\to\infty}\left(\sqrt{4n^2-1}-\sqrt{4n^2+2n}\right)$

27) $\lim_{n\to\infty}\left(\dfrac{n^2+1}{3n+1}-\dfrac{n^2}{3n-1}\right)$

28) $\lim_{n\to\infty}\dfrac{n^2+2n-1}{\sqrt{n^2-n+4}}$

29) $\lim_{n\to\infty}\left(\dfrac{5n^2-1}{5n^2+1}\right)^{\frac{n^2+1}{n-1}}$

30) $\lim_{n\to\infty}\left(\dfrac{2n^2-n+2}{4n^2+2n-3}\right)^{n^3+2n^2-n}$

31) $\lim_{n\to\infty}\left(\sqrt{2n^2-1}-\sqrt{2n^2+2n}\right)$

32) $\lim_{n\to\infty}\left(\dfrac{3n^2-1}{3n^2+1}\right)^{\frac{n^3+1}{n^2-1}}$

33) $\lim_{n\to\infty}\dfrac{3n^5+2n^3-1}{\sqrt{n^6-n^4+4}}$

34) $\lim_{n\to\infty}\left(\sqrt{n^2-3n+2}-n\right)$

35) $\lim_{n\to\infty}\left(\dfrac{3n^2-2}{3n^2+1}\right)^{\frac{2n^2}{n-1}}$

36) $\lim_{n\to\infty}\left(\sqrt{4n^2-2n+1}-2n\right)$

37) $\lim_{n\to\infty}\left(\dfrac{4n-3}{4n}\right)^{\frac{2n^2-1}{2n}}$

38) $\lim_{n\to\infty}\left(\dfrac{2n^2-3n+6}{4n^2+5}\right)^{5n^2-7n-1}$

39) $\lim_{n\to\infty}\left(\sqrt{25n^2-2n}-\sqrt{16n^2-3}\right)$

40) $\lim_{n\to\infty}\dfrac{3n^2-5n}{\sqrt{4n^4-2n+3}}$

41) $\lim_{n\to\infty}\left(\dfrac{2n+7}{2n-3}\right)^{\frac{n^4}{n^3+1}}$

42) $\lim_{n\to\infty}\left(\dfrac{n^3-2n^2-1}{3n^3+4}\right)^{n^2-2n+1}$

43) $\lim_{n\to\infty}\left(\dfrac{n^3+2n^2}{3n^2-n}-\dfrac{2n^3-1}{3n^2-3}\right)$

44) $\lim_{n\to\infty}\sqrt{\dfrac{n^4-2n-3}{2n^5-1}}$

45) $\lim_{n\to\infty}\dfrac{-n^3+2n-1}{n^2+n+2}$

46) $\lim_{n\to\infty}\left(\dfrac{4n^3-2n+1}{3n^3+n-5}\right)^{-n^2+2n-3}$

47) $\lim_{n\to\infty}\left(\sqrt{2n^2+n+1}-\sqrt{2n^2-3n+3}\right)$

48) $\lim_{n\to\infty}\dfrac{-n^4+3}{\sqrt{3n^4-2n}}$

49) $\lim_{n\to\infty}\left(\dfrac{3n^3-n-1}{3n^3-2n}\right)^{\frac{2n^2-1}{n-2}}$

50) $\lim_{n\to\infty}\left(2n^2-1-\dfrac{2n^4+5n^3-n^2}{n^2+2}\right)$

51) $\lim_{n\to\infty}\left(\sqrt{2n^4-1}-2n^2\right)$

52) $\lim_{n\to\infty}\left(\dfrac{3n^2+5}{3n^2-n}\right)^{n^2-2}$

53) $\lim_{n\to\infty}\dfrac{5n^2-n+2}{\sqrt{7n^6+3n^3}}$

54) $\lim_{n\to\infty}\left(\dfrac{2n^4-7n}{5n^4-11}\right)^{n-2n^3+1}$

55) $\lim_{n\to\infty}\left[\sqrt{n^2-10n+8}-(n-3)\right]$

56) $\lim_{n\to\infty}\left(\dfrac{2n+3}{2n-3}\right)^{2n}$

57) $\lim_{n\to\infty}\left[\sqrt{4n^2-5n+2}-(2n-1)\right]$

58) $\lim_{n\to\infty}\left(\dfrac{3n-2}{3n+2}\right)^{3n}$

59) $\lim_{n\to\infty}\left(\sqrt{3n^4-2n^2}-\sqrt{3n^4-4n^2+1}\right)$

60) $\lim_{n\to\infty}\left(\dfrac{4n^2-2n}{3n^2+6}\right)^{\frac{-2n^2+3}{3n-1}}$

61) $\lim_{n\to\infty}\dfrac{-n^3+3n}{\sqrt{9n^4+4}}$

62) $\lim_{n\to\infty}\left(\dfrac{5n^2+n+1}{5n^2+n}\right)^{\frac{7n^2}{n+1}}$

63) $\lim_{n\to\infty}\left(3n+2-\dfrac{7+3n^2}{n+1}\right)$

Review exercises

1. If general term of a progression is $a_n = \dfrac{n^2 + 10}{n+2}$,

   a) Calculate the second and tenth term.

   b) Is there any term whose value is 5? If so, what is its position in the sequence?

   c) Is there any term whose value is 7? If so, indicate its position in the sequence.

2. If the first term of a progression is $a_1 = 3$ and this progression is $a_{n+1} = a_n + 2$, calculate the second and tenth terms.

3. Find out the general term of the following progressions:

   a) −1, 2, 5, 8, 11,...

   b) 1, −2, 4, −8, 16,...

   c) $\dfrac{1}{2}, 1, \dfrac{3}{2}, 2, \dfrac{5}{2}, \dots$

   d) $\dfrac{1}{2}, \dfrac{1}{4}, \dfrac{1}{8}, \dfrac{1}{16}, \dots$

4. Find out the general term of the following progressions:

   a) 0, 3, 8, 15, 24, . . .

   b) $\dfrac{2}{5}, \dfrac{4}{6}, \dfrac{8}{7}, \dfrac{16}{8}, \dfrac{32}{9}, \dots$

   c) 2, 9, 28, 65, 126, . . .

   d) $-\dfrac{3}{2}, \dfrac{4}{3}, -\dfrac{5}{4}, \dfrac{6}{5}, \dots$

   e) $1, -\dfrac{3}{2}, 2, -\dfrac{5}{2}, 3, \dots$

   f) −2 ; −0,5 ; 1; 2,5 ; 4 ; ...

5. Calculate the sum from term a_{15} to term a_{40} (both being included) for the progression with general term $a_n = 2n - 3$.

6. For an arithmetic progression, it is known that $a_1 = 5$ and $d = 2$. Calculate the sum of its 20 first terms.

7. For a geometric progression, it is known that $a_1 = 2$ and $r = 3$. Calculate the sum of its 12 first terms.

8. Calculate the sum:

   $a_7 + a_8 + \ldots\ldots + a_{30}$, if a_n is an arithmetic progression whose general term is $a_n = 3n + 1$

9. Calculate the sum of all the terms of the progression: $2, \dfrac{2}{3}, \dfrac{2}{9}, \dfrac{2}{27}, \dfrac{2}{81}, \ldots\ldots$

10. First term of an arithmetic progression is 12 and the difference is 4. Calculate the sum:

    $a_6 + a_7 + a_8 + \ldots + a_{25}$

11. For the geometric progression whose $a_1 = 3$ and $r = 2$, calculate the sum of its 20 first terms.

12. Calculate the limit of the following progressions and classify them as *convergent* or *divergent*:

    a) $a_n = \dfrac{3n+1}{n+2}$ b) $b_n = (n+1)^2$ c) $b_n = \dfrac{2n+1}{3}$

    d) $a_n = 2 - n^2$ e) $b_n = 2 + \dfrac{1}{n}$ f) $b_n = (-1)^n$ g) $b_n = -n + 2$

13. Calculate the limit of the following progressions:

    a) $\lim \left(\sqrt{n^2 + 4n} - \sqrt{n^2 + 3} \right)$ b) $\lim \dfrac{3n^2 + 4n}{n^3 + 2n^2 + 1}$ c) $\lim \dfrac{n^2 + 3}{\sqrt{4n^4 + n^2}}$

    d) $\lim \left(\sqrt{n^2 + 1} - n \right)$ e) $\lim \left(\sqrt{n^4 + n} - n^3 \right)$ f) $\lim \left(1 + \dfrac{1}{n+5} \right)^n$

    g) $\lim \left(1 - \dfrac{1}{n} \right)^{n+2}$ h) $\lim \left(\dfrac{2n+3}{4n+5} \right)^n$ i) $\lim \left(\dfrac{2n+3}{2n+5} \right)^{3n}$

    j) $\lim \left(\dfrac{2n+3}{n+1} \right)^n$ k) $\lim \left(\dfrac{2n+3}{1-n} \right)^{\frac{n^2+1}{n}}$

- 61 -

Unit 3.- Algebra

1. Brief review of operations with polynomials

1.1. Sum and subtraction of polynomials

Adding or subtracting polynomials consists simply in adding or subtracting their _like_ monomials.

Method 1: Write one below the other, so that we collocate in the same column the like monomials. After that, we add the like monomials.

$$
\begin{array}{l}
A(x) = \quad\quad x^3 - 2x^2 - 7x - 4 \\
B(x) = -x^4 \quad\quad + 8x^2 + 7x + 2 \\
\hline
A(x) + B(x) = -x^4 + x^3 + 6x^2 \quad\quad - 2
\end{array}
$$

Opposite polynomial of a given polynomial is another one that has the same monomials but with opposite coefficients (opposite signs).

If $B(x) = -x^4 + 8x^2 + 7x + 2$, its opposite polynomial is $-B(x) = x^4 - 8x^2 - 7x - 2$.

Do you agree that $3 - 2 = 3 + (-2)$?

So, in the same way, we can calculate $A(x) - B(x) = A(x) + [-B(x)]$:

$$
\begin{array}{l}
A(x) = \quad\quad x^3 - 2x^2 - 7x - 4 \\
-B(x) = x^4 \quad\quad\ - 8x^2 - 7x - 2 \\
\hline
A(x) - B(x) = x^4 + x^3 - 10x^2 - 14x - 6
\end{array}
$$

Method 2: Another method to add or subtract polynomials is the application of the sign's rule for addition or subtraction of like monomials.

First of all, we remove the parentheses:
- If it is preceded by a + sign, we conserve the original signs.
- If it is preceded by a – sign, we change the sign inside the parenthesis.

We are using this method to calculate again $A(x) + B(x)$ and $A(x) - B(x)$:

$A(x) + B(x) = (x^3 - 2x^2 - 7x - 4) + (-x^4 + 8x^2 + 7x + 2) = x^3 - 2x^2 - 7x - 4 - x^4 + 8x^2 + 7x + 2 =$
$= \boxed{-x^4 + x^3 + 6x^2 - 2}$

$A(x) - B(x) = (x^3 - 2x^2 - 7x - 4) - (-x^4 + 8x^2 + 7x + 2) = x^3 - 2x^2 - 7x - 4 + x^4 - 8x^2 - 7x - 2 =$
$= \boxed{x^4 + x^3 - 10x^2 - 14x - 6}$

Exercises

1. Given polynomials $P(x) = 2x^5 - 3x^4 + 3x^2 - 5$ and $Q(x) = x^5 + 6x^4 - 4x^3 - x + 7$, work out: $P(x) + Q(x)$ and $P(x) - Q(x)$

2. Given $P(x) = 4x^3 + 6x^2 - 2x + 3$, $Q(x) = 2x^3 - x + 7$ and $R(x) = 7x^2 - 2x + 1$, work out:

   a) $P(x) + Q(x) + R(x)$ b) $P(x) - Q(x) - R(x)$ c) $P(x) + 3Q(x) - 2R(x)$

1.2. Product of two polynomials

To **multiply** two polynomials, we must multiply each monomial of the first one by all the monomials of the second one, or vice versa.

For example, given the polynomials $A(x) = x^3 + 2x^2 - x + 2$ and $B(x) = 4x - 2$, we calculate $A(x) \times B(x)$:

$$A(x) = \quad x^3 + 2x^2 - x + 2$$
$$\times \quad B(x) = \qquad\qquad 4x - 2$$
$$\overline{\qquad\qquad\qquad -2x^3 - 4x^2 + 2x - 4}$$
$$4x^4 + 8x^3 - 4x^2 + 8x$$
$$\overline{A(x) \cdot B(x) = 4x^4 + 6x^3 - 8x^2 + 10x - 4}$$

As we have already said, we can multiply two polynomials by applying the distributive property. We mean:

$A(x) \times B(x) = (x^3 + 2x^2 - x + 2) \times (4x - 2) = x^3(4x - 2) + 2x^2(4x - 2) - x(4x - 2) + 2(4x - 2) =$

$\qquad = 4x^4 - 2x^3 + 8x^3 - 4x^2 - 4x^2 + 2x + 8x - 4 = 4x^4 + 6x^3 - 8x^2 + 10x - 4$

---

3. Work out the products of the following polynomials:

a) $(3x^2+5x-6)\,(8x^2-3x+4)$

b) $(5x^3-4x^2+x-2)\,(x^3-7x^2+3)$

c) $(2x^4-3x^2+5x)\,(3x^5-2x^3+x-2)$

d) $(ab^2+a^2b+ab)\,(ab-ab^2)$

e) $(-x^6+x^5-2x^3+7)\,(x^2-x+1)$

f) $(x^2y^2-2xy)\,(2xy+4)$

g) $(x^2-4x+3/2)\,(x+2)$

Exercises

1.3. Remarkable identities. Special products

When working on polynomials, there are three identities that are going to be very useful for you to save time. They are named *polynomial identities*, *remarkable identities* or *special products*. They are:

$$(a+b)^2 = a^2 + 2ab + b^2 \qquad (a - b)^2 = a^2 - 2ab + b^2 \qquad (a + b)\cdot(a - b) = a^2 - b^2$$

Example: Work out the following polynomial identities:

a) $(x + 5)^2$
b) $(x - 6)^2$
c) $(x+2)\cdot(x-2)$

Solution:

a) $(x + 5)^2 = x^2 + 2\cdot x\cdot 5 + 5^2 = x^2 + 10x + 25$

b) $(3x - 4)^2 = (3x)^2 - 2\cdot 3x\cdot 4 + 4^2 = 9x^2 - 24x + 16$

c) $(x + 2)\cdot(x - 2) = x^2 - 2^2 = x^2 - 4$.

4. Work out the following products:

a) $(2x^2+3x)^2$ b) $(2x^2-3)^2$ c) $(-x-3)^2$ d) $\left(x+\dfrac{1}{2}\right)^2$

e) $\left(2a-\dfrac{3}{2}\right)^2$ f) $\left(1+\dfrac{x}{2}\right)\left(1-\dfrac{x}{2}\right)$ g) $\left(2x+\dfrac{3}{4}\right)^2$ h) $\left(\dfrac{3}{2}-\dfrac{x}{4}\right)^2$

i) $\left(2+\dfrac{a}{3}\right)\left(-\dfrac{a}{3}+2\right)$ j) $\left(\dfrac{3x}{2}-\dfrac{1}{x}\right)^2$ k) $\left(\dfrac{x^2}{2}-\dfrac{x}{3}\right)\left(\dfrac{x^2}{2}+\dfrac{x}{3}\right)$ l) $\left(\dfrac{3}{2}x+\dfrac{1}{4}\right)^2$

5. Work out and simplify:

a) $(x+1)^2+(x-2)(x+2)$ b) $(3x-1)^2-(2x+5)(2x-5)$ c) $(2x+3)(-3+2x)-(x+1)^2$

d) $(-x+2)^2-(2x+1)^2-(x+1)(x-1)$ e) $-3x+x(2x-5)(2x+5)-(1-x^2)^2$

f) $(3x-1)^2-(-5x^2-3x)^2-(-x+2x^2)(2x^2+x)$

6. Work out and simplify:

a) $(2y + x)(2y - x) + (x + y)^2 - x(y + 3)$ b) $3x(x + y) - (x - y)^2 + (3x + y)y$

c) $(2y + x + 1)(x - 2y) - (x + 2y)(x - 2y)$

7. Work out and simplify:

a) $\dfrac{3x(x + 5)}{5} - \dfrac{(2x + 1)^2}{4} + \dfrac{(x - 4)(x + 4)}{2}$

b) $\dfrac{(8x^2 - 1)(x^2 + 2)}{10} - \dfrac{(3x^2 + 2)^2}{15} + \dfrac{(2x + 3)(2x - 3)}{6}$

1.4. Quotient of two polynomials

Calculating the quotient of two polynomials, named ***dividend, D(x)***, and ***divisor, d(x)***, consists on finding another two polynomials, named **quotient, q(x)**, and **remainder, r(x)**, such that:

$$D(x) = d(x) \times q(x) + r(x); \qquad\qquad \text{degree } r(x) < \text{degree } d(x)$$

Example: Divide polynomials $D(x) = 3x^3 - 2x^2 + 5x - 12$ and $d(x) = x^2 - 3x - 5$.

$$3x^3 - 2x^2 + 5x - 12 \enspace \underline{\big|\, x^2 - 3x - 5}$$

Dividing two polynomials	
Step 1: Write both polynomials in a decreasing order of their terms. If dividend is not a **complete** polynomial, leave gaps in the places of the non-existing terms.	$3x^3 - 2x^2 + 5x - 12 \;\big|\underline{x^2 - 3x - 5}$
Step 2: Calculate the quotient of the first term of dividend divided by the first term of the divisor: $3x^3 : x^2 = 3x$	$3x^3 - 2x^2 + 5x - 12 \;\big|\underline{x^2 - 3x - 5}$ $\qquad\qquad\qquad\qquad\quad 3x$
Step 3: Obtained term of the quotient is multiplied by the whole divisor, and the resulting polynomial is subtracted to the dividend. In this way, we are obtaining the first *partial remainder.*	$3x^3 - 2x^2 + 5x - 12 \;\big|\underline{x^2 - 3x - 5}$ $\underline{-3x^3 + 9x^2 + 15x}\qquad\quad 3x$ $\qquad\quad 7x^2 + 20x$
Step 4: We write down the following term of the dividend after the remainder. We divide the first term of the first partial dividend by the first term of the divisor. We continue this process till obtaining a remainder with degree lower than the divisor's degree.	$3x^3 - 2x^2 + 5x - 12 \;\big|\underline{x^2 - 3x - 5}$ $\underline{-3x^3 + 9x^2 + 15x}\qquad\quad 3x + 7$ $\qquad\quad 7x^2 + 20x - 12$ $\qquad\quad\underline{-7x^2 + 21x + 35}$ $\qquad\qquad\qquad 41x + 23$

Notice that   degree q(x) = degree D(x) – degree d(x)   and it is a general rule.

Exercises

8. Work out the quotient and the reminder of the following divisions:

a) $(7x^2 - 5x + 3) : (x^2 - 2x + 1)$ b) $(2x^3 - 7x^2 + 5x - 3) : (x^2 - 2x)$

c) $(x^3 - 5x^2 + 2x + 4) : (x^2 - x + 1)$ d) $(3x^5 - 2x^3 + 4x - 1) : (x^3 - 2x + 1)$

e) $(x^4 - 5x^3 + 3x - 2) : (x^2 + 1)$ f) $(4x^5 + 3x^3 - 2x) : (x^2 - x + 1)$

9. Divide and check that *Dividend = divisor x quotient + reminder*:

a) $(x^3 - 5x^2 + 3x + 1) : (x^2 - 5x + 1)$ b) $(6x^3 + 5x^2 - 9x) : (3x - 2)$

c) $(x^4 - 4x^2 + 12x - 9) : (x^2 - 2x + 3)$

1.5. Quotient of a polynomial by (x – a). Riffini´s rule

One of the most frequent situation when dividing polynomials is having as divisor a binomial like $(x - a)$, being a a real number. For example, $5x^4 - 4x^2 + 2x - 9) : (x - 4)$, where a = 4.

These division can be calculated in the general way, as above, but there is a more comfortable and quick method, named *Ruffinis´s Rule*. Riffini´s rule uses only the coefficients of the dividend.

Example: Divide $x^4 - 8x^2 + 3x - 1 : (x - 2)$.

Ruffini's Rule

Step 1: Write the coefficients of dividend. You must list them in a decreasing order of degree. Write zero in the places of non-existing terms.
Write the independent term of the divisor, <u>changing its sign</u>.

$$
\begin{array}{r|rrrrr}
 & 1 & 0 & -8 & 3 & -1 \\
2 & & & & & \\
\hline
 & & & & &
\end{array}
$$

Step 2: Write down the first coefficient of the dividend.

$$
\begin{array}{r|rrrrr}
 & 1 & 0 & -8 & 3 & -1 \\
2 & & & & & \\
\hline
 & 1 & & & &
\end{array}
$$

Step 3: Multiply the opposite of the independent term of the divisor, by this number you have just written down, and write it into the box, under the second term of the dividend.

$$
\begin{array}{r|rrrrr}
 & 1 & 0 & -8 & 3 & -1 \\
2 & & 2 & & & \\
\hline
 & 1 & & & &
\end{array}
$$

Step 4: Add in vertical.

$$
\begin{array}{r|rrrrr}
 & 1 & 0 & -8 & 3 & -1 \\
 & & + & & & \\
2 & & 2 & & & \\
\hline
 & 1 & 2 & & &
\end{array}
$$

Step 5: Repeat the product of 2 by the last number you wrote down in the quotient zone.

$$
\begin{array}{r|rrrrr}
 & 1 & 0 & -8 & 3 & -1 \\
2 & & 2 & 4 & & \\
\hline
 & 1 & 2 & & &
\end{array}
$$

Final steps: Continue this process till the end. Last obtained number in the quotient line is "remainder". It is usually written in a box.

$$
\begin{array}{r|rrrrr}
 & 1 & 0 & -8 & 3 & -1 \\
2 & & 2 & 4 & -8 & -10 \\
\hline
 & 1 & 2 & -4 & -5 & \boxed{-11}
\end{array}
$$
$\leftarrow$ Remainder

Important note: Remember that, in general, degree q(x) = = degree D(x) – degree d(x).

In this case, as divisor (x – a) has degree = 1, then:
degree q(x) = degree D(x) – 1.

Example: Determine the value of k that makes $4x^3 + 16x^2 + k$ divisible by $x + 3$.

Solution:

The remainder of the division $(4x^3 + 16x^2 + k) : (x + 3)$ must be zero. Applying the Ruffini's rule, we have:

$$
\begin{array}{r|rrrr}
 & 4 & 16 & 0 & k \\
-3 & & -12 & -12 & 36 \\
\hline
 & 4 & 4 & -12 & \boxed{k + 36 = 0}
\end{array}
$$

From the equation $k + 36 = 0$, we obtain that $\boxed{k = (-36)}$.

Root of a polynomial

Root of a polynomial: We say a is a *root* of a polynomial P(x) if quotient $\dfrac{P(x)}{x-a}$ is an exact division (remainder = 0). In this case, *(x – a)* is said to be a *factor* of P(x).

Example: Check if 2 is a root of the polynomial $P(x) = 3x^2 - 12x + 12$

Solution: We must divide $(3x^2 - 12x + 12) : (x - 2)$, and we do it by using the Ruffini's rule:

$$
\begin{array}{r|rrr}
 & 3 & -12 & 12 \\
2 & & 6 & -12 \\
\hline
 & 3 & -6 & \boxed{0}
\end{array}
$$

As remainder = 0, 2 is a root of the given polynomial.

Exercises	**10.** By uing Ruffini's Rule, work out the quotient and the remainder of the following divisions:
	a) $(5x^3 - 3x^2 + x - 2) : (x - 2)$ b) $(x^4 - 5x^3 + 7x + 3) : (x + 1)$
	c) $(-x^3 + 4x) : (x - 3)$ d) $(x^4 - 3x^3 + 5) : (x + 2)$

1.6. Factor and remainder theorems

> The **remainder** of the division of a polynomial P(x) by a binomial (x – a) is the value of **P(a)**.

11. Work out the quotient and the remainder of $P(x) = 2x^3 - 5x^2 + 7x + 3$ when divided by the following divisors and check the Remainder´s theorem:

a) (x – 2)　　　　b) (x + 3)　　　c) x

12. Use Ruffini´s Rule to work out the quotient and the remainder of $P(x) = x^4 - 3x^2 + 7$ when divided by the following divisors and check the Remainder´s theorem:

a) (x – 2)　　　　b) (x + 3)　　　c) x

13. Check which of the following numbers, 1, –1, 2, –2, 3, –3, are roots of the following polynomials:

a) $P(x) = x^3 - 2x^2 - 5x + 6$　　　　　　b) $Q(x) = x^3 - 3x^2 + x - 3$

14. Check if the following polynomials can be exactly divides by (x – 3) or by (x + 1).

a) $P_1(x) = x^3 - 3x^2 + x - 3$　　　　　　b) $P_2(x) = x^3 + 4x^2 - 11x - 30$

c) $P_3(x) = x^4 - 7x^3 + 5x^2 - 13$

15. The polynomial $x^4 - 2x^3 - 23x^2 - 2x - 24$ is divisible by (x – a) for two integer values of a. Find them out and calculate the quotient in both cases.

1.7. Factorization of a polynomial

> If a, b, c, … are the roots of a polynomial P(x), the **factorization** of P(x) is:
> $$P(x) = (x - a) \cdot (x - b) \cdot (x - c)......$$
> **Note:** Real roots of polynomial $P(x) = ax^n + bx^{n-1} + ... + k$ are divisors of k/a

Example: Factorize the polynomial $P(x) = x^2 + 4x - 5$

Solution: Using either Ruffini´s Rule or quadratic equation expression, we must find its roots. They are x = –5, x = 1.

So, Factorization is $P(x) = x^2 + 4x - 5 = (x + 5)(x - 1)$

Steps to factorize a polynomial are:

Step 1: If possible, extract a common factor.

Example: a) $P(x) = x^4 + 8x^3 + 15x^2 = x^2 \cdot (x^2 + 8x + 15)$

b) $P(x) = 3x^3 + 6x^2 - 15x = 3x \cdot (x^2 + 2x - 15)$

If degree = 2: If the degree of the polynomial is 2, you may use the remarkable products or, in general, solve the quadratic equation.

Example: a) $P(x) = x^2 - 4x^3 + 4 = (x - 2)^2$

b) $P(x) = x^2 - 3x + 2$. Solve the quadratic equation:

$$\left. \begin{matrix} a = 1 \\ b = (-3) \\ c = 2 \end{matrix} \right\} x = \frac{-(-3) \pm \sqrt{(-3)^2 - (4 \cdot 1 \cdot 2)}}{2 \cdot 1} = \frac{+3 \pm \sqrt{9 - 8}}{2} = \frac{+3 \pm 1}{2} \quad \rightarrow \quad \begin{cases} x_1 = \frac{+3+1}{2} = \frac{4}{2} = 2 \\ x_2 = \frac{+3-1}{2} = \frac{2}{2} = 1 \end{cases}$$

So, roots are $x = 2$ and $x = 1$, and factorization is: $P(x) = (x - 2) \cdot (x - 1)$.

If degree > 2: If the degree of the polynomial is bigger than 2, you must factorize it by using Ruffini's Rule. **Note:** When getting a partial reminder with degree = 2, you can use the remarkable products or, in general, solve the quadratic equation.

Example: $2x^4 + 20x^3 + 50x^2 = 2x^2 \cdot (x^2 + 10x + 25) = 2x^2 \cdot (x + 5)^2$

Roots: 0 (double) and 5, (double).

Exercises	**16.** Factorize the following polynomials: a) $x^2 + 8x + 15$ b) $7x^2 - 21x - 280$ c) $3x^2 + 9x - 210$ **17.** By using the remainder's theorem, find out an integer root and, after that, factorize: a) $2x^2 - 9x - 5$ b) $3x^2 - 2x - 5$ c) $4x^2 + 17x + 15$ d) $-x^2 + 17x - 72$ **18.** Extract a common factor and use the remarkable identities to factorize the polynomials: a) $3x^3 - 12x$ b) $4x^3 - 24x^2 + 36x$ c) $45x^2 - 5x^4$ d) $x^4 + x^2 + 2x^3$ e) $x^6 - 16x^2$ f) $16x^4 - 9$ **19.** Complete the factorization of the following expressions: a) $(x^2 - 25)(x^2 - 6x + 9)$ b) $(x^2 - 7x)(x^2 - 13x + 40)$

20. Factorize and indicate the roots:

a) $x^3 + 2x^2 - x - 2$ b) $3x^3 - 15x^2 + 12x$ c) $x^3 - 9x^2 + 15x - 7$

d) $x^4 - 13x^2 + 36$

Mixed

21. Given $P(x) = 4x^5 - 8x^4 + 2x^3 + 2x^2 + 1$ and $Q(x) = 4x^3 - 4x^2 + 2x$, work out:

a) Extract the highest common factor HCF) of $Q(x)$

b) $P(x) - 2x \cdot Q(x)$ c) $Q(x) \cdot Q(x)$ d) $P(x) : Q(x)$

22. Factorize the following polynomials:

a) $2x^4 - 18x^2$ b) $x^4 - x^3 - x^2 - x - 2$ c) $x^3 - 13x^2 + 36x$

d) $2x^3 - 9x^2 - 8x + 15$ e) $x^5 + x^4 - 2x^3$ e) $x^3 - 3x + 2$

(Sol: a) $2x^2 (x + 3) (x - 3)$; b) $(x + 1) (x - 2) (x^2 + 1)$; c) $x (x - 9) (x - 4)$;
d) $2(x - 1) (x - 5) (x + 3/2)$; e) $x^3 (x - 1) (x + 2)$; f) $(x - 1^2 (x + 2))$

23. Find out the value of k so that following division is exact: $(3x^2 + kx - 2) : (x + 2)$

(Sol: $k = 5$)

24. Factorize the following polynomials:

a) $x^6 - 9x^5 + 24x^4 - 20x^3$ b) $x^6 - 3x^5 - 3x^4 - 5x^3 + 2x^2 + 8x$

c) $x^6 + 6x^5 + 9x^4 - x^2 - 6x - 9$ *(Sol.: a) $x^3(x - 2)^2(x - 5)$;*

b) $x(x - 1)(x + 1)(x - 4)(x^2 + x + 2)$; c) $(x + 3)^2(x + 1)(x - 1)(x^2 + 1)$)

2. Algebraic fractions

An ***algebraic fraction*** is a fraction having as numerator and/or denominator an algebraic expression. For example: $\dfrac{x^3 - x^2}{x + 1}$

2.1. Simplification of algebraic fractions

To ***simplify*** an algebraic fraction, you must factorize numerator and denominator and simplify as far as posible.

Example: Simplify $\dfrac{x^3-x^2}{x+1}$:   <u>Solution:</u> $\dfrac{x^4-x^2}{x^2+x}=\dfrac{x^2\cdot(x+1)\cdot(x-1)}{x\cdot(x+1)}=\boxed{x\cdot(x-1)}$

2.2. Common denominator of algebraic fractions. Addition and subtraction

Remember to add or subtract fractions, they must have the same denominator. If not, we need to reduce them to a common denominator. In algebraic fractions, it is the same.

Example: Calculate: a) $\dfrac{x^3-x^2}{x+1}+\dfrac{2}{x}$ b) $\dfrac{2x-1}{x^2-x-2}-\dfrac{x}{x-2}$

<u>Solution:</u>

a) $\dfrac{x-1}{x+1}+\dfrac{2}{x}=\dfrac{x\cdot(x-1)}{x\cdot(x+1)}+\dfrac{(x+1)\cdot2}{x\cdot(x+1)}=\dfrac{x^2-1}{x\cdot(x+1)}+\dfrac{2x+2}{x\cdot(x+1)}=\dfrac{x^2+2x+1}{x\cdot(x+1)}$

Now, we must simplify, if posible: $\dfrac{x^2+2x+1}{x\cdot(x+1)}=\dfrac{(x+1)^2}{x\cdot(x+1)}=\boxed{\dfrac{x+1}{x}}$

b) $x^2-x-2=(x+1)\cdot(x-2)$, so, common denominator is $(x+1)\cdot(x-2)$,

$\dfrac{2x-1}{(x+1)\cdot(x-2)}-\dfrac{x}{x-2}=\dfrac{2x-1}{(x+1)\cdot(x-2)}-\dfrac{x\cdot(x+1)}{(x+1)\cdot(x-2)}=\dfrac{2x-1-x^2-x}{(x+1)\cdot(x-2)}=\dfrac{-x^2+x-1}{(x+1)\cdot(x-2)}$

We must simplify, but numerator has not any real root, so, previous fraction is the solution.

Exercises	**25.** Simplify the following algebraic fractions: a) $\dfrac{x^5+6x^4+9x^3}{x^3+3x^2}$ b) $\dfrac{x^3-x}{x^3+3x^2+2x}$ c) $\dfrac{x^3-x^2-2x}{x^3-3x^2+2x}$ d) $\dfrac{x^3-3x^2+3x-1}{x^3-2x^2+x}$ e) $\dfrac{x^4-2x^3-3x^2}{x^4-9x^2}$ **26.** Calculate and simplify: $\dfrac{x+1}{x-2}+\dfrac{x-2}{x+2}-\dfrac{12}{x^2-4}$

27. Reduce to a common denominator and add the following algebraic fractions:

$$\frac{x+7}{x} \qquad \frac{x-2}{x^2+x} \qquad -\frac{2x+1}{x+1} \qquad Sol.: \quad \frac{-x^2+8x+5}{x^2+x}$$

28. Work out:

$$\frac{1}{x^2-1} + \frac{2x}{x+1} - \frac{x}{x-1} \qquad Sol.: \quad \frac{x^2-3x+1}{x^2-1}$$

2.3. Product and quotient of algebraic fractions

Product and quotient of algebraic expressions have the same rules than for numerical fractions. You must only pay attention to simplifications. In some cases, it may be better to factorize expressions at the beginning.

Example: Calculate: a) $\dfrac{x^3-x^2}{4x+8} \cdot \dfrac{x+2}{x-1}$ b) $\dfrac{x-2}{x^2+x} : \dfrac{x-2}{x+1}$

Solution:

a) $\dfrac{x^3-x^2}{4x+8} \cdot \dfrac{x+2}{x-1} = \dfrac{x^2\cdot(x-1)}{4(x+2)} \cdot \dfrac{x+2}{x-1} = \boxed{\dfrac{x^2}{4}}$

b) $\dfrac{x-2}{x^2+x} : \dfrac{x-2}{x+1} = \dfrac{x-2}{x\cdot(x+1)} : \dfrac{x-2}{x+1} = \dfrac{x-2}{x\cdot(x+1)} \cdot \dfrac{x+1}{x-2} = \boxed{\dfrac{1}{x}}$

29. Work out the following operations:

a) $\dfrac{x^2-2x+3}{x-2} \cdot \dfrac{2x+3}{x+5}$ b) $\dfrac{x^2-2x+3}{x-2} : \dfrac{2x+3}{x+5}$

$Sol.:$ a) $\dfrac{2x^3-x^2+9}{x^2+3x-10}$ b) $\dfrac{x^3+3x^2-7x+15}{2x^2-x-6}$

30. Work out the following operations:

a) $\dfrac{x+2}{x} : \left(\dfrac{x-1}{3} \cdot \dfrac{x}{2x+1}\right)$ b) $\dfrac{x^4-x^2}{x^2+1} \cdot \dfrac{x^4+x^2}{x^4}$

$Sol.:$ a) $\dfrac{6x^2+15x+6}{x^3-x^2}$ b) x^2-1

31. Calculate:

a) $\left(\dfrac{2x-1}{x+1}-\dfrac{3x}{x-1}\right)\cdot\left(\dfrac{x^3-x}{-x^2-6x+1}\right)$

b) $\dfrac{2x}{x-2}+\dfrac{3x-1}{x+2}-\dfrac{1}{x^2-4}$

c) $\dfrac{(x-1)^2}{2}\cdot\dfrac{1}{x^2-1}-\dfrac{3x}{(x+1)^2}$

d) $\dfrac{1}{(x-1)^2}+\dfrac{2}{x-1}+\dfrac{1}{x^2-1}$

e) $\left(\dfrac{3}{x}-\dfrac{2x}{x+1}\right)\cdot\left(\dfrac{x^2+x}{x-1}\right)$

Sol.: a) x b) $\dfrac{-x^2+11x-3}{x^2-4}$ c) $\dfrac{x^2-6x-1}{2(x+1)^2}$

d) $\dfrac{2x^2+2x-2}{(x-1)^2(x+1)}$ e) $\dfrac{-2x^2+3x+3}{x-1}$

3. Quadratic equations

3.1. Review of quadratic equations *(previous years)*

It is named **quadratic equation** an equation that can be transformed into another equivalent equation having this form:

$$ax^2+bx+c=0, \quad \text{with } a\neq 0$$

Solutions of a quadratic equation are the values of x that being substituted in it, make the equation to verify.

Quadratic equations are solved by using this expression: $\quad x=\dfrac{-b\pm\sqrt{b^2-(4\cdot a\cdot c)}}{2\cdot a}$

Example: Solve: $(x+2)^2=8x+1$

Solution: First, develop the bracket, $x^2+4x+4=8x+1$ and arrange all terms in the same member: $x^2-4x+3=0$

$\left.\begin{array}{l}a=1\\b=(-4)\\c=3\end{array}\right\}x=\dfrac{-(-4)\pm\sqrt{(-4)^2-(4\cdot1\cdot3)}}{2\cdot1}=\dfrac{+4\pm\sqrt{16-12}}{2}=\dfrac{+4\pm2}{2}\quad\rightarrow\quad\begin{cases}x_1=\dfrac{+4+2}{2}=\dfrac{6}{2}=3\\[2mm]x_2=\dfrac{+4-2}{2}=\dfrac{2}{2}=1\end{cases}$

So, solutions are $\boxed{x=3}$ and $\boxed{x=1}$.

32. Solve these equations:

a) $3 \cdot (x-1) \cdot (x+2) = 3x - 6$ b) $\dfrac{x}{5}\left(x + \dfrac{1}{6}\right) = x - 1$ c) $(5x-3)^2 - 11(4x+1) = 1$

d) $\dfrac{2x^2 - 1}{2} - \dfrac{x-1}{3} = \dfrac{1-x}{6}$ e) $(x-1)(x+1) + (x-2)^2 = 3$ f) $\dfrac{x(x+3)}{2} - \dfrac{(x+1)^2}{3} + \dfrac{1}{3} = 0$

g) $(x+2)(x-3) + x = 3$ h) $(x+1)^2 - 2x(x+2) + 14 = 0$ i) $x(2x+1) - \dfrac{(x-1)^2}{2} = 3$

j) $(x+1)^2 - (x-1)^2 + 2 = x^2 + 6$ k) $(x+2)^2 = 8x + 1$

33. In equation $x^2 - 5x + c = 0$, one solution is x = 3. a) What is the value of coefficient c? b) What is the other solution?

Incomplete quadratic equations

In a quadratic equation $ax^2 + bx + c = 0$, if parameters b or c, are zero, it is named an *incomplete quadratic equation.*

Incomplete quadratic equations can be solved by using the general expression we have just studied, or by a shorter method:

- If $b = 0$, equation becomes $ax^2 + c = 0$. It is quickly solved.

- If $c = 0$, equation becomes $ax^2 + bx = 0$. It is solved by factorizing. **x(ax + b) = 0.**

Example: Solve: a) $2x^2 - 8 = 0$ b) $x^2 - 12x = 0$
Solution:

a) $2x^2 - 8 = 0$ Adding 8: $2x^2 = 8 \rightarrow$ Dividing by 2: $x^2 = 4 \rightarrow x = \sqrt{4} = \boxed{+2 \text{ and } (-2)}$.

b) $x^2 - 12x = 0$ Factorizing: $x(x - 12) = 0 \rightarrow \begin{cases} x = 0 \\ x - 12 = 0 \rightarrow x = 12 \end{cases}$

For this last step, remember if you have $a \cdot b = 0$, then, there are 2 possible solutions: $\begin{cases} a = 0 \\ b = 0 \end{cases}$

34. Solve these quadratic equations:

a) $x^2 + 2x = 0$ b) $2x^2 + 5x - 3 = 0$ c) $x^2 - 24 = 1$

d) $1 - 4x^2 = -8$ e) $x^2 - 6x = 0$ f) $3x^2 - 39x = 0$

35. Solve these incomplete quadratic equations, <u>without operating</u>:

a) $(3x + 5) \cdot (2x - 1) = 0$ b) $(6x - 3) \cdot (2x - 1) = 0$ c) $(2x - 4) \cdot (x + 6) = 0$

d) $(5x - 8) \cdot (x - 4) = 0$ e) $(x - 4)(x - 6) = 0$ f) $(x + 2)(x - 3) = 0$

g) $x(x + 1)(x - 5) = 0$ h) $(3x + 1)(2x - 3) = 0$

Number of solutions of a quadratic equation

In expression $x = \dfrac{-b \pm \sqrt{b^2 - (4 \cdot a \cdot c)}}{2 \cdot a}$, quantity into the square root is named ***discriminant*** (Δ).

So, discriminant is $\Delta = b^2 - (4 \cdot a \cdot c)$

Value of the discriminant allows to know the number of solutions of a quadratic equation, without having to calculate them:

 ➤ If $\Delta > 0$ equation has **two solutions.**

 ➤ If $\Delta = 0$ equation has **one solution** (named *double solution*).

 ➤ If $\Delta < 0$ equation has **not solution.**

Example : Find the value of k so that equation $x^2 + kx + 36 = 0$ has two identical solutions.

Solution:

If we want the equation to have only one (double) solution, we need $D = 0$.

Substituting: $b^2 - 4ac = k^2 - 4 \cdot 1 \cdot 36 = k^2 - 144 = 0 \rightarrow k^2 = 144 \rightarrow k = \sqrt{144} = \pm 12$

So, equations are: $x^2 + 12x + 36 = 0$ and $x^2 - 12x + 36 = 0$.

Exercises	**36.** Calculate the discriminant of each equation and, without solving it, indicate their number of solutions: a) $5x^2 - 3x + 1 = 0$ b) $x^2 - 4x + 4 = 0$ c) $3x^2 - 6x - 1 = 0$ d) $5x^2 + 3x + 1 = 0$ **37.** Find out the value of m in equation $x^2 - 6x + m = 0$, so that it has two identical solutions. Find also the solutions. **38.** Determine for what values of **m** equation $2x^2 - 5x + m = 0$: a) Has two different solutions. b) Has one solution. c) Has not solution. **39.** Decide for what values of **b** equation $x^2 - bx + 25 = 0$: a) Has two different solutions. b) Has one solution. c) Has not solution.

3.2. Biquadratic equations

Biquadratic equations are those that only have terms in x^4, x^2 and independent term.

For example, $3x^4 + 5x^2 - 2 = 0$.

These equations are solved by means of the change: $\boxed{y = x^2}$

This change is shown in the following example.

Example: Solve the biquadratic equation $x^4 - 5x^2 + 4 = 0$

Solution: First of all, we change $\mathbf{y = x^2} \rightarrow y^2 - 5y + 4 = 0$

$$\left.\begin{array}{l} a = 1 \\ b = (-5) \\ c = 4 \end{array}\right\} y = \frac{-(-5) \pm \sqrt{(-5)^2 - (4 \cdot 1 \cdot 4)}}{2 \cdot 1} = \frac{+5 \pm \sqrt{25-16}}{2} = \frac{+5 \pm 3}{2} \quad \rightarrow \quad \begin{cases} y_1 = \dfrac{+5+3}{2} = \dfrac{8}{2} = 4 \\ y_2 = \dfrac{+5-3}{2} = \dfrac{2}{2} = 1 \end{cases}$$

So, solutions for "y" are $y = 4$ and $y = 1$. Now, we must undo the change we have introduced: $\mathbf{y = x^2}$

$y_1 = 4 \Rightarrow x^2 = 4 \Rightarrow x = \pm 2$

$y_2 = 1 \Rightarrow x^2 = 1 \Rightarrow x = \pm 1$ A biquadratic equation has four solutions, which can be

repeated or not. In our case, solutions are: $\boxed{x_1 = 2; \ x_2 = (-2); \ x_3 = 1; \ x_4 = (-1)}$.

Exercises

40. Solve the following biquadratic equations:

a) $x^4 - x^2 = 0$ b) $x^4 + 20x^2 - 576 = 0$ c) $\dfrac{3 \cdot (x^2 - 11)}{5} - \dfrac{2 \cdot (x^2 - 60)}{7} = 36$

d) $x^4 - 26x^2 + 25 = 0$ e) $4x^4 - 37x^2 + 9 = 0$ f) $x^4 - 10x^2 + 9 = 0$

g) $4x^4 - 17x^2 + 4 = 0$ h) $x^4 - 4x^2 + 3 = 0$ i) $x^4 - 4x^2 = 0$

j) $2x^4 - 5x^2 + 2 = 0$ k) $x^4 + x^2 - 2 = 0$ l) $3x^4 - 2x^2 - 1 = 0$

41. Solve the following equations:

a) $x^4 - x^2 - 12 = 0$ b) $x^4 - 8x^2 - 9 = 0$ c) $x^4 + 10x^2 + 9 = 0$ d) $x^4 - x^2 - 2 = 0$

(Sol.: a) 2 and (–2); b) 3 and (-3); c) No solution; d) $\pm\sqrt{2}$)

3.3. Equations with radicals

Some equations include radicals, usually square roots. In this section, we are learning to solve them.

OBJECTIVE: Isolate the radical in one of the members of the equation.

Once you have got it, you only have to calculate the squares of both members.

Example: Solve this equation: $\sqrt{2x-3} - x = -1$

Solution: First of all, we isolate the radical in the first member → $\sqrt{2x-3} = x - 1$

Now, we only have to calculate the squares of both members: $\left(\sqrt{2x-3}\right)^2 = (x-1)^2$
$2x-3 = x^2 - 2x + 1$ → $x^2 - 4x + 4 = 0$ → Now, we only have to solve this quadratic equation.

Their solutions are: $\boxed{x = 2 \text{ (double)}}$.

Sometimes, it is easier:

Example: Solve this equation: $2\sqrt{x+4} = \sqrt{5x+4}$

Solution: In this case, we can directly calculate the squares of both members:

$\left(2\sqrt{x+4}\right)^2 = \left(\sqrt{5x+4}\right)^2$ $4(x+4) = 5x+4$ → $4x+16 = 5x+4$ →→→→ Its solution is: $\boxed{x = 12}$.

But, sometimes it is more difficult because you have to do it twice:

Example: Solve this equation: $\sqrt{2x+1} + \sqrt{x+4} = 6$

Solution: First of all, we move one root to the right member: $\sqrt{2x+1} = 6 - \sqrt{x+4}$

Now, we calculate the squares of both members: $\left(\sqrt{2x+1}\right)^2 = \left(6 - \sqrt{x+4}\right)^2$

$2x+1 = 36 + x + 4 - 12\sqrt{x+4}$

Now, we can isolate the radical at the left member: $12\sqrt{x+4} = -x + 39$

Again, we calculate the squares of both members:

$144(x+4) = x^2 + 1521 + 78x$ → $x^2 - 66x + 945 = 0$

Now, we only have to solve this quadratic equation.

42. Solve the following equations with radicals:

a) $\sqrt{x} - 3 = 0$ b) $\sqrt{x} + 2 = x$ c) $\sqrt{4x+5} = x+2$ d) $\sqrt{x+1} - 3 = x - 8$

e) $\sqrt{2x^2 - 2} = 1 - x$ f) $\sqrt{3x^2 + 4} = \sqrt{5x+6}$ g) $\sqrt{x^2 + 7} + 2 = 2x$ h) $x - \sqrt{x} = 2$

i) $x - \sqrt{25 - x^2} = 1$ j) $x - \sqrt{169 - x^2} = 17$ k) $x + \sqrt{5x+10} = 8$ l) $\sqrt{x+4} = 7$

m) $x + \sqrt{5x+10} = 8$ n) $x - \sqrt{25 - x^2} = 1$

43. Solve the following equations with radicals:

a) $-\sqrt{2x - 3} + 1 = x$ b) $\sqrt{2x - 3} - \sqrt{x + 7} = 4$ c) $2 + \sqrt{x} = x$

d) $2 - \sqrt{x} = x$ e) $\sqrt{3x + 3} - 1 = \sqrt{8 - 2x}$

(Sol.: a) No solution;   b) 114; c) 4; d) 1; e) 2)

44. To go from A to C I swam at 4 km/h, straight, to P, and I walked, at 5 km/h, form P to C. It took us 99 minutes (99/60 hours). What is the distance, x, from B to P?

(Sol.: 4 km)

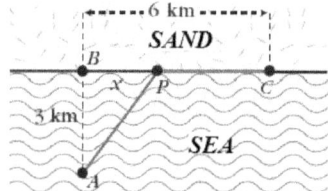

3.4. Equations with algebraic fractions

To solve these equations, you must remember what you have studied about operations and simplification of algebraic fractions, as well as about equations.

Example: Solve: $\dfrac{x+1}{x-1} + \dfrac{3}{x+1} = \dfrac{x-2}{x^2 - 1}$

<u>Solution:</u> Common denominator is LCM $\{(x-1), (x+1), (x^2 - 1)\} = (x^2 - 1)$

$\dfrac{x+1}{x-1} + \dfrac{3}{x+1} = \dfrac{x-2}{x^2 - 1}$ $\Rightarrow$ $\dfrac{(x+1)^2}{x^2 - 1} + \dfrac{3(x-1)}{x^2 - 1} = \dfrac{x-2}{x^2 - 1}$ Now, denominators can be removed:

$(x+1)^2 + 3(x-1) = x - 2$ $\Rightarrow$ $x^2 + 2x + 1 + 3x - 3 = x - 2$ $\Rightarrow$ $x^2 + 4x = 0$

Now, this equation is solved: $x^2 + 4x = 0$ $\Rightarrow$ $x \cdot (x + 4) \Rightarrow \begin{cases} x_1 = 0 \\ x_2 = -4 \end{cases}$

So, solutions are: $\boxed{x_1 = 0}$ and $\boxed{x_2 = (-4)}$.

45. Solve the following equations:

a) $\dfrac{1}{x}+\dfrac{1}{x^2}=\dfrac{3}{4}$ b) $\dfrac{2}{x}+\dfrac{3}{x^2}=1$ c) $\dfrac{1}{x}+\dfrac{2}{x-1}=\dfrac{4}{3}$ d) $\dfrac{1}{x}-\dfrac{1}{x+3}=\dfrac{3}{10}$

e) $\dfrac{5}{x+2}+\dfrac{x}{x+3}=\dfrac{3}{2}$ f) $\dfrac{x-3}{x}+\dfrac{x+3}{x^2}=\dfrac{2}{3}$ g) $\dfrac{(x-2)^2}{x^2}-\dfrac{1}{2x}=\dfrac{8+3x}{2x^2}-\dfrac{2}{x}$

h) $\dfrac{3x+1}{x^3}+\dfrac{x+1}{x}=1+\dfrac{2x+3}{x^2}$

46. Solve the following equations:

a) $\dfrac{1}{x}+\dfrac{1}{x+3}=\dfrac{3}{10}$ b) $\dfrac{4}{x}+\dfrac{2(x+1)}{3(x-2)}=4$ c) $\dfrac{1}{x}+\dfrac{1}{x^2}=\dfrac{3}{4}$

(Sol.: a) $x_1 = 5,489$; $x_2 = -1,822$;   b) $x_1 = 3$; $x_2 = 4/5$;   c) $x_1 = 2$; $x_2 = -2/3$)

47. Solve the following equations:

a) $\dfrac{x}{x-1}+\dfrac{2x}{x+1}=3$ b) $\dfrac{5}{x+2}+\dfrac{x}{x+3}=\dfrac{3}{2}$ c) $\dfrac{x+3}{x-1}-\dfrac{x^2+1}{x^2-1}=\dfrac{26}{35}$

(Sol.: a) $x = 3$;   b) $x_1 = 3$; $x_2 = (-4)$;   c) $x_1 = 6$; $x_2 = -8/13$)

------------------------ **Mixed equations** ---

48. Solve the following equations:

1) $\dfrac{4x^2-4x}{3}-x=x^2-\dfrac{3x+4}{3}$ 2) $x^4-11x^2+28=0$ 3) $x^2+\dfrac{15}{4}=\dfrac{3x^2-x+3}{4}+3$

(Sol.: 1) 2;   2) 2, (-2), $\sqrt{7}$, $-\sqrt{7}$;   3) 0 and (-1))

49. Solve the following equations:

a) $x^4-21x^2-100=0$ b) $x(x+4)-5=\dfrac{x(x-1)}{3}$ c) $x^4-48x^2-49=0$

d) $\sqrt{3x+16}=2x-1$ e) $\sqrt{x+5}-x=3$ f) $\dfrac{4x}{x+2}+\dfrac{x}{x-2}=\dfrac{14}{3}$

(Sol.: a) 5, (-5);   b) 1, (-15/2);   c) 7, (-7);   d) 3;   e) (-1)   f) 14, 4)

50. Solve the following equations:

a) $\dfrac{3}{x}+\dfrac{2}{x+4}=\dfrac{11}{6}$ b) $\dfrac{2}{x-1}+\dfrac{x-2}{x+1}=\dfrac{5}{4}$ c) $x+4=\sqrt{4x+12}$ d) $\dfrac{2x-1}{x}+\dfrac{4}{x-1}=\dfrac{11}{2}$

(Sol.: a) 2, (-36/11);   b) 3 , (-7);   c) (-2);   d) 2, (-1/7))

51. Solve the following equations:

a) $x^4+x^3-9x^2-9x=0$ b) $x^3-2x^2-11x+12=0$ c) $x^4+x^3-4x^2-4x=0$

d) $x^3-2x^2-5x+6=0$ e) $x^3+4x^2-x-4=0$

(Sol.: a) 0, (-1), 3, (-3);   b) 1, 4, (-3);   c) 0, (-1), 2, (-2);
d) 1, 3, (-2);   e) 1, (-1), (-4))

Exercises

4. Systems of equations (simultaneous equations)

4.1. Review of systems of equations *(previous years)*

In order to review methods to solve systems of linear equations and geometrical interpretation of their solutions, you are offered the following systems. Try to use all the methods.

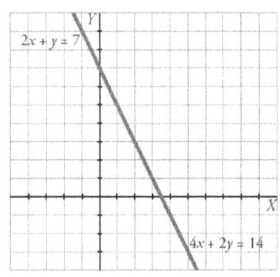

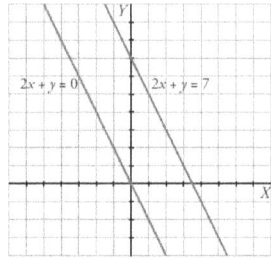

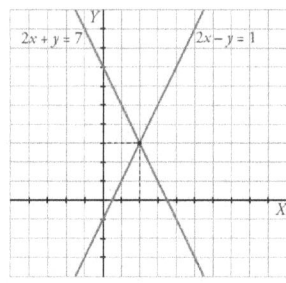

Infinite solutions
↓
The same straight line

Not solution
↓
Parallel straight lines

One solution
↓
Secant straight lines

---

52. Solve and indicate the relative position of the following straight lines:

a) $\left.\begin{array}{l} x+3=y-3 \\ 2(x+3)=6-y \end{array}\right\}$

b) $\left.\begin{array}{l} 3x+2=3(y+2)-4 \\ 5x-2y=5-x-y \end{array}\right\}$

c) $\left.\begin{array}{l} \dfrac{3(x-y)}{2}-\dfrac{2x+y}{3}=-3 \\ 2(x+y-1)+\dfrac{2x-y}{3}=11 \end{array}\right\}$

53. Solve and indicate the relative position of the following straight lines:

a) $\left.\begin{array}{l} 2(x-4)-3(y-7)+22=0 \\ 2(x+1)+4(y+1)-16=0 \end{array}\right\}$

b) $\left.\begin{array}{l} x-7(y+4)=-5 \\ 2x-3y-19=-6 \end{array}\right\}$

c) $\left.\begin{array}{l} x-6(y-2)=-6 \\ 3(x-1)+2y=23 \end{array}\right\}$

54. Solve and indicate the relative position of the following straight lines:

a) $\left.\begin{array}{l} 2x-y=7 \\ \dfrac{4x}{3}-\dfrac{y}{3}=\dfrac{19}{3}-4 \end{array}\right\}$

b) $\left.\begin{array}{l} \dfrac{x}{3}+\dfrac{y}{4}=x-\dfrac{1}{6} \\ \dfrac{y}{3}-\dfrac{x}{5}=\dfrac{x+y+4}{15} \end{array}\right\}$

c) $\left.\begin{array}{l} \dfrac{x-y}{2}-\dfrac{x+y}{10}=\dfrac{3}{5} \\ 3x-\dfrac{5y-4}{2}=\dfrac{25}{2} \end{array}\right\}$

Systems of non-linear equations

Systems of non-linear equations are systems in which, at least one of the equations contain some quadratic terms.

For example, $\left.\begin{array}{l} x - y = 1 \\ x^2 + y^2 - 2x = 31 \end{array}\right\}$

In this section, we are learning to solve these systems. Depending on the system, we can use *substitution* or *reduction* methods. We are seeing it with some examples.

Example: Solve the following system: $\left.\begin{array}{l} x - y = 1 \\ x^2 + y^2 - 2x = 31 \end{array}\right\}$

<u>Solution:</u> When one of the equations contain quadratic terms but the other one does not, we will use substitution method: We can isolate "x" in the first equation: $x = y + 1$ and substitute it in the second one:

$$(y+1)^2 + y^2 - 2(y+1) = 31 \implies y^2 + 2y + 1 + y^2 - 2y - 2 = 31 \implies 2y^2 = 32 \implies y^2 = 16 \implies \boxed{y \pm 4}.$$

Now, we must find out the corresponding values for "x". We must substitute, for example, in the first one:

$y_1 = 4;$ Como $x - y = 1;$ $x = y + 1 \rightarrow x_1 = 4 + 1 = 5.$

$y_2 = (-4);$ $x = y + 1 \rightarrow x_2 = -4 + 1 = (-3).$ So, solutions are $\boxed{(5, 4) \text{ and } (-3, -4)}$.

Example: Solve the following system: $\left.\begin{array}{l} x^2 - 2y = 0 \\ 2x^2 + y = 0 \end{array}\right\}$

<u>Solution:</u> Notice if we multiply by (-2) the first equation and add both equations, x unknown will disappear.

$\left\{\begin{array}{l} x^2 - 2y = 0 \ \underline{\textit{Multiply by }(-2)} \\ 2x^2 + y = 0 \ \underline{\textit{Leave}} \end{array}\right.$ $\left\{\begin{array}{l} -2x^2 + 4y = 0 \\ 2x^2 + y = 0 \end{array}\right.$ If we add them, we have $5y = 0 \ \rightarrow \ y = 0.$

Later, we substitute $y = 0$ in the first equation, for example, and we obtain $x = 0$. Sol.: $\boxed{x = y = 0}$.

55. Solve the following systems:

a) $\left.\begin{array}{l} x - y = 2 \\ x^2 - y^2 = 4 \end{array}\right\}$
b) $\left.\begin{array}{l} x + 2y = 6 \\ xy = -8 \end{array}\right\}$
c) $\left.\begin{array}{l} x + y = 4 \\ x^2 - xy = 6 \end{array}\right\}$

56. Solve the following systems:

a) $\left.\begin{array}{l} x - y + 3 = 0 \\ x^2 + y^2 = 5 \end{array}\right\}$
b) $\left.\begin{array}{l} x + y = 1 \\ xy + 2y = 2 \end{array}\right\}$
c) $\left.\begin{array}{l} 2x + y = 3 \\ xy - y^2 = 0 \end{array}\right\}$
d) $\left.\begin{array}{l} 3x + 2y = 0 \\ x(x - y) = 2y^2 - 8 \end{array}\right\}$

57. Solve the following systems:

a) $\left.\begin{array}{l} x^2 + y^2 = 41 \\ x^2 - y^2 = 9 \end{array}\right\}$
b) $\left.\begin{array}{l} 3x^2 + 2y^2 = 35 \\ x^2 - 2y^2 = 1 \end{array}\right\}$
c) $\left.\begin{array}{l} x^2 + y^2 + x + y = 32 \\ x^2 - y^2 + x - y = 28 \end{array}\right\}$
d) $\left.\begin{array}{l} x^2 + 2y^2 + x + 1 = 0 \\ x^2 - 2y^2 + 3x + 1 = 0 \end{array}\right\}$

58. Solve the following systems:

a) $\left\{\begin{array}{l} 2x - y - 1 = 0 \\ x^2 - 7 = y + 2 \end{array}\right.$
b) $\left\{\begin{array}{l} \dfrac{1}{x} + \dfrac{1}{y} = 1 - \dfrac{1}{xy} \\ xy = 6 \end{array}\right.$
c) $\left\{\begin{array}{l} x = 2y + 1 \\ \sqrt{x + y} - \sqrt{x - y} = 2 \end{array}\right.$

(Sol.: a) $x_1 = 4$; $y_1 = 7$; $x_2 = -2$; $y_2 = -5$;
b) $x_1 = 2$; $y_1 = 3x_2 = 3$; $y_2 = 2$;   c) $x = 17$; $y = 8$)

59. Solve the following systems:

a) $\left.\begin{array}{l} \dfrac{x}{3} + \dfrac{y}{2} = 3 \\ \dfrac{x}{2} + \dfrac{y}{2} = 4 \end{array}\right\}$
b) $\left.\begin{array}{l} y - 4x - 2 = 0 \\ y = x^2 + 3x \end{array}\right\}$
c) $\left.\begin{array}{l} y = x^2 - 2x \\ y + x - 6 = 0 \end{array}\right\}$

d) $\left.\begin{array}{l} \dfrac{x - 1}{3} + \dfrac{y}{2} = 2 \\ 3x + y = 7 \end{array}\right\}$
e) $\left.\begin{array}{l} y = x^2 - 3x \\ y - 2x + 6 = 0 \end{array}\right\}$

(Sol.: a) $x = 6$; $y = 2$;   b) $x_1 = 2$, $y_1 = 10$, $x_2 = 1$; $y_2 = 2$;
c) $x_1 = 3$, $y_1 = 3$, $x_2 = (-2)$; $y_2 = 8$;   d) $x = 1$, $y = 4$;
e) $x_1 = 3$, $y_1 = 0$, $x_2 = 2$; $y_2 = (-2)$)

60. Solve the following systems:

a) $\left.\begin{array}{l} y = 3x + 1 \\ \sqrt{x+y+4} = y - x \end{array}\right\}$

b) $\left.\begin{array}{l} \dfrac{3}{x} - \dfrac{x}{y} = 0 \\ 2x - y = 3 \end{array}\right\}$

c) $\left.\begin{array}{l} \dfrac{2}{x} + \dfrac{3}{y} = 3 \\ x + y = 4 \end{array}\right\}$

d) $\left.\begin{array}{l} 2x + y = 6 \\ \sqrt{x} - y = -3 \end{array}\right\}$

e) $\left.\begin{array}{l} \dfrac{1}{x+y} = \dfrac{2}{5} \\ \dfrac{1}{x} + \dfrac{1}{y} = \dfrac{5}{2} \end{array}\right\}$

(Sol.: a) x = 1; y = 4;   b) x = 3; y = 3;   c) x_1 = 8/3, y_1 = 4/3, x_2 = 1; y_2 = 3;
d) x = 1, y = 4;   e) x_1 = 2, y_1 = 1/2, x_2 = 1/2; y_2 = 2)

Gauss method for bigger linear systems

Till now, you have solved systems with 2 unknowns and 2 equations. In this part of the unit, you will learn to solve systems of linear equations having more unknowns and equations, as:

$$\begin{cases} x + 2y - z = -5 \\ 2x + 3y + 2z = 8 \\ 4x + y - z = 6 \end{cases}$$

This method starts transforming the system into a matrix, a chart of values:

$$\left(\begin{array}{ccc|c} 1 & 2 & -1 & -5 \\ 2 & 3 & 2 & 8 \\ 4 & 1 & -1 & 6 \end{array}\right)$$

The next step is operating linear combinations of the rows in order to convert into "0" all the terms below the main diagonal,

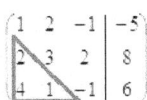

Allowed operations are:

➢ Sum of rows.
➢ Product of a row by a number.
➢ Product of 2 rows by 2 numbers and sum of the results.
➢ Change the position of rows.

Step 1: Leave the first row unchanged:

$$\begin{pmatrix} 1 & 2 & -1 & | & -5 \\ & & & | & \\ & & & | & \end{pmatrix}$$

Step 2: Notice you can eliminate "2" by multiplying "1" above by (-2) and adding our "2".

We will name rows as R_1, R_2 and R_3.

$(-2)\cdot R_1 \quad \rightarrow \quad -2 \quad -4 \quad 2 \quad | \quad 10$

$\underline{\quad\quad R_2 \quad \rightarrow \quad 2 \quad\quad 3 \quad\quad 2 \quad | \quad 8 \quad}$

Add: $\quad\quad\quad\quad 0 \quad -1 \quad 4 \quad | \quad 18$ Write this row as the 2nd one:

$$\begin{pmatrix} 1 & 2 & -1 & | & -5 \\ 0 & -1 & 4 & | & 18 \end{pmatrix}$$

Step 3: Notice you can eliminate "4" by multiplying "1" above by (-4) and adding our "4".

$(-4)\cdot R_1 \quad \rightarrow \quad -4 \quad -8 \quad 4 \quad | \quad 20$

$\underline{\quad\quad R_3 \quad \rightarrow \quad 4 \quad\quad 1 \quad -1 \quad | \quad 6 \quad}$

Add: $\quad\quad\quad\quad 0 \quad -7 \quad 3 \quad | \quad 26$ Write this row as the 3rd one:

$$\begin{pmatrix} 1 & 2 & -1 & | & -5 \\ 0 & -1 & 4 & | & 18 \\ 0 & -7 & 3 & | & 26 \end{pmatrix}$$ We are finishing. We only must eliminate (-7)   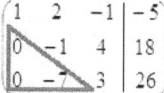

Step 4: As 1st and 2nd rows are correct, you can copy them:

$$\begin{pmatrix} 1 & 2 & -1 & | & -5 \\ 0 & -1 & 4 & | & 18 \end{pmatrix}$$

Step 5: Now, eliminate "(-7)" by multiplying "(-1)" above by (-7) and adding our "(-7)".

$(-7)\cdot R_2 \quad \rightarrow \quad 0 \quad 7 \quad -28 \quad | \quad -126$

$\underline{\quad\quad R_3 \quad \rightarrow \quad 0 \quad -7 \quad\quad 3 \quad | \quad 26 \quad}$

Add: $\quad\quad\quad\quad 0 \quad 0 \quad -25 \quad | \quad -100$ Write this row as the 3rd one:

$$\begin{pmatrix} 1 & 2 & -1 & -5 \\ 0 & -1 & 4 & 18 \\ 0 & 0 & -25 & -100 \end{pmatrix}$$ Now, you can rewrite the system and solve it just in a second:

$$\begin{cases} x + 2y - z = -5 \\ -y + 4z = 18 \\ -25z = -100 \ \rightarrow \ z = 4 \end{cases}$$

Substitute "z" by 4 in the 2nd equation: $-y + 4 \cdot 4 = 18 \ \rightarrow \ y = 16 - 18 = -2$

Substitute "z" by 4 and y by (-2) in the 1st equation: $x - 4 - 4 = -5 \ \rightarrow \ x = 3$.

So, solution is: $\boxed{x = 3, \ y = (-2), \ z = 4}$.

Exercises

61. Use Gauss method to solve the following systems:

a) $\begin{cases} x + 3y - 2z = 4 \\ 3x + 2y + z = 5 \\ 4x + 4y + 2z = 6 \end{cases}$ b) $\begin{cases} -x - 2y + z = -1 \\ 2x + 3y - z = 1 \\ 3x + 3y + 9z = 6 \end{cases}$

(Sol.: a) (2, 0, (-1)); b) ((-11/3), 5/3, 2/3)).

62. Use Gauss method to solve the following systems:

a) $\begin{cases} 2x + 3y - z = 15 \\ 2x - y + z = -3 \\ x - y = 0 \end{cases}$ b) $\begin{cases} 2x - 5y + 12z = 9 \\ 4x - y - 2z = -2 \\ 2x + 4y + 10z = -11 \end{cases}$ c) $\begin{cases} x - 4y = -5 \\ 2x + y = -1 \\ 2x - 8y = -10 \end{cases}$

d) $\begin{cases} x + y - z = 36 \\ x - y + z = 13 \\ -x + y + z = 7 \end{cases}$ e) $\begin{cases} x + 2y + z = 4 \\ 2x + 7y - z = 8 \\ 3x - 5y + 3z = 1 \end{cases}$ f) $\begin{cases} 2x + y = 5 \\ x + 3z = 16 \\ 5y - z = 10 \end{cases}$

(Sol.: a) (2, 2, (-5)); b) (-19/18, -20/9, 0); c) (-1, 1);
d) (49/2, 43/2, 10); e) (1, 1, 1); f) (1, 3, 5)).

63. Use Gauss method to solve the following systems:

a) $\begin{cases} -x - 2y - z = -4 \\ x + 3y + z = 5 \\ 4x + 2y + 2z = 8 \end{cases}$ b) $\begin{cases} x + y + z = 3 \\ 2x - y + 2z = 0 \\ 2x + 3z = 3 \end{cases}$ c) $\begin{cases} x + y = 2 \\ y + z = 3 \\ x - y - z = 5 \end{cases}$

(Sol: a) x = 1; y = 1; z = 1; b) x = 0; y = 2; z = 1; c) x = 8; y = (-6); z = 9).

64. Use Gauss method to solve the following systems:

a) $\begin{cases} 3x + 2y + z = 7 \\ 2x - 2y - z = 8 \\ x + 5y + z = -2 \end{cases}$ b) $\begin{cases} 3x + y - 2z = -6 \\ 2x - y + 3z = -8 \\ x + y - z = 4 \end{cases}$ c) $\begin{cases} -2x - y + z = -4 \\ 3x + y - 2z = 6 \\ 2x + y + z = 6 \end{cases}$

d) $\begin{cases} 2x - y + 2z = 2 \\ x + 2y - z = 3 \\ 2x - y + 3z = 1 \end{cases}$ e) $\begin{cases} x + 2y - 2z = 6 \\ x - 3y + z = -7 \\ 2x - y + z = -3 \end{cases}$ f) $\begin{cases} x + y - z = 2 \\ 2x - 2y + 3z = 1 \\ x + 2y - z = 4 \end{cases}$

g) $\begin{cases} x - 2y + z = 6 \\ 3x + y - z = 7 \\ x - y + 2z = 6 \end{cases}$ h) $\begin{cases} x - y + 2z = 7 \\ x + y - 3z = -5 \\ 2x - y + 2z = 9 \end{cases}$ i) $\begin{cases} x + y + 2z = 6 \\ x - 3y - z = 1 \\ x - y - z = -1 \end{cases}$

(Sol.: a) (3, (-1), 0); b) (0, 2, (-2)); c) (3, (-1), 1); d) (2, 0, (-1)); e) (0, 2, (-1));
f) (1, 2, 1); g) (3, (-1), 1); h) (2, (-1), 2); i) (1, (-1), 3).

5. Inequalities and systems of inequalities

Inequalities are expressions quite similar to equations, but instead of expressing the equality of both members by the symbol "=", their inequality is expressed by $<, >, \leq$ or $\geq$.

		symbol	Graphically
$\geq$ bigger or equal than $\leq$ lower or equal than	Values are included	•	[ ]
$>$ bigger than $<$ lower than	Values are NOT included	o	( )

Linear inequalities

Inequalities are solved by the same steps than equations. We must only be careful in one aspect:

When **multiplying** or **dividing** both members of an inequality by a **negative** number, the sense of the inequality must be **changed**.

For example: $2x + 5 < 5x + 20 \rightarrow 2x - 5x < 20 - 5 \rightarrow (-3)x < 15 \rightarrow x > \dfrac{15}{(-3)} \rightarrow \boxed{x > (-5)}$.

Now, some inequalities are solved, step by step, for you to understand the process:

Example: Solve: $2\,(x + 1) - 3\,(x - 2) < x + 6$

Solution: $2x + 2 - 3x + 6 < x + 6$

   $2x - 3x - x < -2 - 6 + 6$

   $-2x < -2$ → $\boxed{x > 1}$ or $(1, +\infty)$ or

Example: Solve: $\dfrac{3x+1}{7} - \dfrac{2-4x}{3} \geq \dfrac{-5x-4}{14} + \dfrac{7x}{6}$

Solution: $LCM(7, 3, 14, 6) = 42.$ So, we multiply all the terms by 42:

   $6(3x + 1) - 14(2 - 4x) \geq 3(-5x - 4) + 49x$

   $18x + 6 - 28 + 56x \geq -15x - 12 + 49x$

   $18x + 56x + 15x - 49x \geq -12 - 6 + 28$

   $40x \geq 10 \rightarrow 4x \geq 1$ → $\boxed{x \geq 1/4}$ or $[1/4, +\infty)$

Exercises

65. Solve the following linear inequalities:

   a) $x + 2x + 3x < 5x + 1$ b) $5x + 10 > 12x - 4$ c) $4x + 2 - 2x < 8x$

   d) $2x + 4 > x + 6$ e) $-x + 1 > 2x + 4$ f) $5x + 10 < 12x - 4$

66. Solve the following linear inequalities:

   a) $x + 2x + 3x > 5x + 1$ b) $5x + 10 < 12x - 4$ c) $4x + 2 - 2x > 8x$

   d) $2x + 4 > x + 6$ e) $-x + 1 < 2x + 4$ f) $x + 51 > 15x + 9$

67. Find out the numbers whose triple minus 20 units is lower than its double plus 40 units.

68. The price for mobile calls at the firm A is fix 20 euros/month plus 7 cents/min, and in the firm B it is fix 11 euros/month plus 12 cents/min. For more than how minutes of conversation is more convenient for A?

69. Solve the following linear inequalities:

   a) $2(x - 3) > 1 - 3(x - 1)$ b) $2(x + 1) + 4 < -2(x + 3)$

   c) $(x - 20) / 8 < (1 - 2x) / 10$ d) $\dfrac{x-1}{4} < \dfrac{x+3}{3} - \dfrac{x-5}{2}$

70. A man is 25 years older than his son. Find out the period of their lives in which father's age exceeds in more than 5 years the double of the age of his son.

71. Solve the following inequalities:

a) $3x - 2 \le 10$ b) $x - 2 > 1$ *Sol.: a) $(-\infty, 4]$; b) $(3, +\infty)$;*

c) $2x + 5 \ge 6$ d) $3x + 1 \le 15$ *c) $[1/2, +\infty)$; $(-\infty, 14/3]$*

72. Solve the following inequalities:

a) $\dfrac{(x-2)^2}{2} + \dfrac{5x+6}{6} < \dfrac{(x+3)(x-3)}{3} + 6$ b) $\dfrac{x+3}{x-7} \le \dfrac{1}{2}$ *Sol.: a) $x \in (0,7)$;*

b) $x \in [(-13),7)$

Quadratic inequalities

Quadratic inequalities are solved following the same steps than quadratic equations. The only difference is, at the end, when looking for the intervals that are solution of the inequality. We are seeing it with some examples:

Example: Solve: $x^2 + 3x - 4 > 0$

<u>Solution:</u> We are solving it as though it were a quadratic equation: $x^2 + 3x - 4 = 0$

$$x = \frac{(-3) \pm \sqrt{9+16}}{2} = \frac{-3 \pm 5}{2} \quad \rightarrow \quad \begin{cases} x_1 = 1 \\ x_2 = (-4) \end{cases}$$

We have three zones in the Real Line:

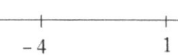

-4 1

Now, we must decide which of these zones fulfil the condition $x^2 + 3x - 4 > 0$

We use one value included in each zone:

$P(-6) = 36 - 18 - 4 = 14 > 0 \rightarrow$ This zone fulfils the inequality.

$P(0) = -4 < 0 \rightarrow$ This zone does not fulfil the inequality.

$P(10) = 100 + 30 - 4 = 126 > 0 \rightarrow$ This zone fulfils the inequality.

IMPORTANT: See that (-4) and 1 do NOT fulfil the inequality, as $P(-4) = P(1) = 0$

Solution is: $(-\infty; -4)\ U\ (1; +\infty)$ or

Example: Solve: $-x^2 + 4x - 7 < 0$

Solution: We are solving it as though it were a quadratic equation: $x^2 - 4x + 7 = 0$

$$x = \frac{4 \pm \sqrt{16 - 28}}{2} = \frac{4 \pm \sqrt{-12}}{2} \quad \rightarrow$$

As you know, this equation has NOT any solution. So, Real Line is not divided into several

zones, we must consider it as an only zone $(-\infty; +\infty)$.

So, we will consider only one value: For example, $P(0) = -7 < 0$.

So, the whole Real Line is solution of this inequality.   Sol.: $x =$ **R**.

Exercises

73. Solve the following quadratic inequalities:

   a) $4x^2 - 2x < 2$ b) $5x^2 - 6x + 1 \geq 0$ c) $3x^2 < -4x + 4$ d) $(4x - 8)(x + 3) < 0$

   e) $(x + 1)(2x + 1) \geq 0$ f) $-x^2 - x + 3 < 0$ g) $\dfrac{x^2 + x}{3} - 1 > -\dfrac{1 - 2x^2}{6}$

74. Solve the following quadratic inequalities:

   a) $\dfrac{2x^2}{3} - x < \dfrac{8x}{3}(1 + x) + 1$ b) $\dfrac{x - 4}{3} < \dfrac{x^2}{x + 42}$ c) $\dfrac{x + 2}{3} < \dfrac{x^2}{3x + 4}$

   d) $x^2 + 2x + 3 \leq -1$ e) $(x + 5)(x - 4) \geq 0$ f) $x^2 - 9 < 0$

75. Solve the following quadratic inequalities:

   a) $x^2 - 3x - 4 < 0$ b) $x^2 - 3x - 4 \geq 0$ c) $x^2 + 7 < 0$ d) $x^2 - 4 \leq 0$

   Sol.: a) (–1, 4);   b) (–∞, –1] U [4, +∞);   c) No solution;   d) [–2, 2]

Systems of inequalities with one unknown

Systems of inequalities are formed by two inequalities and their solution is the zone that fulfils both of them at the same time.

Example: Solve the following system of inequalities:

$$\begin{cases} 10(x+1) + x \le 6(2x+1) \\ 4(x-10) < -6(2-x) - 6x \end{cases}$$

Solution: We will solve both inequalities separately:

$10(x+1) + x \le 6(2x+1)$ $4(x-10) < -6(2-x) - 6x$

$10x + 10 + x \le 12x + 6$ $4x - 40 < -12 + 6x - 6x$

$10x - 12x \le -10 + 6$ $4x < 40 - 12 \rightarrow 4x < 28 \rightarrow x < 7$

$-2x \le -4 \rightarrow 2x \ge 4 \rightarrow x \ge 2$

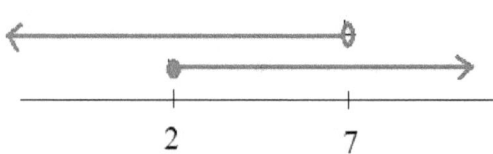

In the following figure, you can see there is a zone that has been marked twice.

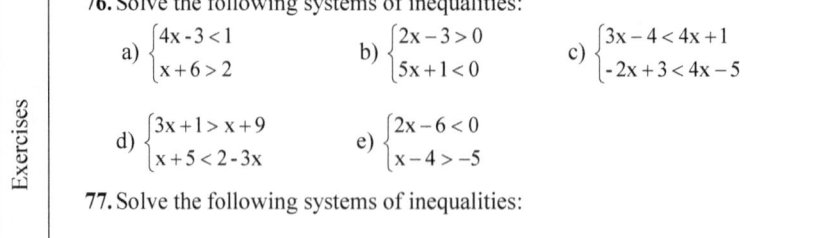

And this is the solution of this system: $[2, 7)$.

---

76. Solve the following systems of inequalities:

a) $\begin{cases} 4x - 3 < 1 \\ x + 6 > 2 \end{cases}$ b) $\begin{cases} 2x - 3 > 0 \\ 5x + 1 < 0 \end{cases}$ c) $\begin{cases} 3x - 4 < 4x + 1 \\ -2x + 3 < 4x - 5 \end{cases}$

d) $\begin{cases} 3x + 1 > x + 9 \\ x + 5 < 2 - 3x \end{cases}$ e) $\begin{cases} 2x - 6 < 0 \\ x - 4 > -5 \end{cases}$

77. Solve the following systems of inequalities:

a) $\begin{cases} 2x + 2 < 6 \\ 3x - 12 \ge -7 \end{cases}$ b) $\begin{cases} x - 2 \ge 0 \\ 2x \le 10 \end{cases}$ c) $\begin{cases} 5 - x < -12 \\ 16 - 2x < 3x - 3 \end{cases}$ d) $\begin{cases} 3x - 2 > -7 \\ 5 - x < 1 \end{cases}$

Exercises

78. Solve the following systems of inequalities:

a) $\begin{cases} x^2 - 3x - 4 \geq 0 \\ 2x - 7 > 5 \end{cases}$ b) $\begin{cases} x^2 - 4 \leq 0 \\ x - 4 > 1 \end{cases}$ *Sol.: a) (6, +∞); b) No solution*

79. Solve the following systems of inequalities:

a) $\begin{cases} \dfrac{5 - 3x}{4} - 3(x + 4) \leq \dfrac{3(x + 2)}{2} + 2 \\ \dfrac{2(2x + 1) - (x - 1)}{3} - \dfrac{2x + 1}{5} < 2 \end{cases}$ b) $\begin{cases} 2x - 10 > -x + 2 \\ 12 - 4x > -3x + 2 \\ 3(x + 2) \geq 2(x + 6) \end{cases}$

Sol.: a) [(-3), 2); b) (6, 10)

6. Logarithmic and exponential equations

When logarithms are included in an equation or a system, you must start applying the rules for logarithms. <u>At the end, you must check solutions are valid, because **argument of logarithms must be positive**</u>. Look at the following examples.

Example: Solve the following equations:

a) $\log 2 + \log(11 - x^2) = 2\log(5 - x)$ b) $2\log x = 3 + \log \dfrac{x}{10}$ c) $5^{2x} - 5^{x+1} = -4$

Solution:

a) $\log[2 \cdot (11 - x^2)] = \log(5 - x)^2$ → $2 \cdot (11 - x^2) = (5 - x)^2$ → $22 - 2x^2 = 25 - 10x + x^2$

$3x^2 + 10x + 3 = 0$ → $\begin{cases} x = 3 \\ x = \frac{1}{3} \end{cases}$ $\begin{array}{ll} \to 11 - 3^2 > 0 & 5 - 3 > 0 \\ \to 11 - \frac{1}{3} > 0 & 5 - \frac{1}{3} > 0 \end{array}$

So, both solutions are valid: $\boxed{x = 3}$ and $\boxed{x = 1/3}$.

b) $\log x^2 = \log 1000 + \log \dfrac{x}{10}$ → $\log x^2 = \log 1000 \cdot \dfrac{x}{10}$ → $\log x^2 = \log 100x$ →

$x^2 = 100x$ → $x^2 - 100x = 0$ → $x \cdot (x - 100) = 0$ → $\begin{cases} x = 0 \\ x = 100 \end{cases}$

As log0 is not defined, the only valid solution is $\boxed{x = 100}$.

c) You must write all the terms as powers "5^x": $5^{2x} - 5^{x+2} = -4 \rightarrow (5^x)^2 - 5^x \cdot 5^1 = -4 \rightarrow$

Let $t = 5^x \rightarrow t^2 - 5t + 4 = 0 \rightarrow \begin{cases} t = 4 \\ t = 1 \end{cases}$

Now, you must undo previous change, $(t = 5^x)$:

$$\begin{cases} t = 4 \rightarrow 5^x = 4 \rightarrow Ln5^x = Ln4 \rightarrow x \cdot Ln5 = Ln4 \rightarrow x = \dfrac{Ln4}{Ln5} = 0.86 \\ t = 1 \rightarrow 5^x = 1 \rightarrow x = 0 \end{cases}$$

So, solutions are $\boxed{x = 0.86}$ and $\boxed{x = 0}$.

Exercises

80. Solve the following logarithmic and exponential equations:

a) $\log 4x = 3\log 2 + 4\log 3$ b) $\log (2x-4) = 2$ c) $4\log (3 - 2x) = -1$

d) $\log (x + 1) + \log x = \log (x + 9)$ e) $\log (x + 3) = \log 2 - \log (x + 2)$

f) $\log (x^2 + 15) = \log (x + 3) + \log x$ g) $2\log (x + 5) = \log (x + 7)$

h) $\log \sqrt{x-1} = \log(x+1) - \log \sqrt{x+4}$ i) $\dfrac{\log(7 + x^2)}{\log(x - 4)} = 2$

j) $2\log (3x - 4) = \log 100 + \log (2x + 1)^2$ k) $\log_2 (x^2 - 1) - \log_2 (x + 1) = 2$

l) $\log^2 x - 3\log x = 2$ m) $2^{3x-1} = 3^{x+2}$ n) $5^{2x-3} = 2^{2-4x}$

ñ) $\log (x - a) - \log (x + a) = \log x - \log (x - a)$

81. Solve the following logarithmic and exponential equations:

a) $2 \log x - \log (x - 16) = 2$ b) $2^{2(x+1)} + 2^{x+2} - 80 = 0$

c) $\log (3x+5) - \log (2x+1) = 1 - \log 5$ d) $\log \sqrt{x+3} + \dfrac{1}{2}\log(x-2) = 1 + \log 0'6$

e) $5^{2x+1} - 5^{x+2} = 2.500$ f) $\log(2x+4) + \log(3x+1) - \log 4 = 2\log(8 - x)$

g) $\log \sqrt{3x+4} + \dfrac{1}{2}\log(5x + 1) = 1 - \log 3$ h) $\dfrac{\log 2 + \log(11 - x^2)}{\log(5 - x)} = 2$

i) $\log \sqrt{3x+2} + \dfrac{1}{2}\log(2x - 3) = 1 + \log 0,8$ j) $3^{x+1} + 3^{2-x} - 28 = 0$

k) $3^x + 3^{2-x} = 20$ l) $\log(2x + 2) + \log(x + 3) = \log 6$

Review exercises

Factorization of polynomials

1. Calculate the roots of:

   a) $x^3 + 6x^2 - x - 6$ b) $x^3 + 3x^2 - 4x - 12$ c) $x^4 - 5x^2 + 4$

2. Factorize the polynomials:

   a) $x^4 + 2x^3 - 13x^2 - 14x + 24$ b) $x^4 + 4x^3 + 4x^2$ c) $x^4 - 5x^2 + 4$

   d) $x^3 + 2x^2 + 4x$ e) $2x^3 + 11x^2 + 2x - 15$ f) $3x^4 - 3x^3 - 18x^2$

   g) $4x^2 + 12x + 9$ h) $25x^2 - 4$

3. Find out the HCF and LCM of the following polynomials:

   $P(x) = x^4 + 7x^3 + 12x$ $Q(x) = x^5 + 2x^4 - 3x^3$

Remainder theorem

4. Find out the value of *m* so that $5x^3 - 12x^2 + 4x + m$ is divisible by $x - 2$.

5. Calculate *a* so that polynomial $x^3 + ax + 10$ is divisible by $x + 5$.

6. Given polynomial $x^4 + 6x^3 - 3x^2 + 5x + m$, find out *m* so that when divided by $x + 3$ the reminder is 100.

Algebraic fractions

7. Simplify the following algebraic fractions:

   a) $\dfrac{x+3}{x^2-1} \cdot \dfrac{x-1}{x+2}$ b) $\dfrac{x^2+4x+4}{x^2-1} : \dfrac{x+2}{x+1}$ c) $\dfrac{x^3-3x+2}{x^3+x^2-2x}$ d) $\dfrac{x^2+2x-3}{x^3+2x^2-x-2}$

   e) $\dfrac{x^3-3x^2+4}{x^3+5x^2+8x+4}$ f) $\dfrac{x^3-7x^2+15x-9}{x^3-5x^2+3x+9}$ g) $\dfrac{x^2+6x+9}{x^2-1} \cdot \dfrac{x+1}{x+3}$

   h) $\dfrac{x^2+10x+25}{x^2-4} \cdot \dfrac{x+2}{x+5}$ i) $\dfrac{x^2-4}{x+6} : \dfrac{x^2-5x+6}{x^2-36}$ j) $\left(\dfrac{1}{x} - \dfrac{2}{x+1}\right) : \left(\dfrac{x^2+2}{x^2} + \dfrac{3}{x}\right)$

8. Calculate and simplify:

   a) $\dfrac{x}{x^2-4x+3} - \dfrac{3}{x^2-5x+6}$ b) $\dfrac{x}{x+1} + \dfrac{1+x}{x^2+2x+1}$ c) $\dfrac{x-1}{x^2-5x+6} + \dfrac{x-2}{x^2-4x+3}$

   d) $\dfrac{x-3}{x^2+x+1} - \dfrac{3x^2}{x^3-1}$ e) $\dfrac{2}{x^2-2x+1} + \dfrac{x+1}{x^2-1}$ f) $\dfrac{1}{x^2-9x+20} - \dfrac{11}{x^2-11x+30}$

   g) $\dfrac{1-x}{x^2-4x+3} - \dfrac{1+2x}{x^2-6x+9} - \dfrac{x+1}{x^2-9}$ h) $\dfrac{1+2x}{x^2+3x+2} - \dfrac{1-x}{x^2+5x+6} - \dfrac{1+x}{x^2+4x+3}$

Equations

9. Solve the following equations:

a) $\dfrac{x^2}{2} - 4x = 3 + \dfrac{x^2 - 12}{4}$

b) $x^4 - 4x^2 + 3 = 0$

c) $\dfrac{2x+1}{x+3} + \dfrac{x-3}{x} = \dfrac{1}{2}$

d) $x^4 + 2x^2 - 3 = 0$

e) $\sqrt{x+4} + \sqrt{2x-1} = 6$

f) $-x.(x-1).(x^2-2) = 0$

g) $\dfrac{2x^3 - x^2 - 2x + 25}{x^2 - 1} = 2x$

h) $2x^4 + 4x^3 - 18x^2 - 36x = 0$

i) $\dfrac{x^2 - 16}{3} - x = \dfrac{2 - 3x}{3} - \dfrac{x^2}{3}$

j) $x^4 - 5x^2 - 36 = 0$

k) $\sqrt{3x-3} + x = 7$

l) $\dfrac{2}{x-1} + \dfrac{x-2}{x+1} = \dfrac{5}{4}$

m) $x + \sqrt{3x+10} = 6$

n) $x^4 - 5x^2 + 4 = 0$

ñ) $\sqrt{x^2 + 3x} = \sqrt{2x}$

o) $\dfrac{x+1}{x-1} - 1 = \dfrac{1}{x}$

p) $\sqrt{2x+8} - \sqrt{x} = 2$

q) $\dfrac{3}{x} + \dfrac{2}{x^2} = 1 + \dfrac{4}{x^2}$

Systems

10. Solve the following systems and indicate the relative position of the following straight lines:

a) $\left.\begin{array}{l} \dfrac{x+y}{3} - \dfrac{2(x-3)}{5} = \dfrac{11}{5} \\ \dfrac{3x+y}{2} + 2(x-y) = \dfrac{-9}{2} \end{array}\right\}$

b) $\left.\begin{array}{l} \dfrac{5(x+1)}{3} - \dfrac{x+y}{2} = \dfrac{7}{2} \\ 2(x-y+5) - \dfrac{x+y}{3} = 11 \end{array}\right\}$

c) $\left.\begin{array}{l} \dfrac{2(x+1)}{5} - \dfrac{3(y-2)}{2} = 0 \\ \dfrac{x+y}{4} = \dfrac{1}{4} \end{array}\right\}$

d) $\left.\begin{array}{l} \dfrac{3x}{5} - \dfrac{2y}{3} = 7 \\ \dfrac{5x}{3} - 2y = 2 \end{array}\right\}$

e) $\left.\begin{array}{l} \dfrac{x}{3} + \dfrac{y}{4} = x - \dfrac{1}{6} \\ \dfrac{y}{3} - \dfrac{x}{5} = \dfrac{x+y+4}{15} \end{array}\right\}$

f) $\left.\begin{array}{l} 2x - y = 7 \\ \dfrac{4}{3}x - \dfrac{1}{8}y = \dfrac{19}{3} - 4 \end{array}\right\}$

g) $\left.\begin{array}{l} 3y - 2x - 16 = 0 \\ 2(x-5) + 6(y-2) + 20 = 0 \end{array}\right\}$

h) $\left.\begin{array}{l} \dfrac{x-2}{3} + \dfrac{y-1}{4} - 1 = x \\ 4x + 5y = -18 \end{array}\right\}$

i) $\left.\begin{array}{l} \dfrac{x-y}{2} - \dfrac{x+y}{10} = \dfrac{3}{5} \\ 3x - \dfrac{5y-4}{2} = \dfrac{25}{2} \end{array}\right\}$

11. Solve the following systems and indicate the relative position of the following straight lines:

a) $\left\{\begin{array}{l} x+y=1 \\ 2x-y=2 \end{array}\right.$

b) $\left\{\begin{array}{l} x+y=1 \\ x+y=2 \end{array}\right.$

c) $\left\{\begin{array}{l} x+y=1 \\ 2x+2y=2 \end{array}\right.$

d) $\left\{\begin{array}{l} y=x^2+4x+2 \\ x+y+2=0 \end{array}\right.$

e) $\left\{\begin{array}{l} y=x^2+4x+2 \\ 4x-y+2=0 \end{array}\right.$

f) $\left\{\begin{array}{l} y=x^2+4x+2 \\ x-y-2=0 \end{array}\right.$

12. Solve the following systems:

a) $\begin{cases} 2x^2 - y = 4 \\ 4x + 3y = -2 \end{cases}$

b) $\begin{cases} x + y = 2 \\ 3x - 3y = -4 \end{cases}$

c) $\begin{cases} x + 2y = 0 \\ x^2 + y^2 = 5 \end{cases}$

d) $\begin{cases} \dfrac{x}{2} - \dfrac{y}{4} = 1 \\ 3x - y = 8 \end{cases}$

e) $\begin{cases} 2x - 1 = y \\ \dfrac{x-1}{2} = y^2 - 1 \end{cases}$

f) $\begin{cases} x + 2y = \dfrac{3}{x} \\ x + y = \dfrac{2}{y} \end{cases}$

g) $\begin{cases} 2x + y = 52 \\ \sqrt{x} + y = 7 \end{cases}$

h) $\begin{cases} \sqrt{x} + \sqrt{y} = 20 \\ 3x - y = 122 \end{cases}$

Gauss method for linear systems

13. Solve the following method by mean of Gauus method:

a) $\begin{cases} 2x + y - z = 0 \\ x - y + 2z = 5 \\ x + y + z = 3 \end{cases}$

b) $\begin{cases} x + 2y + z = 4 \\ 2x + 5y + z = -3 \\ 4x + 9y + 3z = 2 \end{cases}$

c) $\begin{cases} x + 2y - 3z = 5 \\ 2x - 3y + z = 3 \\ 4x + y - 5z = 13 \end{cases}$

d) $\begin{cases} x + z = 4 \\ y + z = 3 \\ x + y = 5 \\ x - 2y - z = -2 \end{cases}$

e) $\begin{cases} x + y + z = 3 \\ x + y = 2 \\ y + z = 3 \end{cases}$

f) $\begin{cases} x - y + 2z = 7 \\ 2x + y + 5z = 10 \\ x + y - 4z = -9 \end{cases}$

g) $\begin{cases} 3x + 4y - z = 3 \\ 6x - 6y + 2z = -16 \\ x - y + 2z = -6 \end{cases}$

h) $\begin{cases} x - 2y + 3z = 5 \\ 2x - y + z = 3 \\ x - y + 3z = 6 \\ 3x + y - 2z = 0 \end{cases}$

One unknown inequalities

14. Solve the following inequalities:

a) $-2x + 4 \le -2$

b) $x^2 + x - 6 \le 0$

c) $2x + 1 > -5$

d) $-3x + 1 > -5$

e) $x^2 - 4 \le 0$

f) $2x - 3 < 5$

g) $3x - 1 \le 4x$

h) $x^2 - 3x > -2$

i) $\dfrac{x-1}{3} \le 2x + 1$

j) $\dfrac{2(x-1)}{3} > x - 1$

k) $x^2 - 4 \ge 0$

l) $3(x-1) + 1 \le 2(x+1)$

m) $2 - 3x < 2(x+1)$

n) $-x^2 + 4x - 4 \le 0$

ñ) $\dfrac{x-2}{3-x} > 0$

o) $\dfrac{x+3}{x^2 - x} > 0$

p) $\dfrac{x^2 + 2}{x - 3} \le 0$

q) $\dfrac{x^2 + x - 6}{x^2 - 2x + 1} \ge 0$

r) $x^3 - 4x \ge 0$

s) $x^3 + 3x^2 - x - 3 < 0$

t) $3x^2 - 6x > 0$

One unknown systems of inequalities

15. Solve the following systems of inequalities:

a) $\begin{cases} 3x + 8 \le x + 14 \\ 2x > \dfrac{3}{2}x - 1 \end{cases}$
b) $\begin{cases} x^2 - 3x - 4 > 0 \\ 2x - 3 < 0 \end{cases}$
c) $\begin{cases} 2x - 3\left(\dfrac{x}{2} + 1\right) \ge x - 8 \\ x + \dfrac{x}{3} - \dfrac{1}{2} + 2 > 2x - \dfrac{5}{6} \end{cases}$
d) $\begin{cases} 10 - 3x - x^2 < 0 \\ 3x + 5 > -16 \end{cases}$

Problems

16. The sum of the 3 digits of a number is 9. If the order of the digits is inverted, the number decreases in 99 units. Digit of tens is double than the units. Find out this number.

17. Area of an isosceles triangle is 60 m^2 and each of the similar sides measures 13 m. Find out the base and the height of this triangle.

18. Difference of the ages of two brothers is 4 years. In 8 years, the sum of both ages will be 40. What are their current ages?

19. About a rectangle, it is known its area, 192 cm^2 and its diagonals, 20 cm. Calculate the length of their sides.

20. For 2 pens, a pencil and a marker I have paid 6 €. For 4 pens and 2 markers, 10 €. And for 5 pencils and 3 markers, 11 €. What is the price of each of the articles?

21. Find out four consecutive numbers whose sum is 366.

22. Find out two numbers knowing their sum is 7 and the sum of their inverses is 7/12.

23. Calculate the length of the sides of a rectangle if its perimeter is 20 cm and its diagonal, 58 cm.

24. If each of the edges of a cubic deposit is increased in 2 dm, its volume increases in 98 litres. Find out its initial volume.

25. A group of friends are meeting for a trip. They are 20 people, men, women and children. Men and women altogether are three times the quantity of children. Moreover, if there were one more woman, the number of women would be the same as men. How many men, women and children are there?

26. Ana is investing 100.000 €. The bank offers her two financial products. Product A offers a annual interest of 4 % and product B, 6 %. She invests a portion of her 100,000 €. In each product and one year later, she receives 4,500 € as interests. How much did she invest in each product?

27. The difference of sides of a rectangle is 2 m. If each side were increased in 2 m, its area would get increased in 40 m^2. Find out dimensions of the rectangle.

28. Renting a tent costs 90 € per day. Inés is preparing a camp weekend and she thinks: "If we were three more friends, we would have to pay 6 € each. How many people are they?

29. Now, the age of a man is 3 times the age of his son, and in 15 years the age of the father will be twice the age of his son. How old are both of them now?

Logarithmic and exponential equations and systems

30. Solve the following logarithmic equations:

a) $\log x = \log 2 + \log(x-3)$ b) $\log(3x+1) - \log(2x-3) = 1 - \log 5$

c) $\log(20x) + \log(2x) = 3$ d) $\log(x+2) + \log(10x+20) = 3$

e) $\log x + \log 50 = 3$ f) $5\log(x+3) = \log 32$

g) $2\log x = \log(10-3x)$

31. Solve the following systems:

a) $\begin{cases} \log \dfrac{x^2}{y} = 3 \\ \log x + 3\log y = 5 \end{cases}$ b) $\begin{cases} 2x^2 + y = 75 \\ 2\log x - \log y = 2\log 2 + \log 3 \end{cases}$

c) $\begin{cases} \log \dfrac{x^2}{y} = 3 \\ \log x + 3\ \log y = 5 \end{cases}$ d) $\begin{cases} x + y = 52 \\ \log x + \log y = 2 \end{cases}$

e) $\begin{cases} \log(x+y) - \log(x-y) = \log 5 \\ \dfrac{2^x}{2^y} = 2 \end{cases}$

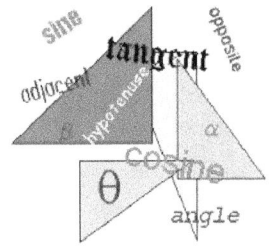

Unit 4.- Trigonometry

1. Degrees and radians

Angles are measured using two different systems of units, radians and degrees, some of whose main relationships are given.

$$360° = 2\pi \, rad \qquad 180° = \pi \, rad$$

$$90° = \frac{\pi}{2} rad$$

Example: Transform the following angles: a) $30°$ b) $\frac{3\pi}{2} rad$

Solution:

a) $30° \cdot \dfrac{2\pi \, rad}{360°} = \dfrac{60\pi \, rad}{360} = \boxed{\dfrac{\pi}{60} rad}$

b) $\dfrac{3\pi}{2} rad \cdot \dfrac{360°}{2\pi \, rad} = \dfrac{1080\pi}{4\pi} = \boxed{270°}$

1. Transform into radians the following angles:

    a) 45°   b) 60°   c) 90°   d) 120°   e) 150°

    f) 180°   g) 210°   h) 225°   i) 270°   j) 300°

2. Transform into degrees the following angles:

    a) $\dfrac{7\pi}{6}\,rad$ b) 2,5 rad   c) $\dfrac{7\pi}{3}\,rad$ d) $\dfrac{3\pi}{4}\,rad$ e) $\dfrac{5\pi}{6}\,rad$

2. Right triangles. Trigonometric ratios of an acute angle

When working in a right angled triangle, the longest side is known as the **hypotenuse**, and the other two sides are known as the **opposite** and the **adjacent**. The adjacent is the side next to a marked angle, and the opposite side is opposite this angle.

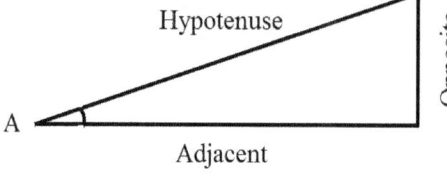

For an angle A ($0° \leq A \leq \frac{\pi}{2}$), its *trigonometric ratios* are defined as:

$$Sin\,A = \frac{opposite}{hypotenuse} \qquad Cos\,A = \frac{adjacent}{hypotenuse} \qquad tg\,A = \frac{opposite}{adjacent}$$

$$Co\sec A = \frac{1}{SinA} \qquad Sec\,A = \frac{1}{CosA} \qquad Cotg\,A = \frac{1}{TgA}$$

Example: Write down the values of sinA , cosA and tgA for the triangle shown. Then use a calculator to find the angle in each case.

Solution:

$$Sin\,A = \frac{8}{10} = 0,8 \qquad Cos\,A = \frac{6}{10} = 0,6 \qquad tg\,A = \frac{8}{6} = 1,33...$$

➢ SinA = 0,8. With you calculator, press the following keys:

| SHIFT | | Sin | | 0 | | · | | 8 | | = |, and look at the screen: 53,13.....

If you need this angle in degrees, then push | ° ′ ″ |, and you will have **53° 7′ 48.37″**.

➢ CosA = 0,6. With you calculator, press the following keys:

| SHIFT | | Cos | | 0 | | · | | 6 | | = |, and look at the screen: 53,13.....

If you need this angle in degrees, then push | ° ′ ″ |, and you will have **53° 7′ 48.37″**.

➢ tgA = 0,8. With you calculator, press the following keys:

| SHIFT | | tan | | 1 | | · | | 3 | | 3 | | 3 | | 3 | | 3 | | = |, At the screen: 53,13.....

If you need this angle in degees, then push | ° ′ ″ |, and you will have **53° 7′ 48.37″**.

Exercises

3. Calculate sine, cosine and tangent of the acute angles of the following right triangles:

a)
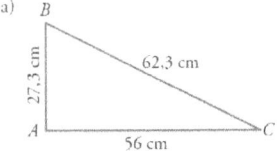

b)
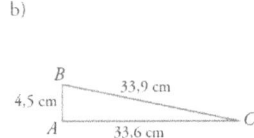

c)
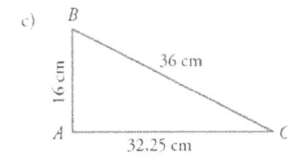

4. For each triangle, write sinθ, cosθ and tanθ as fractions and demonstrate they fulfil the relationship: $sin^2θ + cos^2θ = 1$.

(a)
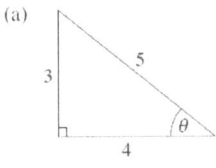

(b)
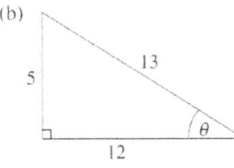

(c)
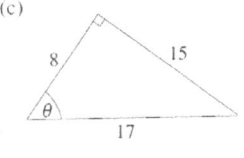

(d)
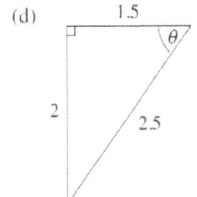

(e)
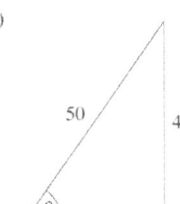

(f)
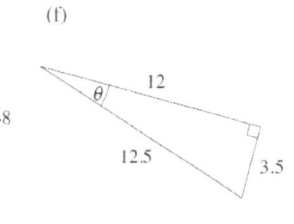

5. Given the following trigonometric ratios, find out their acute angle:

    a) senA = 0,25   b) cosA = 0,74   c) tgA = 3

    d) secA = 1,18   e) cotgA = 1,5

6. Given cosA = 0,2, use your calculator to calculate tg A.

7. Given senA = 0,56, use your calculator to calculate cosA.

8. Given tgA = 2, use your calculator to calculate senA.

9. Given cosecA = 3, use your calculator to calculate cosA.

10. Given secA = 1,5, use your calculator to calculate tgA.

11. Given cotgA = 3, use you calculator to calculate cosecA.

3. Calculations with right triangles

Naming angles and sides of a triangle

In a triangle, whether right or not,

> ➢ angles are named by using capital letters, A, B, C, counterclockwise,
> ➢ sides are named by the same letters, a, b, c, locating each side opposite the corresponding angle.

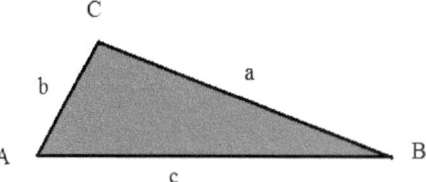

Now, we are going to remember how right triangles are solved. This means we are going to calculate all their sides and angles. We are seeing it with some examples.

Example: Find the length of the side marked x in the triangle shown.

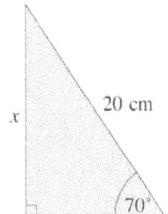

Solution: In this triangle, opposite = x and hypotenuse = 20.

Using $Sin\,70 = \dfrac{x}{20} = 0,94 \quad \rightarrow \quad x = 20 \cdot 0,94 = 18,8\,cm$

Example: Find the length of the hypotenuse, marked x, in the triangle.

Solution: In this triangle, opposite = 10 and hypotenuse = x.

Using $Sin\,28 = \dfrac{10}{x} = 0,47 \quad \rightarrow \quad x = \dfrac{10}{0,47} = 21,3\,cm$

Exercises

12. Draw and solve the following triangles, (where A is the right angle), by using the trigonometric ratios (you cannot use the Pythagoras´ Theorem). Calculate their areas.

a) a = 320 m, B = 47°   *(Sol: C = 43°; b = 234,03 m; c = 218,24 m; A = 25537,64 m²)*

b) a = 42,5 m, b = 35,8 m

(Sol: B = 57°23'22''; C = 32°36'38''; c = 22,90 m; A = 410 m²)

c) b = 32,8 cm, B = 22°   *(Sol: C = 68°; a = 87,56 cm; c = 1,18 cm; A = 331,40 cm²)*

d) b = 8 mm, c = 6 mm   *(Sol: B = 53°7'48''; C = 36°52'12''; a = 10 mm; A = 24 mm²)*

e) a = 8 km, b = 6 km   *(Sol: B = 48°35'; C = 41° 25'; c = 5,30 km; A = 15,87 km²)*

f) a = 13 m, c = 5 m   *(Sol: B = 67°22'48''; C = 22°37'12''; b = 12 m; A = 30 m²)*

g) c = 42,7 dam, C = 31°   *(Sol: B = 59°; a = 82,91 dam; b = 71,06 dam; A = 1517 dam²)*

h) c = 124 dm, B = 67° 21'   *(Sol: C = 22°39'; a = 322 dm; b = 297 dm; A = 18424 dm²)*

13. The diagram below, represents one face of the roof of a house in the shape of a parallelogram *EFGH*. Angle *EFI* = 40° and *EF* = 8 m. *EI* represents a rafter placed perpendicular to *FG* such that *IG* = 5 m.

Calculate, giving your answers to 3 significant figures:

a) the length of *FI*   b) the length of *EI*   c) the area of *EFGH*.

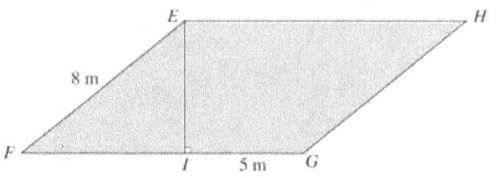

14. Find the length of the side marked *x* in each triangle:

(a)   (b)   (c)

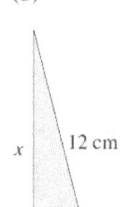   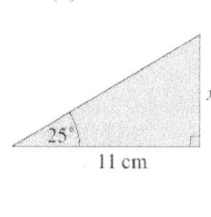

15. Find the angle in the following right triangles:

(a)   (b)   (c)

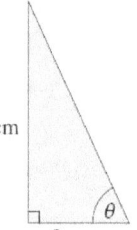   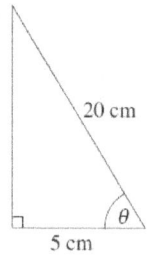

16. A soldier runs 500 metres east and then 600 metres north. If he had run directly from his starting point to his final position, what bearing should he have run on?

17. A ship is 50 km south and 70 km west of the port that it is heading for. What bearing should it sail on to reach the port? *(Bearing = rumbo)*.

Exercises

18. A ladder leans against a wall. The length of the ladder is 4 metres and the base is 2 metres from the wall. Find the angle between the ladder and the ground.

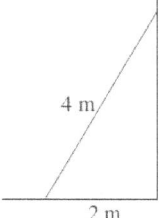

19. As cars drive up a ramp at a multi-storey car park, they go up 2 metres. The length of the ramp is 10 metres. Find the angle between the ramp and the horizontal.

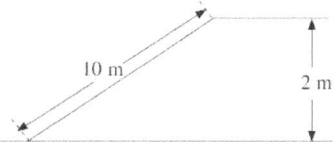

20. Calculate the trigonometric ratios of the angles *A, C, ABD* and *CBD*.

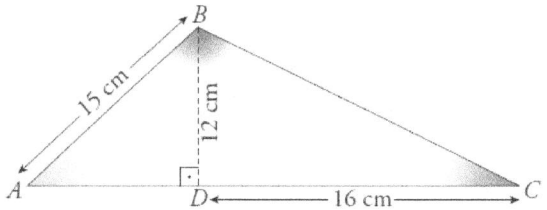

21. **(Height method)** Solve the following *not right* triangle *(Clue: calculate the height by mean of tangents of angles you know and notice both heights are equal)*.

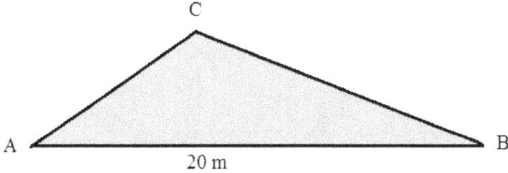

22. From my position, I can see the top of a tree under a 30° angle. If I get 40 m further, I can see its top under a 10° angle. Answer: a) what is the height of the tree? b) what was my distance from it at the beginning?

4. An important relationship: $\sin^2 A + \cos^2 A = 1$

Exercises

23. Write the algebraic expressions of sinA and cosA and demonstrate the relationship between them: $\sin^2 A + \cos^2 A = 1$.

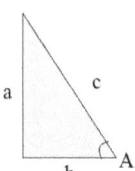

24. Knowing A is an acute angle and that $\cos A = 1/5$, calculate *senA* and *tgA*.

25. Complete the following chart by using the previous relationship. A, B and C are acute angles.

	A	B	C
Sin			0,3
cos	0,25		
Tg		0,6	

26. Knowing that $0° < A < 90°$, complete the following chart by using the previous relationship.

	A	B
sin		0,8
cos		
tg	0,75	

27. Calculate the missing trigonometric ratios or angles, without using the calculator. $0° < A < 90°$:

sinA			1	
cosA	1/2			
tgA		$\sqrt{3}/2$		
A				45°

28. Calculate sinA and cosA, knowing that tgA= ¾. A is an acute angle.

29. Calculate the missing trigonometric ratios or the angles, without using the calculator. 0° <A<90°:

sinA	$\sqrt{3}/2$			
cosA				$\sqrt{2}/2$
tgA		0		
A			30°	

30. Calculate the missing trigonometric ratios or the angle, without using the calculator. 0° <A<90°:

sinA		1/2		
cosA				0
tgA			1	
A	0			

31. If $sin\ A = 0,28$, calculate $cosA$ and tgA (A < 90°).

32. Find out the exact value (with radicals) of $sinA$ and tgA knowing that $cosA = 2/3$ (A < 90°).

33. If $tgA = \sqrt{5}$, calculate $sinA$ and $cosA$ (A < 90°).

5. Another important relationships: sinA = cos(90–A) cosA = sin(90–A)

34. Given the following right triangle, calculate sine and cosine of both acute angles and check they fulfil: $sinA = cos(90–A)$ $cosA = sin(90–A)$

Notice that $B = 90° - A$

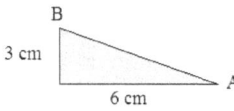

3 cm

6 cm

Exercises

35. Given the following right triangle, calculate sine and cosine of both acute angles and check they fulfil: $sinA = cos(90–A)$ $cosA = sin(90–A)$

Notice that $B = 90° - A$

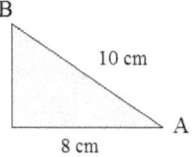

B

10 cm

8 cm

A

36. For each triangle, write $sin\theta$, $cos\theta$ and $tan\theta$ as fractions and demonstrate they fulfil the relationship: $sin^2\theta + cos^2\theta = 1$.

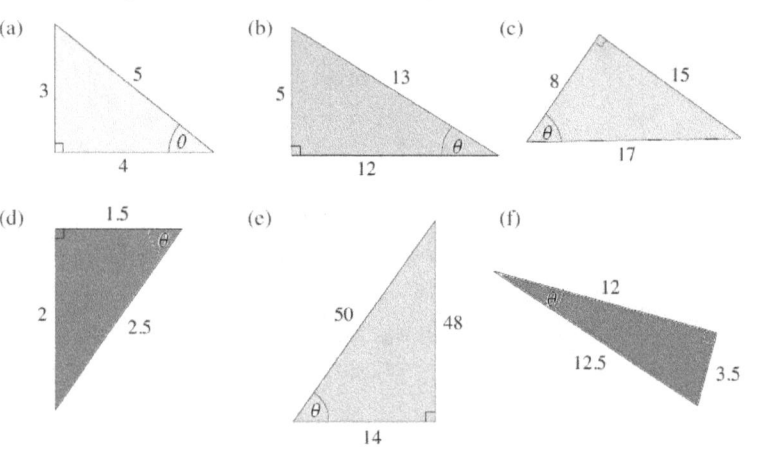

(a)

3

5

4

θ

(b)

5

13

12

θ

(c)

8

15

17

θ

(d)

1.5

2

2.5

θ

(e)

50

48

14

θ

(f)

12

12.5

3.5

θ

6. Trigonometric ratios in the other quadrants

Till now, you have worked with trigonometric ratios in the first quadrant ($\alpha < \frac{\pi}{2}$). Now, you will learn to work with larger angles, $\alpha > \frac{\pi}{2}$.

The goniometric circumference is divided into four quadrants, called:

First	I	$0 < \alpha < \dfrac{\pi}{2}$	
Second	II	$\dfrac{\pi}{2} < \alpha < \pi$	
Third	III	$\pi < \alpha < \dfrac{3\pi}{2}$	
Fourth	IV	$\dfrac{3\pi}{2} < \alpha < 2\pi$	

Sign of the trigonometric ratios

➢ In the goniometric circumference, angles are measured from the positive horizontal semi-axis of the first quadrant. So:

- Opposite side is located as a vertical line.

- Adjacent side is located as a horizontal line.

➢ Remember in Maths and Physics, criteria for signs are:

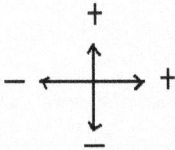

So, notice that signs for trigonometric ratios depend on the quadrant:

	I	II	III	IV
Sin	+	+	−	−
cos	+	−	−	+
Tg	+	−	+	−

Their inverse ratios, *cosecA, secA* and *cotgA*, have the same signs.

Example: If $\sin \alpha = \dfrac{\sqrt{5}}{3}$ and $90° < \alpha < 180°$, what are the values of $\cos\alpha$ and $tg\alpha$?

<u>Solution:</u> Using the trigonometric relationship $\sin^2 \alpha + \cos^2 \alpha = 1$, we have:

$$\left(\frac{\sqrt{5}}{3}\right)^2 + \cos^2 \alpha = 1 \;\rightarrow\; \frac{5}{9} + \cos^2 \alpha = 1 \;\rightarrow\; \cos^2 \alpha = 1 - \frac{5}{9} = \frac{4}{9} \;\rightarrow\; \cos\alpha = \sqrt{\frac{4}{9}} = \frac{2}{3}$$

But, as we are at the second quadrant, $\cos\alpha < 0$, so $\boxed{\cos \alpha = -\dfrac{2}{3}}$

Now, $tg\alpha$ is: $tg\alpha = \dfrac{\sin \alpha}{\cos \alpha} = \dfrac{\sqrt{5}/3}{-2/3} = \boxed{-\dfrac{\sqrt{5}}{2}}$ As you already knew, in quadrant II, $tg < 0$.

We also need their inverses *(having the same signs)*: Remember rationalization of fractions:

$$\sin \alpha = \frac{\sqrt{5}}{3} \qquad \cos \alpha = -\frac{2}{3} \qquad tg\alpha = -\frac{\sqrt{5}}{2}$$

$$\cos ec\alpha = \frac{3\sqrt{5}}{5} \qquad \sec\alpha = -\frac{3}{2} \qquad \cot g\alpha = -\frac{2\sqrt{5}}{5}$$

Example: Calculate $\sin\alpha$ and $\cos\alpha$ knowing that $tg\alpha = -\sqrt{5}$ and $90° < \alpha < 180°$.

<u>Solution:</u> You cannot use the trigonometric relationship $\sin^2 \alpha + \cos^2 \alpha = 1$, because you do not know sine or cosine. So, you must do first: $tg\alpha = \dfrac{\sin \alpha}{\cos \alpha} = -\sqrt{5} \;\rightarrow\; \sin \alpha = -\sqrt{5}\cdot\cos \alpha$ *(Eq. 1)*

At this moment, you can use the trigonometric relationship $\sin^2 \alpha + \cos^2 \alpha = 1$:

$$\left(-\sqrt{5}\cdot\cos \alpha\right)^2 + \cos^2 \alpha = 1 \;\rightarrow\; 5\cos^2 \alpha + \cos^2 \alpha = 1 \;\rightarrow\; 6\cos^2 \alpha = 1 \;\rightarrow\; \cos \alpha = \sqrt{\frac{1}{6}} = \frac{1}{\sqrt{6}} = \frac{\sqrt{6}}{6}$$

But, as we are at the second quadrant, $\cos\alpha < 0$, so $\boxed{\cos \alpha = -\dfrac{\sqrt{6}}{6}}$

Now, *sinα* is calculated by substituting in Equation 1 $\rightarrow\; \sin \alpha = -\sqrt{5}\cdot\cos \alpha = -\sqrt{5}\cdot\left(-\dfrac{\sqrt{6}}{6}\right) = \boxed{\dfrac{\sqrt{30}}{6}}$

As you already knew, in quadrant II, sine > 0

We also need their inverses *(having the same signs)*: Remember rationalization of fractions:

$$\sin \alpha = \frac{\sqrt{30}}{6} \qquad \cos \alpha = -\frac{\sqrt{6}}{6} \qquad tg\alpha = -\sqrt{5}$$

$$\cos ec\alpha = \frac{\sqrt{30}}{5} \qquad \sec\alpha = -\sqrt{6} \qquad \cot g\alpha = -\frac{\sqrt{5}}{5}$$

Example: If $\cos\alpha = -\dfrac{\sqrt{7}}{4}$ and $180° < \alpha < 270°$, find out the other trigonometric ratios of α:

<u>Solution:</u> Using the trigonometric relationship $\sin^2\alpha + \cos^2\alpha = 1$, we have:

$$\sin^2\alpha + \left(-\frac{\sqrt{7}}{4}\right)^2 = 1 \quad \rightarrow \quad \sin^2\alpha + \frac{7}{16} = 1 \quad \rightarrow \quad \sin^2\alpha = 1 - \frac{7}{16} = \frac{9}{16} \quad \rightarrow \quad \sin\alpha = \sqrt{\frac{9}{16}} = \frac{3}{4}$$

But, as we are at the third quadrant, sinα < 0, so $\boxed{\sin\alpha = -\dfrac{3}{4}}$

Now, tangent is calculated: $tg\alpha = \dfrac{\sin\alpha}{\cos\alpha} = \dfrac{-\frac{3}{4}}{-\frac{\sqrt{7}}{4}} = \dfrac{3}{\sqrt{7}} = \boxed{\dfrac{3\sqrt{7}}{7}}$

As you already knew, in quadrant III, tg < 0.

Now, we only need their inverses:

$$\sin\alpha = -\frac{3}{4} \qquad \cos\alpha = -\frac{\sqrt{7}}{4} \qquad tg\alpha = \frac{3\sqrt{7}}{7}$$

$$\cos ec\,\alpha = -\frac{4}{3} \qquad \sec\alpha = -\frac{4\sqrt{7}}{7} \qquad \cot g\alpha = \frac{\sqrt{7}}{3}$$

<table>
<tr><td rowspan="1" style="writing-mode:vertical">Exercises</td><td>

37. If sinα = -2/3 and α is in the third quadrant, calculate all its trigonometric ratios.

38. Calculate sinα, knowing that tgα = 3/2 and α is in the third quadrant.

39. Calculate α knowing that sinα = 1/2 and 90° < α < 270°.

40. If cosα = 1/3 and Л < α < 2Л, find out all its trigonometric ratios.

41. If secα = 2 and 3Л/2 < α < 2Л, calculate all its trigonometric ratios.

42. Knowing that cotgα = -1/2 and that 0 < α < Л, find out all its trigonometric ratios.

43. Knowing that cosecα = -5 and that Л < α < 3Л/2, calculate all its trigonometric ratios.

44. Find out the other trigonometric ratios:

    a) sinα = 0.6   α ∈ II   b) cosα = −0.6   α ∈ III

    c) tgα = −1   α ∈ II   d) cosecα = −2   α ∈ IV

</td></tr>
</table>

45. Knowing that cosA = -3/5 and 180°<a<270°, calculate the other trigonometric ratios. *(Sol: senA = -4/5, tgA = 4/3)*

46. Knowing that tgA = -3/4 and that A Є IV, calculate the other trigonometric ratios. *(Soluc: sen a=-3/5, cos a=4/5)*

47. The same for secA = 2 and 0 < A < Л/2. *(Sol: senA = 3/2, cosA = 1/2, tgA = 3)*

48. The same for tgA = -3 and Л/2 < A < Л. $\left(Sol: senA = \dfrac{3\sqrt{10}}{10}, cosA = -\dfrac{\sqrt{10}}{10}\right)$

49. The same for cosA = 0,2 and 3Л/2 < A < 2Л. $\left(Sol: senA = -\dfrac{2\sqrt{6}}{5}, tgA = -2\sqrt{6}\right)$

50. The same for senA = -0,3 and Л < A < 3Л/2. *(Sol: cosA = -0,95, tgA = 0,32)*

51. The same for tgA = 4/3 and Л < A < 3Л/2. *(Sol: senA = -4/5, cosA = -3/5)*

52. Calculate the other trigonometric ratios knowing that:

a) cosA = 4/5 and 270° < A < 360°   b) tgA = 3/4 and 180° < A < 270°

c) senA = 3/5 and 90° < A < 180°   d) cotgA = -2 and 90° < A < 180°

e) senA = 1/4 and A Є I   f) cosA = -1/3 and A Є II

g) cosecA = -2 and 180° < A < 270°   h) secA = 1 and 0° < A < 90°

i) tgA = 3/4 and 0° < A < 90°   j) tgA = 3/4 and A Є III

53. Solve the following trigonometric equations:

a) $\operatorname{sen} x = \dfrac{1}{2}$ b) $\cos x = -\dfrac{\sqrt{3}}{2}$ c) $\operatorname{tg} x = 1$

d) $\operatorname{sen} x = -\dfrac{\sqrt{2}}{2}$ e) $\cos x = -\dfrac{1}{2}$ f) $\operatorname{tg} x = -\sqrt{3}$

7. Reduction to the first quadrant

Usually, trigonometric ratios of some angles in the first quadrant are known. In this part of the unit, you will learn to calculate trigonometric ratios of angles belonging to other quadrants.

Although there are expressions to calculate this, it is not necessary for you to learn them by heart, as it is easier to draw a scheme. We are seeing this in several examples.

Angles whose difference is 90°: B = 90° + α

cos β = cos (90 + α) = - sen α

sen β = sen (90 + α) = cos α

tag β = tag (90 + α) = - ctg α

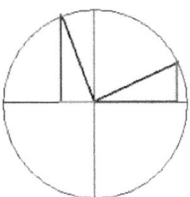

Supplementary angles: α + B = 180°

cos β = cos (180 - α) = - cos α

sen β = sen (180 - α) = sen α

tag β = tag (180 - α) = - tg α

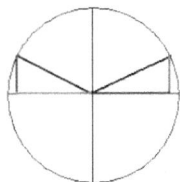

Angles whose difference is 180°: B = 180° + α

cos β = cos (180 + α) = - cos α

sen β = sen (180 + α) = - sen α

tag β = tag (180 + α) = tg α

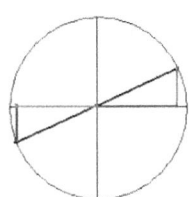

Angles whose addition is 270°: α + B = 270°

cos β = cos (270 - α) = - sen α

sen β = sen (270 - α) = - cos α

tag β = tag (270 - α) = ctg α

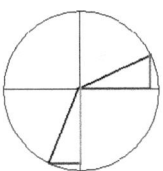

Angles whose difference is 270°: B = 270° + α

$\cos \beta = \cos (270 + \alpha) = \text{sen } \alpha$

$\text{sen } \beta = \text{sen } (270 + \alpha) = - \cos \alpha$

$\text{tag } \beta = \text{tag } (270 + \alpha) = - \text{ctg } \alpha$

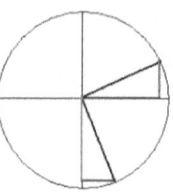

Opposite angles: α + B = 360°   or   α + B = 0°

$\cos (-\alpha) = \cos (360 - \alpha) = \cos \alpha$

$\text{sen } (-\alpha) = \text{sen } (360 - \alpha) = - \text{sen } \alpha$

$\text{tag } (-\alpha) = \text{tag } (360 - \alpha) = - \text{tg } \alpha$

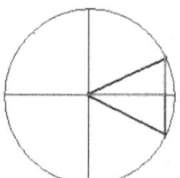

Angles whose difference is a number of whole rounds: B = k·360° + α

$\cos \beta = \cos (\alpha + 360°k) = \cos \alpha$

$\text{sen } \beta = \text{sen } (\alpha + 360°k) = \text{sen } \alpha$

$\text{tag } \beta = \text{tag } (\alpha + 360°k) = \text{tag } \alpha$

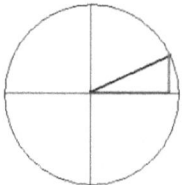

Example: Express sin 2130° as a trigonometric ratio of an angle α, so that $0 \leq \alpha \leq 90°$.

<u>Solution:</u> If you divide 2130: 360, you will obtain an integer quotient of 5.

$5 \cdot 360 = 1800 \quad \rightarrow \quad 2130 - 1800 = 330 \quad \rightarrow \quad \sin 2130° = \sin 330°$

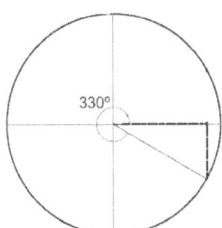

Now, draw a 330° angle in the goniometric circumference:

Notice the little angle under the horizontal axis. It is 360 – 330 = 30°

Now, add a positive 30° angle in the same drawing:

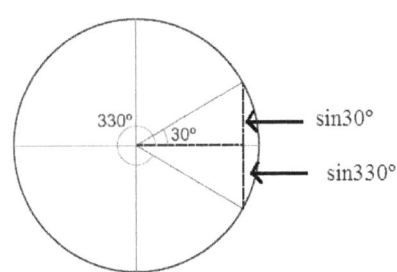

And notice they have equal sine, although with opposite signs.

So sin330° = $\boxed{-\sin30°}$.

Exercises

54. Calculate, without calculator:

    a) sen 570°   b) cos 14520°   c) sen (-120°)   d) cos (-240°)

    e) tg 2565°   f) cos 15Л/2 rad   g) sen 55Л/6 rad   h) tg 79Л rad

                (Sol: a) -1/2; b) -1/2; c) - 3 /2; d) -1/2; e) 1; f) 0; g) -1/2; h) 0)

55. Calculate, without calculator:

    a) cos 225°   b) cos(-60°)   c) tg 120°   d) sen (-1470°)

    e) tg 900°   f) sen 19Л/6 rad   g) cos 11Л rad

                (Sol: a) - 2 /2; b)1/2; c) - 3 ; d) -1/2; e) 0; f) -1/2; g) -1)

56. Express the following trigonometric ratios as a function of an angle at the first quadrant: a) sen 1485°   b) cos 1560°   c) sen 1000°

                (Soluc: sen 45°; -cos 60°; -sen 80°)

57. Express the following trigonometric ratios as a function of an angle at the first quadrant:

    a) sen 1300°   b) cos (-690°)   c) tg 170°   d) sen (-1755°)

    e) sen (-120°)   f) ctg (-150°)   g) sen 2700°   h) sec (-25°)

    i) cos (-30°)   j) cosec 4420°

8. Expressions for the addition of angles (A+B)

In this part of the unit, we are going to learn to calculate the trigonometric ratios of the addition of 2 angles.

> $sin(A+B) = sinA \cdot cosB + sinB \cdot cosA$ $\qquad cos(A+B) = cosA \cdot cosB - sinA \cdot sinB$
>
> $sin(A–B) = sinA \cdot cosB - sinB \cdot cosA$ $\qquad cos(A–B) = cosA \cdot cosB + sinA \cdot sinB$
>
> $tg(A+B) = \dfrac{tgA + tgB}{1 - tgA \bullet tgB}$ $\qquad\qquad tg(A-B) = \dfrac{tgA - tgB}{1 + tgA \bullet tgB}$

Example: Knowing sinA = 1/2 and sin B = 1/5, calculate sine, cosine and tangent of (A + B).

<u>Solution:</u> First, we need cosA and cosB. We will apply the relationship $\sin^2 \alpha + \cos^2 \alpha = 1$:

$$\left(\frac{1}{2}\right)^2 + \cos^2 A = 1 \quad \rightarrow \quad \frac{1}{4} + \cos^2 A = 1 \quad \rightarrow \quad \cos^2 A = 1 - \frac{1}{4} = \frac{3}{4} \quad \rightarrow \quad \cos A = \sqrt{\frac{3}{4}} = \frac{\sqrt{3}}{2}$$

$$\left(\frac{1}{5}\right)^2 + \cos^2 B = 1 \quad \rightarrow \quad \frac{1}{25} + \cos^2 B = 1 \quad \rightarrow \quad \cos^2 B = 1 - \frac{1}{25} = \frac{24}{25} \quad \rightarrow \quad \cos B = \sqrt{\frac{24}{25}} = \frac{\sqrt{24}}{5}$$

$$sin(A+B) = sinA \cdot cosB + sinB \cdot cosA = \frac{1}{2} \bullet \frac{\sqrt{24}}{5} + \frac{1}{5} \bullet \frac{\sqrt{3}}{2} = \frac{\sqrt{24}}{10} + \frac{\sqrt{3}}{10} = \boxed{0.663}$$

$$cos(A+B) = cosA \cdot cosB - sinA \cdot sinB = \frac{\sqrt{3}}{2} \bullet \frac{\sqrt{24}}{5} - \frac{1}{2} \bullet \frac{1}{5} = \frac{\sqrt{72}}{10} - \frac{1}{10} = \boxed{0.749}$$

To calculate *tg(A+B)*, we need *tgA* and *tgB*:

$$tgA = \frac{\sin A}{\cos A} = \frac{1/2}{\sqrt{3}/2} = \frac{\sqrt{3}}{3} \qquad\qquad tgB = \frac{\sin B}{\cos B} = \frac{1/5}{\sqrt{24}/5} = \frac{\sqrt{24}}{24} = \frac{2\sqrt{6}}{24} = \frac{\sqrt{6}}{12}$$

$$tg(A+B) = \frac{tgA + tgB}{1 - tgA \bullet tgB} = \frac{\dfrac{\sqrt{3}}{3} + \dfrac{\sqrt{6}}{12}}{1 - \dfrac{\sqrt{3}}{3} \bullet \dfrac{\sqrt{6}}{12}} = \boxed{0.886}$$

Exercises

58. Knowing that $sinx = 12/13$ and $siny = 4/5$, with x and y in the 1st quadrant, calculate:

a) sen (x+y)   b) sen (x-y)   c) cos (x+y)   d) cos (x-y)

(Sol: a) 56/65; b) 16/65; c) -33/65; d) 63/65)

59. Knowing that $tgA = 3/4$, calculate $(A+30°)$ and $tg (45°- A)$.

60. SinA $= 1/3$ and $0 < A < 90°$. Calculate:

a) $sin(A+30°)$ b) $sin(A + 45°)$ c) $cos(A+60°)$ d) $tan(60° - A)$

---

9. Expressions for the double angle (2·A)

Expressions for the trigonometric ratios of double angle, $2 \cdot A$, is obtained from the expressions for the addition of angles:

$$sin(2A) = sin(A+A) = sinA \cdot cosA + sinA \cdot cosA = 2 \cdot sinA \cdot cosA$$

$$cos(2A) = cos(A+A) = cosA \cdot cosA - sinA \cdot sinA = cos^2A - sin^2A$$

$$tg(2A) = tg(A+A) = \frac{tgA + tgA}{1 - tgA \bullet tgA} = \frac{2 \cdot tgA}{1 - tg^2A}$$

> $sin(2A) = 2 \cdot sinA \cdot cosA$
> $cos(2A) = cos^2A - sin^2A$
> $tg(2A) = \dfrac{2 \cdot tgA}{1 - tg^2A}$

Exercises

61. Calculate sine, cosine and tagent of $20°$ as a function of $sin10°$, and check your results with your calculator.

62. Find out *sen 2x, cos 2x* and *tg 2x*, with angle x in the first quadrant:

a) sen x $= 1/2$ b) cos x $= 3/5$ c) sen x $= 5/13$

(Soluc: a)3/2; 1/2;3;   b) 24/25; -7/25; -24/7;   c) 120/169; 119/169; 120/119)

63. If A is an angle of the 3rd quadrant and $tgA = \dfrac{\sqrt{3}}{3}$, calculate the trigonometric ratios of angle $2A$.

64. If $cosA = 1/5$ and $0 < A < 90°$, calculate trigonometric ratios of $(90° - 2A)$.

65. If $cotgA = 4/3$, calculate cos2A.

66. Knowing that $tg2A = 3$, Find out $sinA$ and $cosA$, with A < 90°. What is the value of angle A?

10. Expressions for the half angle (A/2)

Expressions for the trigonometric ratios of half angle, A/2, is obtained from the expressions for the double angle. We will use these relationships:

$$cos(2A) = cos^2A - sin^2A \quad \text{and} \quad cos^2A + sin^2A = 1.$$

<table>
<tr><td>Adding:</td><td>Subtracting:</td></tr>
<tr><td>$1 = cos^2A + sin^2A$</td><td>$1 = cos^2A + sin^2A$</td></tr>
<tr><td>$+ \quad cos(2A) = cos^2A - sin^2A$</td><td>$- \quad cos(2A) = cos^2A - sin^2A$</td></tr>
<tr><td>$1 + cos(2A) = 2cos^2A$</td><td>$1 - cos(2A) = 2sin^2A$</td></tr>
</table>

$\cos A = \sqrt{\dfrac{1+\cos(2A)}{2}}$ Naming A = a/2:

$\sin A = \sqrt{\dfrac{1-\cos(2A)}{2}}$ Naming A = a/2:

$$\boxed{\cos\left(\frac{a}{2}\right) = \sqrt{\frac{1+\cos a}{2}}} \qquad \boxed{\sin\left(\frac{a}{2}\right) = \sqrt{\frac{1-\cos a}{2}}}$$

Finally, $tg\left(\dfrac{a}{2}\right) = \dfrac{\sin\left(\frac{a}{2}\right)}{\cos\left(\frac{a}{2}\right)} = \dfrac{\sqrt{\dfrac{1+\cos a}{2}}}{\sqrt{\dfrac{1-\cos a}{2}}} = \boxed{\sqrt{\dfrac{1+\cos a}{1-\cos a}}}$

$$sin(2A) = sin(A+A) = sinA \cdot cosA + sinA \cdot cosA = 2 \cdot sinA \cdot cosA$$

$$cos(2A) = cos(A+A) = cosA \cdot cosA - sinA \cdot sinA = cos^2A - sin^2A$$

$$tg(2A) = tg(A+A) = \frac{tgA + tgA}{1 - tgA \bullet tgA} = \frac{2 \cdot tgA}{1 - tg^2A}$$

$$\boxed{\cos\left(\frac{a}{2}\right) = \sqrt{\frac{1+\cos a}{2}}} \qquad \boxed{\sin\left(\frac{a}{2}\right) = \sqrt{\frac{1-\cos a}{2}}} \qquad \boxed{tg\left(\frac{a}{2}\right) = \sqrt{\frac{1+\cos a}{1-\cos a}}}$$

67. Calculate *tgЛ/8*.

68. Given A $\in$ IV, if *secA* = 2, find out *cosA/2*.

69. A is an angle of the 2nd quadrant and *tg A* = -3/4. Calculate trigonometric ratios of angle A/2.

70. Given A $\in$ III, and *sinA* = -1/2, calculate trigonometric ratios of A/2.

71. A is an angle of the *4th* quadrant and *tgA* = - 3. Calculate trigonometric ratios of angle A/2.

Mixed

72. *CosA* = - ½ (A $\in$ III). Calculate (rationalized and simplified results, not decimals):

    a) sen2A   *b) cosA/2*   *c) sen(A-30°)*   *d) tg(A+60°)*

73. *CosA* = - ½ (A $\in$ IV). Calculate (rationalized and simplified results, not decimals):

    a) cos(A+30°)   *b) tg(a-45°)*   *c) sen(A+1650°)*   *d) senA/2*   *e) cos2A*

74. *SecA* = - 3 (A $\in$ III). Calculate (rationalized and simplified results, not decimals):

    a) sen(A-60°)   *b) tg(A+45°)*   *c) cos(A -2640°)*   *d) cosA/2*   *e) sen2A*

11. Use of trigonometric relationships to simplify expressions

75. Simplify the following expressions:

a) $sen\alpha \cdot \dfrac{1}{tg\,\alpha}$ b) $sen\alpha.cos\alpha \cdot \left(tg\alpha + \dfrac{1}{tg\alpha} \right)$ c) $sen^3\,\alpha + sen\,\alpha \cdot cos^2\,\alpha$

d) $\dfrac{cos^2\,\alpha - sen^2\alpha}{cos^4\,\alpha - sen^4\alpha}$ e) $\dfrac{cosec\alpha}{1 + cot\,g^2\alpha}$ f) $\dfrac{cos^2\,\alpha}{1 - sen\alpha}$

g) $2 \cdot \sqrt{3} \cdot cos30° + tg^2 60° - cot\,g\dfrac{\pi}{4}$

h) $cos^3\,\alpha + cos^2\,\alpha \cdot sen\,\alpha + cos\,\alpha \cdot sen^2\,\alpha + sen^3\,\alpha$

76. Simplify the following expressions:

a) $\dfrac{\text{sen}^2 x.(1+\cos x)}{1-\cos x}$

b) $\dfrac{\cos x}{\text{tag} x.(1-\text{sen} x)}$

77. Check the following equality:

$$\dfrac{\sec^2 x}{\cos ec^2 x - \sec^2 x} + \dfrac{\text{ctg}^2 x}{\text{ctg}^2 x - 1} = \dfrac{1}{\cos^2 x - \text{sen}^2 x}$$

78. Check the following equalities:

a) $\dfrac{\text{sen } 4\alpha + \text{sen } 2\alpha}{\cos 4\alpha + \cos 2\alpha} =$ *(Soluc: tg 3α)*

b) $\dfrac{\text{sen } 2\alpha}{1-\cos^2\alpha} =$ *(Soluc: 2 ctgα)*

c) $\dfrac{2\cos(45°+\alpha)\cos(45°-\alpha)}{\cos 2\alpha} =$ *(Soluc: 1)*

d) $2\,\text{tg } x\cos^2\dfrac{x}{2} - \text{sen } x =$ *(Soluc: tg x)*

e) $2\,\text{tg }\alpha\,\text{sen}^2\dfrac{\alpha}{2} + \text{sen }\alpha =$ *(Soluc: tg α)*

f) $\dfrac{\cos(a+b)+\cos(a-b)}{\text{sen}(a+b)+\text{sen}(a-b)} =$ *(Soluc: ctg a)*

79. Check the following equalities:

a) $\dfrac{1-\cos 2\alpha}{\text{sen}^2\alpha+\cos 2\alpha} = 2\,\text{tg}^2\alpha$

b) $\text{sen } 2\alpha\cos\alpha - \text{sen }\alpha\cos 2\alpha = \text{sen }\alpha$

c) $\cos\alpha\cos(\alpha-\beta) + \text{sen }\alpha\,\text{sen}(\alpha-\beta) = \cos\beta$

d) $\text{sen }\alpha + \cos\alpha = \sqrt{2}\cos\left(\dfrac{\pi}{4}-\alpha\right)$

e) $\dfrac{2\,\text{sen }\alpha - \text{sen } 2\alpha}{2\,\text{sen }\alpha + \text{sen } 2\alpha} = \dfrac{1-\cos\alpha}{1+\cos\alpha} = \text{tg}^2\dfrac{\alpha}{2}$

f) $\text{sen}^2\dfrac{\alpha+\beta}{2} - \text{sen}^2\dfrac{\alpha-\beta}{2} = \text{sen }\alpha\,\text{sen }\beta$

12. Trigonometric equations

80. Solve the following trigonometric equations:

a) sen x = 0   *b) sen (x + Π/4) =* $\dfrac{\sqrt{3}}{2}$ *c) 2tagx – 3 cotag x -1 = 0*

d) 3sen²x – 5senx + 2 = 0 :   *e) cos² x – 3 sen²x = 0*

f) 2cosx = 3 tag x

81. Solve the following trigonometric equations:

a) $\text{sen}\,x = \dfrac{\sqrt{3}}{2}$ **b)** $\cos x = -\dfrac{\sqrt{2}}{2}$ **c)** $\text{ctg}\,x = -\sqrt{3}$

d) $\text{sen}\,x = \dfrac{1}{3}$ **e)** $\cos x = -\dfrac{4}{5}$ **f)** $\text{sen}\,x = 0$

g) $\cos x = -1$ **h)** $\text{cosec}\,x = -2$ **i)** $\sec x = -\dfrac{2\sqrt{3}}{3}$

j) $\text{tg}\,x = \sqrt{3}$ **k)** $\text{cosec}\,x = \dfrac{1}{2}$ **l)** $\text{sen}^2x + \cos^2x = 1$

m) $\cos 3x = \dfrac{\sqrt{3}}{2}$ **n)** $\text{sen}\left(x + \dfrac{\pi}{4}\right) = \dfrac{\sqrt{2}}{2}$

82. Solve the following trigonometric equations:

a) cos2x+3senx=2   b) tg2x tgx=1

c) cosx cos2x+2cos²x=0   d) 2sen x=tg 2x

e) $\sqrt{3}\,\text{sen}\dfrac{x}{2} + \cos x = 1$ f) sen2x cosx=6sen³x

g) $\text{tg}\left(\dfrac{\pi}{4} - x\right) + \text{tg}\,x = 1$ h) sen3x-senx=cos2x

i) $\dfrac{\text{sen }5x + \text{sen }3x}{\cos x + \cos 3x} = 1$ j) $\dfrac{\text{sen }3x + \text{sen }x}{\cos 3x - \cos x} = \sqrt{3}$

k) sen3x-cos3x=senx-cosx

Exercises

13. Solving non-right triangles

Although sometimes we can cut a non-right triangle, converting it into 2 right ones, there are two very useful expressions, known as *sine* and *cosine theorems*, we can use.

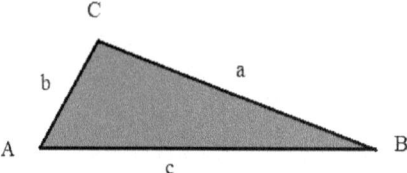

> Sine theorem: $\dfrac{a}{\sin A} = \dfrac{b}{\sin B} = \dfrac{c}{\sin C}$

> Cosine theorem: $a^2 = b^2 + c^2 - 2 \cdot b \cdot c \cdot \cos A$
> $$b^2 = a^2 + c^2 - 2 \cdot a \cdot c \cdot \cos B$$
> $$c^2 = a^2 + b^2 - 2 \cdot a \cdot b \cdot \cos C$$

Exercises

83. This stick is held to the floor by a wire as shown in the figure. Calculate the height of the stick and the length of the wire.

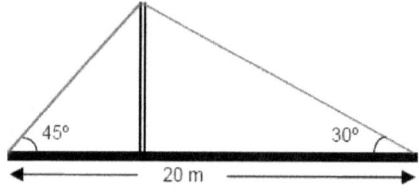

84. This globe is held to the floor by 2 wires. Calculate the length of the longest wire and the height of the globe.

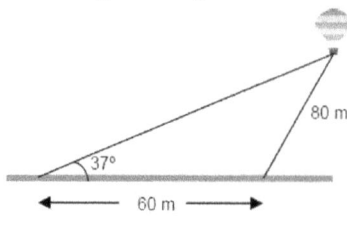

85. Three towns, A, B and C, are connected by straight roads. AB distance is 6 km, BC is 9 km and the angle formed by AB and BC is 120°. What is the distance between A and C?

86. A group have decided to climb a mountain whose height is unknown by them. When they left the town, they determined the elevation angle, 30°. After that, they have advanced 100 m towards the mountain base and now the elevation angle is 45°. Calculate the height of the mountain.

87. Solve the following triangles and calculate their areas:

a) a = 6 m, B = 45°, C = 105°   *(Sol: A =30°, b = 8,49 m, c = 11,59 m, S = 24,60 m²)*

b) a = 10 dam, b = 7 dam, C = 30°   *(Sol: c = 5,27 dam, B = 41° 38', A = 108° 22')*

c) b = 35,42 dm, A = 49° 38', B = 70° 21'

(Sol: C = 60° 1', a = 28,66 dm, c = 32,58 dm, S = 439,94 dm²)

d) a = 13 m, b = 14 m, c = 15 m

(Sol: A = 53° 7' 48", B = 59° 29' 23", C = 67° 22' 48", S = 84m²)

e) a = 15 m, b = 22 m, c = 17 m   *(Sol: A = 42° 54', B = 86° 38', C = 50° 28')*

f) a = 10 mm, b = 7 mm, C = 60°   *(Sol: 8,89 mm, A = 77°, B = 43°, S = 30,31mm²)*

g) a = 10 m, b = 9m, c = 7m   *(Sol: A = 76° 13', B = 60° 57, C = 42° 50')*

88. Solve the triangle ABC knowing that its perimeter is 24 cm, A is a right angle and sinB = 3/5   *(Sol: a = 10 cm, b = 6 cm, c = 8 cm)*

89. Calculate the area of the triangle that is a = 8 m, B = 30°, C = 45°.

90. Fing out the sides of a triangle knowing that its area is 18 cm² and two of its angles are A = 30° and B = 45°.

91. From the door of a house, A, cinema B, situated at 120 m, and school C, at 85 m, are observed under an angle *BAC* = 40°. What is the distance betwwen the cinema and the school?   *(Sol: 77,44 m)*

Exercises

Review exercises

Operating with angles

1. Express in radians the following angles:

   a) 45°   b) 120°   c) 690°   d) 1470°

2. Express in degrees the following angles:

   a) 3 rad   b) 2,5 rad   c) $\dfrac{7\pi}{2}$ rad   c) $\dfrac{\pi}{5}$ rad

3. Find out, without calculator, the quadrant and the trigonometric ratios of the following angles:

   a) 135°   b) 450°   c) 210°   d) –60°

4. Calculate the trigonometric ratios of the following angles:

   a) 1035°   b) –3400°   c) 10.000°   d) 2700°

5. Calculate, without calculator, the values of the following expressions:

   a) 2.tag 30° + 5.tag 240° - cos 270°

   b) cos 60° + sen 150° + sen 210° + cos 240°

6. Calculate the trigonometric ratios of 120°.

7. Knowing that sin25° = 0.42, cos25° = 0.91 and tg25° = 0.47, find out, without using the trigonometric keys of your calculator, the trigonometric ratios of 155° and 205°.

8. Calculate the trigonometric ratios of 130° and 230° knowing that: sen40° = 0,64; cos40° = 0,77 and tg40° = 0,84.

Change of quadrants

9. Knowing that secA = -4 and 0 < A < 90°, calculate:

   a) cosec (270° + A)   b) sen (90° - A)   c) tg (630° - A)

10. Knowing that senA = 2/3 and Л/2 < A < 3Л/2, calculate:

    a) cos (3Л/2 + A)   b) tg (Л - A)

11. Knowing that senA = - 2/3 and Л < A < 2Л, calculate:

    a) cos (3Л/2 - A)   b) tg (Л + A)

12. Knowing that cos 53° = 0,6. Calculate:

    a) cos 37°   b) sen 143°   c) tag 127°   d) cotag 233°   e) sec (-53°)

13. Knowing that tg A = ½ and that Л < A < 3Л/2, calculate:

   a) sen (Л/2 + A)   b) cos (Л/+ A)   c) tag (Л/2 - A)   d) sec (360° - A)

14. Knowing that cotg A = -2 and that Л < A < 2Л, calculate:

   a) cos(Л /2 + A)   b) sen (Л +A)   c) cotag (Л /2 -A)

15. Knowing that sen (Л /2 + A) = -1/3, calculate senA and cos A (A ∈ II)

16. Calculate the value of the expressions:

   a) $\dfrac{sen(\pi/2+x)+\cos(\pi-x)+sen(\pi-x)}{\cos(-x)+sen(-x)}$

   b) $\dfrac{\cot ag(\pi/2-x).sen(\pi/2+x)}{2.\tan(\pi-\alpha)}$

   c) $\dfrac{\tan(\pi-x).\cos(-x)}{\cot ag(\pi+x).\cos(\pi/2-x)}$

Trigonometric formulae

17. If Л /2 < x < 2 Л and being tg x = 3/4, calculate the value and the quadrant of:

   a) sen (x/2)   b) tg (x + 3Л /4)

18. If cos x = -4/5 and Л < x < 2 Л, calculate the value and the quadrant of cos (x/2) and sen (2x).

19. Knowing sen x = - 3/5 and knowing Л /2 < x < 3 Л /2, calculate the value and the quadrant of: a) tag (x - Л /4)   b) sen (x/2)

20. If cos A = -5/13 and Л < A < 2 Л, calculate the value and the quadrant of the following angles: a) sen(2A)   b) tag (A/2)

21. If x is an angle between Л /2 and 3 Л /2 and its sine is 3/5, calculate sen (2x) and cos(x/2).

22. If sen x = -3/5 (90° < x < 270°), calculate and indicate the quadrant of:

   a) sen (x/2)   b) cos (2x)

23. Knowing that Л /2 < A < 3 Л /2 and sen A = 1/3:

   a) Find out the quadrant and the other trigonometric ratios of A.

   b) Find out the quadrant and the other trigonometric ratios of 2A.

   c) Find out the quadrant and the value of sen (A/2).

24. Knowing that 90° < x < 270° and sen x = -2/5, find out the quadrant and the value of:

   a) sen (2x)   b) cos (x/2)   c) ctg (x + 45°)

Simplifications and identities

25. Simplify the following expressions:

a) $\dfrac{\left(1-\tan^2 x\right)\sin x.\sec^2 x}{\left(\cos^2 x-\sin^2 x\right)\tan x}$

b) $\dfrac{\sin(\pi+x)\tan\left(\dfrac{\pi}{2}+x\right)}{\sec^2\left(\dfrac{\pi}{2}+x\right)\left(1-\cos^2 x\right)\cos x}-\cos^2\left(\dfrac{\pi}{2}+x\right)$

c) $\dfrac{1}{1-\sin x}+\dfrac{1}{1+\sin x}-2$

d) $\left[\dfrac{\sec x}{1+\tan^2 x}\right]:\left[(\sin x+\cos x)^2-(\sin x-\cos x)^2\right]$

26. Check **if** the following identities are true:

a) $\dfrac{1-\sin^2\alpha}{\cos\alpha}=\cos\alpha$

b) $\tan x+\dfrac{1}{\tan x}=\tan x.\dfrac{1}{1-\cos^2 x}$

c) $\cos^2 x+\sin^2 x+\tan^2 x=\dfrac{1}{\cos^2 x}$

d) $1+\dfrac{1}{\tan^2 x}=\dfrac{1}{\sin^2 x}$

27. Check the following identities are true:

a) $\dfrac{\sin x\cos x}{\cos^2 x-\sin^2 x}=\dfrac{1}{2}\tan 2x$

b) $\sin(x+y)\cdot\sin(x-y)=\sin^2 x-\sin^2 y$

c) $\cos(x+45°)\cdot\cos(x-45°)=\dfrac{1}{2}\cos 2x$

d) $\dfrac{\sin 2x}{\sin x}+\cos^2\dfrac{x}{2}=\dfrac{5\cos x+1}{2}$

e) $\cos x+2\sin^2\dfrac{x}{2}=1$

f) $\dfrac{\sin x}{1+\cos x}+\dfrac{1+\cos x}{\sin x}=\dfrac{4+4\cos x}{2\sin x+\sin 2x}$

Trigonometric equations

28. Solve the following trigonometric equations:

a) $\tan^2 x-\tan x=0$

b) $2\cos^2 x-\sin^2 x+1=0$

c) $2\sin x.\cos^2 x-6\sin^3 x=0$

d) $\cos(2x+20°)=-\dfrac{\sqrt{3}}{2}$

e) $3\sec x-2\sin x.\tan x=-3$

f) $\sin^2 x+\dfrac{1}{\sec x}=\dfrac{5}{4}$

29. Solve the following trigonometric equations:

a) $\cos 2x=3\sin x-1$

b) $\sin^2 x=1+\cos^2 x$

c) $(\sin^2 x)-1=2\cos^2 x$

d) $\sin x\sin 2x+2\sin^2 x=0$

e) $2-4\cos^2 x=2\sin x$

f) $\cos 2x+\sin^2 x-\dfrac{1}{2}=0$

g) $\sin(x+45°)+\sin(x-45°)=1$

h) $\cos^2\dfrac{x}{2}\cos x=\dfrac{1}{4}$

i) $\cos 2x+\cos^2 x=2$

j) $\cos(6x)+\cos(8x)=-\sqrt{3}.\cos x$

k) $\cos 5x+\cos 3x=\sqrt{2}.\cos 4x$

Unit 5.- Complex numbers

1. Complex numbers. Standard form

When solving equations in the form $x^2 + 9 = 0$

$$x^2 = -9$$

$$x = \pm \sqrt{-9}$$

previous years you have answered *"This equation has not any solution because there are not squared root for negative numbers"*

From now, we will answer: *"This equation has not any REAL solution because there are not REAL squared root for negative numbers"*

At this moment, the set of the Complex numbers (C) is being defined.

The basic unit of the complex numbers is $\sqrt{-1}$, which is called the ***Imaginary Unit***. $i = \sqrt{-1}$

For the example above, $x = \pm \sqrt{-9} = \pm \sqrt{9 \cdot (-1)} = \pm \sqrt{9} \sqrt{(-1)} = \pm 3 \cdot i$)

Exercises

1. Solve the following equations, using complex numbers:

   a) $x^2 - 2x + 2 = 0$ b) $x^2 + 3 = 0$ c) $x^2 - 2x + 4$

   d) $x^2 + x + 1 = 0$ e) $x^4 - 1 = 0$

   (Sol: a) $1 \pm i$; b) $\pm \sqrt{3}i$; c) $1 \pm \sqrt{3}i$; d) $-\dfrac{1}{2} \pm \dfrac{\sqrt{3}}{2}i$; e) $\pm 1, \pm i$)

Standard form of a complex number

A *complex number* z is a number of the form $a + bi$ where a, b are real numbers and $i^2 = -1$.

The set **C** of all complex numbers is defined by $\mathbf{C} = \{a + bi : a, b \in R \text{ and } i^2 = -1\}$ where:

➤ *a* is called the ***real part***.

➤ *b* is called the ***imaginary part***.

Notice that:

- If a = 0, the complex number is said to be ***purely imaginary***.
- If b = 0, the complex number z is ***real***.

Opposite and conjugate of a complex number

Given a complex number, $z = a + b \cdot i$,

➤ ***Opposite of z:*** $-z = -a - b \cdot i$

➤ ***Conjugate of z:*** $\overline{z} = a - b \cdot i$

Notice: $z + Op(z) = 0$ and $z \cdot \overline{z}$ = Real number

2. Indicate real part (Re(z)), imaginary part (Im(z)), opposite and conjugate of the following complex numbers:

a) $z = 2 + 3i$ b) $z = 3 - i$ c) $z = 1 + i$ d) $z = 3 - 3\sqrt{3}\,i$

e) $z = 3$ f) $z = -2i$ g) $z = i$

Operations with complex numbers in their standard form

Let $z_1 = a + bi$ and $z_2 = c + di$, then:

- Addition: $z_1 + z_2 = (a+c) + (b+d)i$

- Subtraction: $z_1 - z_2 = (a-c) + (b-d)i$

- Product: $z_1 z_2 = (a+bi)(c+di) = (ac - bd) + (ad + bc)i$

- Quotient: $\dfrac{z_1}{z_2} = \dfrac{a+bi}{c+di} \cdot \dfrac{c-di}{c-di} = \dfrac{ac+bd}{c^2+d^2} + \dfrac{(bc-ad)}{c^2+d^2}i$,   where $z_2 \neq 0$.

Do not memorize the rules for product and quotient. Look at the following examples:

Example: If $z_1 = -2 + 3i$ and $z_2 = 1 - 4i$, find: a) $z_1 + 2z_2$ b) $z_1 - z_2$ c) $z_1 z_2$ d) $\dfrac{z_1}{z_2}$

Solution: For additions and subtractions, add or subtract, separately, real and imaginary parts:

a) $z_1 + 2z_2 = (-2 + 3i) + 2\cdot(1 - 4i) = (-2 + 3i) + (2 - 8i) = (-2 + 2) + (3 - 8)\cdot i = \boxed{(-5)\cdot i}$

b) $z_1 - z_2 = (-2 + 3i) - (1 - 4i) = (-2 - 1) + (3 + 4)\cdot i = \boxed{-3 + 7i}$

c) In products, we will apply the distributive property:

$z_1 z_2 = (-2 + 3i)\cdot(1 - 4i) = (-2)\cdot1 + (-2)\cdot(-4)\cdot i + 3\cdot i\cdot1 + 3\cdot(-4)\cdot i^2$ Remember, $i^2 = -1$

$= -2 + 8i + 3i + 3\cdot(-4)\cdot(-1) = -2 + 8i + 3i + 12 = \boxed{10 + 11i}$

d) In quotients, we will multiply numerator and denominator by the *conjugate* of the denominator. The conjugate of a complex number is another complex number obtained by changing the sign of the imaginary part. For example, conjugate of (5 – 3i) is (5 + 3i).

But, why do we do this?

For example, multiply complex number (3 – 2i) by its conjugate:

$$(3-2i)\cdot(3+2i) = 3\cdot3 + 3\cdot2i - 2i\cdot3 - 2\cdot2\cdot i^2 = 9 + 6i - 6i + 4 = 9 + 4 = 13$$

When we multiply a complex number by its conjugate, we are obtaining a real number, and its value in the sum of the squares of the real and imaginary parts.　　$(a+bi)\cdot(a-bi) = a^2 + b^2$

$$\frac{z_1}{z_2} = \frac{(-2+3i)}{(1-4i)} = \frac{(-2+3i)\cdot(1+4i)}{(1-4i)\cdot(1+4i)} = \frac{-2-8i+3i-12}{1+16} = \frac{-14-5i}{17} = \boxed{-\frac{14}{17} - \frac{5}{17}i}$$

3. Given complex numbers $z_1 = 2 + 3i$, $z_2 = -1 + 4i$ and $z_3 = 2 - 5i$, calculate:

　　a) z_1+z_2　　　b) z_1+z_3　　　　c) z_1-z_2　　　d) z_3-z_2　　　e) $3z_2+2z_3$

　　f) $2z_1$-$3z_2$　　g) z_3-$3z_1+4z_2$　　h) $z_1 + \overline{z_2}$　　i) $z_3 - \overline{z_3}$　　j) $2\overline{z_1} - z_1$

　　　　　　　　(Sol: a) 1+7i; b) 4-2i; c) 3-i; d) 3-9i; e) 1+2i; f) 7-6i;

　　　　　　　　　　　　　　g) -8+2i; h) 1-i; i) -10i; j) 2-9i)

4. Calculate the following additions and subtractions of complex numbers:

　　a) $(-4+7i)+(5-10i)$　　　　b) $(4+12i)-(3-15i)$　　　c) $5i-(-9+i)$

　　　　　　　　(Sol.: a) 1 – 3i;　　b) 1 + 27i;　　c) 9 + 4i)

5. Calculate x and y so that $(2 + xi) + (y + 3i) = 7 + 4i$.　　　　*(Sol: x=1, y=5)*

6. Work out each of the following products and write the answers in standard form:

　　a) $7i\cdot(-5+2i)$　　b) $(1-5i)\cdot(-9+2i)$　　c) $(4+i)\cdot(2+3i)$　　d) $(1-8i)\cdot(1+8i)$

　　　　　　　　(Sol.: a) -14 – 35i;　　b) 1 + 47i;　　c) 5 + 14i;　　d) 65)

7. Calculate:

a) $(2+5i)(3+4i)$ *(Sol: -14+23i)*   b) $(1+3i)(1+i)$ *(Sol: -2+4i)*

c) $(1+i)(-1-i)$ *(Sol: -2i)*   d) $(2-5i) \cdot i$ *(Sol: 5+2i)*

e) $(2+5i)(2-5i)$ *(Sol: 29)*   f) $(1+i)(1-i)$ *(Sol: 2)*

g) $(5+2i)(3-4i)$ *(Sol: 23-14i)*   h) $(3+5i)^2$ *(Sol: -16+30i)*

i) $(1+3i)(1-3i)$ *(Sol: 10)*   j) $(-2-5i)(-2+5i)$ *(Sol: 29)*

k) $(2+3i)\cdot 3i$ *(Sol: -9+6i)*   l) $(3i)\cdot(-3i)$ *(Sol: 9)*

m) $(2+3i)^2$ *(Sol: -5+12i)*   n) $(6-3i)^2$ *(Sol: 27-36i)*

o) $(2+3i)\cdot(1-i)$ *(Sol: 5+i)*   p) $(1-3i)\cdot 2i$ *(Sol: 6+2i)*

q) $(1+i)\cdot(2-3i)$ *(Sol: 5-i)*   r) $(5+i)\cdot(5-i)$ *(Sol: 26)*

8. Given complex numbers $(2 - mi)$ and $(3 - ni)$, find out m and n so that their product is $(8 + 4i)$.   *(Sol: m_1=-2, n_1=1; m_2=2/3, n_2=-3)*

9. Solve equation: $(a + i)\cdot(b - 3i) = (7 - 11i)$ *(Sol: a_1=4, b_1=1; a_2=-1/3 y n_2=-12)*

10. Calculate the following quotients:

a) $\dfrac{1+3i}{1+i}$ *(Sol : 2+i)*   h) $\dfrac{1+i}{i}$ *(Sol : 1-i)*

b) $\dfrac{2+5i}{3+4i}$ $\left(Sol : \dfrac{26}{25}+\dfrac{7}{25}i \right)$ i) $\dfrac{1+2i}{2-i}$ *(Sol : i)*

c) $\dfrac{1+i}{1-i}$ *(Sol : i)*   j) $\dfrac{1-i}{2+3i}$ $\left(Sol : -\dfrac{1}{13}-\dfrac{5}{13}i \right)$

d) $\dfrac{3+5i}{1-i}$ *(Sol : -1+4i)*   k) $\dfrac{19-4i}{2-5i}+\dfrac{3+2i}{i}$ *(Sol : 4)*

e) $\dfrac{2-5i}{i}$ *(Sol : -5-2i)*   l) $\dfrac{2-i}{3+i}-\dfrac{1}{2i}$ $\left(Sol : \dfrac{1}{2} \right)$

f) $\dfrac{20+30i}{3+i}$ *(Sol : 9+7i)*   m) $\dfrac{(5-3i)(1+i)}{1-2i}$ $\left(Sol : \dfrac{12}{5}-\dfrac{14}{5}i \right)$

g) $\dfrac{i}{3-2i}$ $\left(Sol : -\dfrac{2}{13}+\dfrac{3}{13}i \right)$ n) $\dfrac{(3+2i)^2 + 3-2i}{(5+i)^2}$ $\left(Sol : \dfrac{73}{169}+\dfrac{40}{169}i \right)$

Inverse of a complex number

Given a complex number, $z = (a + b \cdot i)$, its *inverse*, z^{-1}, is another complex number that satisfy:

$$z \cdot z^{-1} = 1$$

Example: Calculate the inverse of the following complex numbers:

        a) 5i   b) 2 + 3i

<u>Solution:</u> Inverse is unknown, so, we can name it as $z^{-1} = (a + b \cdot i)$:

a) $5i \cdot (a + b \cdot i) = 1 \quad \rightarrow \quad 5 \cdot a \cdot i + 5 \cdot b \cdot i^2 = 5a \cdot i + 5b \cdot (-1) = 5a \cdot i - 5b = 1 + 0 \cdot i$

Comparing: $\begin{cases} 5a = 0 & \rightarrow \ a = 0 \\ -5b = 1 & \rightarrow \ b = -\frac{1}{5} \end{cases}$ So, inverse is $\boxed{z^{-1} = -\frac{1}{5} i \cdot i}$

b) $(2 + 3i) \cdot (a + b \cdot i) = 1 \quad \rightarrow \quad 2a + 2b \cdot i + 3a \cdot i - 3b = (2a - 3b) + (2b + 3a) \cdot i = 1 + 0 \cdot i$

Comparing: $\begin{cases} 2a - 3b = 1 \\ 3a + 2b = 0 \end{cases}$ Solving the system, we have: $\left(a = \frac{2}{13}, b = -\frac{3}{13} \right)$.

So, inverse $\boxed{z^{-1} = \left(\dfrac{2}{3} - \dfrac{3}{13} \cdot i \right)}$

➤ You should check your results satisfy that: $z \cdot z^{-1} = 1$

Exercises

11. Calculate the inverse of the following complex numbers:

a) 3i $\left(Sol: -\frac{1}{3} i \right)$ **c)** 2+3i $\left(Sol: \frac{2}{13} - \frac{3}{13} i \right)$ **e)** -2+i $\left(Sol: -\frac{2}{5} - \frac{1}{5} i \right)$

b) 1+i $\left(Sol: \frac{1}{2} - \frac{1}{2} i \right)$ **d)** 1-i $\left(Sol: \frac{1}{2} + \frac{1}{2} i \right)$ **f)** i $(Sol: -i)$

Powers of imaginary unit, i^n

Observe:

$i^1 = 1$ $i^2 = (-1)$ $i^3 = i^2 \cdot i = (-1)\cdot i = -i$ $i^4 = i^2 \cdot i^2 = (-1)\cdot(-1) = 1$

$i^5 = i^4 \cdot i^1 = i$ $i^6 = i^4 \cdot i^2 = (-1)$ $i^7 = i^4 \cdot i^3 = -i$ $i^8 = (i^4)^2 = 1$

$i^9 = i^8 \cdot i^1 = i$ $i^{10} = i^8 \cdot i^2 = (-1)$ $i^{11} = i^8 \cdot i^3 = -i$ $i^{12} = (i^4)^3 = 1$

$i^{13} = i^{12} \cdot i^1 = i$ $i^{14} = i^{12} \cdot i^2 = (-1)$ $i^{15} = i^{12} \cdot i^3 = -i$ $i^{16} = (i^4)^4 = 1$

....

$i^{4n+1} = i$ $i^{4n+2} = (-1)$ $i^{4n+3} = -i$ $i^{4n} = 1$

Exercises

12. Calculate the following sucesive powers of i:

a) i^{12} b) i^{77} c) i^{125} d) i^{723} e) i^{2344} f) $1/i$ g) $1/i^2$

h) $1/i^3$ i) i^{-4} j) $1/i^5$ k) i^{-6} l) i^{544} m) i^{6254} n) i^{-1}

o) i^{-527} *(Sol: 1, i, i, -i, 1, -i, -1, i, 1, -i, -1, 1, -1, -i, i)*

13. Calculate:

a) $(2+i)^3$ *(Sol: 2+11i)*   f) $\dfrac{2i-1}{i^{45}} + \dfrac{4-3i}{1+2i}$ *(Sol: 4+2i)*

b) $(1+i)^3$ *(Sol: -2+2i)*

c) $(2-3i)^3$ *(Sol: -46-9i)*   g) $\dfrac{(3-2i)^2 + (2-3i)^2}{i^{12} + i^{-5}}$ *(Sol: 12-12i)*

d) i^{-131} *(Sol: i)*   h) $\dfrac{(2+3i)(1-i) - (3+4i)^2}{2i^{14} - i^{-7}}$ $\left(Sol:\ -\dfrac{1}{5} + \dfrac{58}{5}i\right)$

e) $\dfrac{i^7 - 1}{1+i}$ *(Sol: -1)*   i) $\dfrac{(3-2i)(3+i) - (2i-3)^2}{i^{23} - i^{13}}$ $\left(Sol:\ -\dfrac{9}{2} + 3i\right)$

14. Calculate the value of m that makes complex number z = (m-2i) (2+4i) be a real number. What is the real number obtained?

15. Determine x so that product z = (2-5i) (3+xi) is:

a) A real number. What number? *(Sol: x=15/2; z=87/2)*

b) A purely imginary number. What number? *(Sol: x=-6/5; z=-87i/5)*

16. a) Find out x that makes $(x-2i)^2$ be a purely imaginary number. *(Sol: $x = \pm 2$)*

    b) Idem with $(3x-2i)^2$ *(Sol: $x = \pm 2/3$)*

    c) Idem with $(2+xi)^2$ *(Sol: $x = \pm 2$)*

17. Calculate x and y so that $\dfrac{3-xi}{1+2i} = y + 2i$ *(Sol: $x = (-16)$; $y = 7$)*

18. Calculate x so that $\dfrac{x+3i}{3+2i}$ is a purely imaginary. What number? *(Sol: $x = (-2)$; i)*

19. Determine k so that quotient $z = \dfrac{-2+ki}{k-i}$ is:

    a) A real number. What number?   *(Sol: $k = \pm \sqrt{2}$; $z = \pm \sqrt{2}$)*

    b) A purely imaginary complex number. What number?   *(Sol: $k = 0$; $z = -2 \cdot i$)*

20. Check the following equality, discovered, by chance by the German mathematician Gottfried Leibniz (1646-1716):

$$\sqrt{1+\sqrt{3}\,i} + \sqrt{1-\sqrt{3}\,i} = \sqrt{6}$$

21. Find out two complex numbers knowing that their difference is a real number and their sum has Re(z) = 1 and their product is -7+i.   *(Sol: $(3+i)$ and $(-2+i)$)*

22. Check that complex numbers $(2\pm3i)$ verify equation $x^2 - 4x + 13 = 0$.

2. Geometrical representation of a complex number

A complex number $z = a + bi$ is defined by the two real numbers a and b. Hence, if we consider the real part a as the $x-coordinate$ in the rectangular coordinates system and the imaginary part b as the $y-coordinate$, then the complex number z can be represented by the point (a,b) on the plane. This plane is called the ***complex plane***. On this plane, real

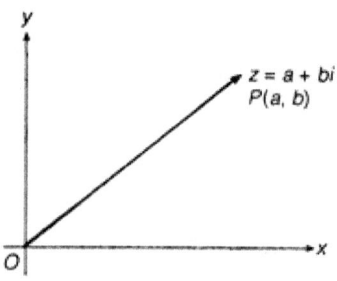

numbers are represented by points on $x-axis$, called the ***real axis***; imaginary numbers are represented by points on the $y-axis$, called the ***imaginary axis***.

Any point (a,b) on this plane can be used to represent a complex number $z = a + bi$. For example, as shown in Figure, the z_1, z_2, z_3 represent respectively the complex numbers

$z_1 = -3, \quad z_2 = 2i \quad and \quad z_3 = 4 - 3i$

These points are called *affix*.

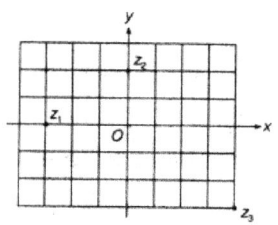

Exercises

23. Represent the following complex numbers, their opposite and their conjugates:

a) $z_1 = 3 + 4i$ b) $z_2 = 1 - i$ c) $z_3 = -3 + i$ d) $z_4 = -2 - 5i$ e) $z_5 = 7i$

3. Polar form of a complex number. Norm and argument

Complex numbers can be also defined by the vector from origin to the point P. So, we should know the *modulus* of this vector and its orientation.

- Norm, modulus or absolute value of a complex number (r): Remember that the absolute value of a number is simply defined as the distance the number is from the origin. As we graph complex numbers on a rectangular coordinate system, their norm is the distance from point P(a, b) to the origin. Real and imaginary coordinates along with the distance from the point to the origin make a right triangle. So, we may use the Pythagorean Theorem to calculate it.

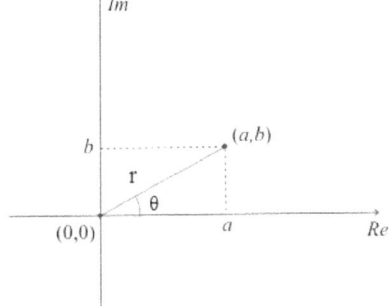

$$a^2 + b^2 = r^2 \rightarrow r = \sqrt{a^2 + b^2}$$

- Argument of a complex number (θ): Orientation is indicated by the angle, θ, between vector **OP** and the positive OX semiaxis.

$$tg\,\theta = \frac{b}{a} \rightarrow \theta = acrtg(b/a)$$

- 137 -

$$a+b\cdot i \leftrightarrow r_\theta \quad where \quad \begin{cases} r = \sqrt{a^2+b^2} \\ \theta = acrtg(b/a) \end{cases}$$

Example: Given complex $\sqrt{3}-i$, give its graphical representation and polar form.

<u>Solution:</u>

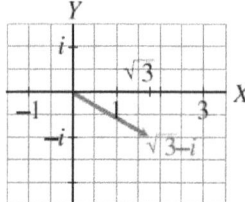

$$r = \sqrt{\sqrt{3}^2+(-1)^2} = \sqrt{3+1} = \sqrt{4} = 2$$

$$\left. \theta = acrtg(-1/\sqrt{3}) = 330° \ (\textit{it must be at the 3rd quadrant}) \right\} \rightarrow 2_{330}$$

Example: Given complex number 4_{135}, give its standard form and its graphical representation.

<u>Solution:</u>

$$\left. a = r\cdot\cos\theta = 4\cdot\cos 135° = 4\left(-\frac{\sqrt{2}}{2}\right) = -2\sqrt{2} \\ \ \\ b = r\cdot\sin\theta = 4\cdot\sin 135° = 4\left(\frac{\sqrt{2}}{2}\right) = 2\sqrt{2} \right\} \rightarrow -2\sqrt{2}+2\sqrt{2}\cdot i$$

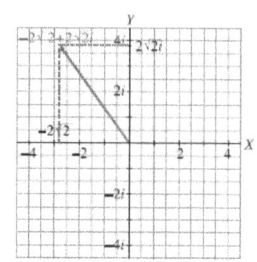

Exercises	**24.** Give the graphical representation of the complex number $-1-i$, give its polar form and find its opposite and its conjugate. *(Sol: $\sqrt{2}_{225}$; $1+i$; $-1+i$)*
	25. Give the graphical representation of the complex number $2-2\sqrt{3}i$, give its polar form and find its opposite and its conjugate. *(Sol: 4_{300}; $-2+2\sqrt{3}i$; $2+2\sqrt{3}i$)*

26. Give the graphical representation of the complex number $-1-i$, give its polar form and find its opposite and its conjugate. *(Sol:* $\sqrt{2}_{225}$; $\quad 1+i$; $\quad -1+i$ *)*

27. Give the graphical representation of the complex number $2-2\sqrt{3}i$, give its polar form and find its opposite and its conjugate. *(Sol:* 4_{300}; $\quad -2+2\sqrt{3}i$; $\quad 2+2\sqrt{3}i$ *)*

28. Give the polar form of the following complex numbers. You should, previously, represent them in order to determine the argument:

a) $4+4\sqrt{3}\,i$ *(Sol:* $8_{60°}$*)*

b) $3-3\sqrt{3}\,i$ *(Sol:* $6_{300°}$*)*

c) $-\sqrt{2}+i$ *(Sol:* $\sqrt{3}_{144°\,44'}$ *)*

d) $-\sqrt{2}-\sqrt{2}\,i$ *(Sol:* $2_{225°}$*)*

e) $\sqrt{3}-i$ *(Sol:* $2_{330°}$*)*

f) $1+i$ *(Sol:* $\sqrt{2}_{45°}$ *)*

g) $1-i$ *(Sol:* $\sqrt{2}_{315°}$ *)*

h) $-1-i$ *(Sol:* $\sqrt{2}_{225°}$ *)*

i) i *(Sol:* $1_{90°}$*)*

j) $-i$ *(Sol:* $1_{270°}$*)*

k) $3+4i$ *(Sol:* $5_{53°\,8'}$*)*

l) $3-4i$ *(Sol:* $5_{306°}$*)*

m) $-3+4i$ *(Sol:* $5_{126°\,52'}$*)*

n) $-5+12i$ *(Sol:* $13_{112°\,37'}$*)*

o) $-8i$ *(Sol:* $8_{270°}$*)*

p) 8 *(Sol:* $8_{0°}$*)*

q) -8 *(Sol:* $8_{180°}$*)*

r) $3+2i$ *(Sol:* $\sqrt{13}_{33°41'}$ *)*

s) $-2-5i$ *(Sol:* $\sqrt{29}_{248°12'}$ *)*

29. a) Find out m so that complex number $(m+3i)$ has norm 5. *(Sol: $m = \pm 4$)*

b) Find out m so that its argument is 60°. *(Sol: $m = \sqrt{3}$)*

30. Find out a complex number at the second quadrant that has norm 2 and such that $Re(z) = (-1)$. Express it in its polar form. *(Sol: $2_{120°}$)*

31. Find out a complex number with argument 45° that, when added to $(1 + 2i)$, you obtain another complex number with norm 5. *(Sol: $2+2i$)*

32. Express in its standard form:

a) $4_{30°}$ b) $4_{90°}$ c) $2_{0°}$ d) 5_{π} e) $2_{3\pi/2}$ f) $1_{90°}$ g) $1_{30°}$

Exercises

33. Express in its standard form:

a) $2_{60°}$ (*Sol:* $1+\sqrt{3}i$)

b) $6_{225°}$ (*Sol:* $-3\sqrt{2}-3\sqrt{2}i$)

c) $4_{120°}$ (*Sol:* $-2+2\sqrt{3}i$)

d) $2_{150°}$ (*Sol:* $-\sqrt{3}+i$)

e) $3_{60°}$ (*Sol:* $\dfrac{3}{2}+\dfrac{3\sqrt{3}}{2}i$)

f) $3_{50°}$ (*Sol:* $1,929+2,298i$)

g) $2_{180°}$ (*Sol:* -2)

h) $1_{210°}$ (*Sol:* $-\dfrac{\sqrt{3}}{2}-\dfrac{i}{2}$)

34. Find out the complex numbers that correspond to the vertex of these hexagons:

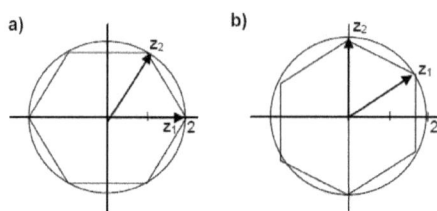

a)   b)

4. Operations with complex numbers in their polar form

4.1. Product and quotient

➤ **Product:** $r_{1_{\theta_1}} \cdot r_{2_{\theta_2}} = (r_1 \cdot r_2)_{(\theta_1+\theta_2)}$

➤ **Quotient:** $r_{1_{\theta_1}} : r_{2_{\theta_2}} = (r_1 : r_2)_{(\theta_1-\theta_2)}$

Exercises

35. Given complex numbers $3_{30°}$ and $5_{60°}$, check that you obtain the same results when multiplying them in their polar and standard forms.   *(Sol: 15i)*

36. Idem with 3i and (2 − 2i).   *(Sol: (6 + 6i) = $6\sqrt{2}_{45°}$)*

37. Given complex numbers $3_{30°}$ and $5_{60°}$, check that you obtain the same results when multiplying them in their polar and standard forms. *(Sol: 15i)*

38. Idem with 3i and (2 − 2i). *(Sol: (6 + 6i) = $6\sqrt{2}_{45°}$)*

39. Work out and express your results in their standard form:

a) $3_{45°} \cdot 2_{15°}$ (*Sol:* $6_{60°} = 3 + 3\sqrt{3}\,i$)

b) $3_{150°} \cdot 4_{45°}$ (*Sol:* $12_{195°} \cong -11,59 - 3,11i$)

c) $1_{33°} \cdot 2_{16°} \cdot 3_{41°}$ (*Sol:* $6_{90°} = 6i$)

d) $3_{12°} \cdot 4_{17°} \cdot 2_{1°}$ (*Sol:* $24_{30°} = 12\sqrt{3} + 12i$)

e) $2_{106°} : 1_{61°}$ (*Sol:* $2_{45°} = \sqrt{2} + \sqrt{2}\,i$)

f) $9_{37°} : 3_{97°}$ (*Sol:* $3_{300°} = \dfrac{3}{2} - \dfrac{3\sqrt{3}}{2}i$)

g) $(2_{40°})^3$ (*Sol:* $8_{120°} \cong -4 + 4\sqrt{3}i$)

h) $1_{33°} : 2_{16°} \cdot 3_{41°}$ (*Sol:* $\left(\dfrac{3}{2}\right)_{58°} \cong 0,79 + 1,27i$)

i) $3_{12°} : 4_{17°} : 2_{1°}$ (*Sol:* $\left(\dfrac{3}{8}\right)_{354°} \cong 0,37 - 0,04i$)

40. Complex number with argument 80° and norm 12 is the product of two complex numbers, one of them with norm 3 and argument 50°. Write, in its standard form, the other one. *(Sol: $2\sqrt{3} + 2i$)*

41. Work out in polar form and express the results in standard form:

a) $\dfrac{2_{15°} \cdot 4_{135°}}{8_{170°}}$ (*Sol:* $1_{340°} \cong 0,94 - 0,34i$)

b) $\dfrac{2_{15°} \cdot (1+i)}{2_{-15°} \cdot (1-i)}$ (*Sol:* $1_{120°} = -\dfrac{1}{2} + \dfrac{\sqrt{3}}{2}i$)

c) $(1 + \sqrt{3}\,i)(1+i)(\sqrt{3} - i)$ (*Sol:* $4\sqrt{2}_{75°} \cong 1,46 + 5,46i$)

42. Calculate the value of α so that product $3_{\pi/2} \cdot 1_{\alpha}$ is:

a) A positive real number. *(Sol: α = 3Π/2)*

b) A negative real number. *(Sol: α = Π/2)*

43. Calculate the value of α so that product $5_{\pi} : 3_{\alpha}$ is:

a) A positive real number. *(Sol: α = Π)*

b) A negative real number. *(Sol: α = 0)*

c) A purely imaginary number with positive argument. *(Sol: α = Π/2)*

d) A purely imaginary number with positive argument. *(Sol: α = 3Π/2)*

44. Without calculating in its standard form, find the value of *m* so that the complex number **z** = (m-2i)·(2+4i) has norm 10. *(Sol: m = ±1)*

Exercises

45. Without calculating in its standard form, find the value of a so that quotient $\dfrac{a+2i}{1-i}$

has norm 2.   *(Sol: m = ±2)*

46. Find out 2 complex numbers knowing their product is (-8) and quotient of one of them by squared of the other one is 1. Use polar form.   *(Sol: $z_1 = 4_{120°}$; $z_2 = 2_{60°}$)*

47. Find out 2 complex numbers knowing their product is 4 and quotient of one of them by squared of the other one is 2. Use polar form.

(Sol: $z_1 = 2\sqrt[3]{4}\ _{0°}$; $z_2 = \sqrt[3]{2}\ _{0°}$)

48. Give the graphical interpretation of the result of multiplying a complex number by imaginary unit, i.

4.2. Powers

Power:	$\left(r_\theta\right)^n = \left(r^n\right)_{(n\cdot\theta)}$
Moivre's expression	$\left(\cos\alpha + sen\alpha\cdot i\right)^n = \cos(n\alpha) + sen(n\alpha)\cdot i$

Notice: $\left(r_\theta\right)^n = r_\theta \cdot r_\theta \cdot r_\theta \cdot r_\theta \cdot \ldots \cdot r_\theta = \left(r^n\right)_{(n\cdot\theta)}$

Exercises

49. Calculate in the most suitable way and express the results in their standard form:

a) $(1+i)^2$ *(Sol: 2i)*

b) $(2-2i)^2$ *(Sol: -8i)*

c) $(1+i)^3$ *(Sol: -2+2i)*

d) $(2+3i)^3$ *(Sol: -46+9i)*

e) $(1-i)^4$ *(Sol: -4)*

f) $(-2+i)^5$ *(Sol: 38+41i)*

g) $\dfrac{(1+i)^2}{4+i}$ $\left(Sol:\ \dfrac{2}{17}+\dfrac{8}{17}i\right)$

h) $\dfrac{2+i}{(1+i)^2}$ $\left(Sol:\ \dfrac{1}{2}-i\right)$

i) $(i^4+i^{-13})^3$ *(Sol: -2-2i)*

j) $(1+i)^{20}$ *(Sol: -1024)*

k) $(-2+2\sqrt{3}\ i)^6$ *(Sol: 4096)*

l) $\dfrac{i^7-i^{-7}}{2i}$ *(Sol: -1)*

- 142 -

50. Calculate in the most suitable way and express the results in their standard form:

a) $(4 - 4\sqrt{3}\ i)^3$ (*Sol:* *-512*)

b) $(-2 + 2\sqrt{3}\ i)^4$ (*Sol:* $-128 + 128\sqrt{3}i$)

c) $(\sqrt{3} - i)^5$ (*Sol:* $-16\sqrt{3} - 16i$)

d) $\left(\dfrac{3\sqrt{3}}{2} + \dfrac{3i}{2}\right)^3$ (*Sol:* *27i*)

e) $(-1 + i)^{30}$ (*Sol:* $2^{15}i$)

f) $\dfrac{(-1+i)^2}{(1+i)^3}$ $\left(\textit{Sol:}\ -\dfrac{1}{2} + \dfrac{i}{2} \right)$

g) $(2 + 2\sqrt{3}\ i)^4$ (*Sol:* $-128 - 128\sqrt{3}i$)

h) $(4 + 4\sqrt{3}\ i)^4$ (*Sol:* $-2048 - 2048\sqrt{3}i$)

i) $(2 + 2\sqrt{3}\ i)^2$ (*Sol:* $-8 + 8\sqrt{3}i$)

j) $(1 + i)^5$ (*Sol:* *-4-4i*)

k) $(1 + 2i)^3$

l) $(2 + i^5)^3$ (*Sol:* *2+5i*)

m) $(3 + 3i)^5$ (*Sol:* *-972-972i*)

n) $\dfrac{\left(2\sqrt{3} - 2i\right)^8}{\left(-4\sqrt{2} + 4\sqrt{2}\ i\right)^6}$ $\left(\textit{Sol:}\ \left(\dfrac{1}{4}\right)_{30^\circ} = \dfrac{\sqrt{3}}{8} + \dfrac{1}{8}i \right)$

ñ) $\dfrac{(1 - \sqrt{3}\ i)^3 \cdot (\sqrt{2} + \sqrt{2}\ i)}{(-2\sqrt{2} + 2\sqrt{2}\ i)^2}$ $\left(\textit{Sol:}\ \dfrac{\sqrt{2}}{2} - \dfrac{\sqrt{2}}{2}i \right)$

51. Use the Moivre´s expression to obtain expressions for:

a) sen4α b) cos4α   c) sen2α d) cos2α   e) sen2α b) cos2α

4.3. Roots

First of all, you must know that complex numbers have n roots, where n is the index of the root.

For example, they have 2 squared roots, 3 cubic roots, ...

$$\textbf{Roots:}\quad \sqrt[n]{r_\theta} = r'_{\theta'}\ \ where:\ \begin{cases} r' = \sqrt[n]{r} \\ \theta' = \dfrac{\theta + k \cdot 2\pi}{n} \end{cases}$$

Notice: If $\sqrt[n]{r_\theta} = r'_{\theta'}\ \Rightarrow\ \left(r'_{\theta'}\right)^n = r_\theta\ \Rightarrow\ (r')^n_{n\cdot\theta'} = r_\theta \Rightarrow \begin{cases} (r')^n = r \rightarrow r' = \sqrt[n]{r} \\ n\cdot\theta' = \theta + k\cdot 2\pi \rightarrow \theta' = \dfrac{\theta + k\cdot 2\pi}{n} \end{cases}$

Example: Calculate: $\sqrt[4]{81_{40°}}$.

Solution: $\sqrt[4]{81_{40°}} = \sqrt[4]{81}\left(\frac{40°+2k180°}{4}\right) = 3_{(10°+k90°)}$

By substituting k by values 0, 1, 2 and 3, you will obtain the 4 roots:

$k = 0 \;\rightarrow\; z_0 = 3_{(10°+0\cdot90°)} = 3_{10°}$

$k = 1 \;\rightarrow\; z_1 = 3_{(10°+1\cdot90°)} = 3_{100°}$

$k = 2 \;\rightarrow\; z_2 = 3_{(10°+2\cdot90°)} = 3_{190°}$

$k = 3 \;\rightarrow\; z_3 = 3_{(10°+3\cdot90°)} = 3_{280°}$

Example: Find out the solutions of $x^5 + \sqrt{2} - \sqrt{2}.i = 0$.

Solution: $x^5 + \sqrt{2} - \sqrt{2}.i = 0 \;\rightarrow\; x^5 = -\sqrt{2} + \sqrt{2}.i \;\rightarrow\; x = \sqrt[5]{-\sqrt{2} + \sqrt{2}.i}$

First, write convert complex number into its polar form: $2_{135°}$

$\sqrt[5]{2_{135°}} = \sqrt[5]{2}\left(\frac{135°+k360°}{5}\right) = \sqrt[5]{2}_{(27°+k72°)}$

By substituting k by values 0, 1, 2 and 3, you will obtain the 5 roots:

$k = 0 \;\rightarrow\; z_0 = \sqrt[5]{2}_{(27°+0\cdot72°)} = \sqrt[5]{2}_{27°}$

$k = 1 \;\rightarrow\; z_1 = \sqrt[5]{2}_{(27°+1\cdot72°)} = \sqrt[5]{2}_{99°}$

$k = 2 \;\rightarrow\; z_2 = \sqrt[5]{2}_{(27°+2\cdot72°)} = \sqrt[5]{2}_{171°}$

$k = 3 \;\rightarrow\; z_3 = \sqrt[5]{2}_{(27°+3\cdot72°)} = \sqrt[5]{2}_{243°}$

$k = 4 \;\rightarrow\; z_4 = \sqrt[5]{2}_{(27°+4\cdot72°)} = \sqrt[5]{2}_{315°}$

52. Calculate the following roots:

a) $\sqrt[4]{1+i}$ $\qquad$ ($Sol:$ $\sqrt[8]{2}_{11,25°}$ $\sqrt[8]{2}_{101,25°}$ $\sqrt[8]{2}_{191,25°}$ $\sqrt[8]{2}_{281,25°}$)

b) $\sqrt[3]{1-i}$ $\qquad$ ($Sol:$ $\cdot\sqrt[6]{2}_{105°}$ $\sqrt[6]{2}_{225°}$ $\sqrt[6]{2}_{345°}$)

c) $\sqrt[4]{\dfrac{-4}{1-\sqrt{3}\,i}}$ $\qquad$ ($Sol:$ $\cdot\sqrt[4]{2}_{60°}$ $\sqrt[4]{2}_{150°}$ $\sqrt[4]{2}_{240°}$ $\sqrt[4]{2}_{330°}$)

d) $\sqrt[3]{\dfrac{-1+3i}{2-i}}$ $\qquad$ ($Sol:$ $\cdot\sqrt[6]{2}_{45°}$ $\sqrt[6]{2}_{165°}$ $\sqrt[6]{2}_{285°}$)

e) $\sqrt[3]{-i}$ $\qquad$ $\left(Sol:\ i;\ -\dfrac{\sqrt{3}}{2}\pm\dfrac{1}{2}i \right)$

f) $\sqrt[3]{-\dfrac{\sqrt{2}}{2}+\dfrac{\sqrt{2}}{2}i}$ $\left(Sol:\ \dfrac{\sqrt{2}}{2}+\dfrac{\sqrt{2}}{2}i;\ -0,97+0,26i;\ 0,26-0,97i \right)$

g) $\sqrt{i}$ $\qquad$ $\left(Sol:\ \dfrac{\sqrt{2}}{2}+\dfrac{\sqrt{2}}{2}i;\ -\dfrac{\sqrt{2}}{2}-\dfrac{\sqrt{2}}{2}i \right)$

h) $\sqrt[3]{\dfrac{1-i}{\sqrt{3}+i}}$ $\qquad$ ($Sol:$ $0,89_{95°}$ $0,89_{215°}$ $0,89_{335°}$)

i) $\sqrt[3]{8i}$ $\qquad$ (. $Sol:$ $2i;\ \pm\sqrt{3}+i$)

j) $\sqrt[4]{-1}$ $\qquad$ $\left(Sol:\ \pm\dfrac{\sqrt{2}}{2}\pm\dfrac{\sqrt{2}}{2}i \right)$

k) $\sqrt[3]{8}$ $\qquad$ ($Sol:$ $2;\ -1\pm\sqrt{3}\,i$)

l) $\sqrt[4]{-2+2\sqrt{3}\,i}$ $\qquad$ $\left(Sol:\ \dfrac{\sqrt{6}}{2}+\dfrac{\sqrt{2}}{2}i;\ -\dfrac{\sqrt{6}}{2}-\dfrac{\sqrt{2}}{2}i;\ -\dfrac{\sqrt{2}}{2}+\dfrac{\sqrt{6}}{2}i;\ \dfrac{\sqrt{2}}{2}-\dfrac{\sqrt{6}}{2}i \right)$

m) $\sqrt[3]{4-4\sqrt{3}\,i}$ $\qquad$ ($Sol:$ $2_{100°}$ $2_{220°}$ $2_{340°}$)

n) $\sqrt[3]{\dfrac{8+8i}{1-i}}$ $\qquad$ $\left(Sol:\ \cdot -2i;\ \pm\dfrac{\sqrt{3}}{2}+i \right)$

o) $\sqrt[4]{-2+2i}$ $\qquad$ (. $Sol:$ $\sqrt[8]{8}_{33,75°}$ $\sqrt[8]{8}_{123,75°}$ $\sqrt[8]{8}_{213,75°}$ $\sqrt[8]{8}_{303,75°}$)

53. Number 4+3i is the fourth root of a certain complex number. Find out the other three roots.

54. Can 2+i, -2+i, -1-2i y 1-2i be the fourth roots of a complex number? Justify your answer.

55. Can $2_{28°}$, $2_{100°}$, $2_{172°}$, $2_{244°}$ and $2_{316°}$ be the roots of a complex number? If so, what id this number?

Exercises

56. Complex number $3_{40°}$ is one of the vertices of a regular pentagon. Find out the other vertices and the complex number whose roots are these vertices.

57. One of the cubic roots of a complex number is i+i. Find out the other cubic roots.

58. a) Find out the cubic roots of the unit in their standard form and draw them.

$$\left(Sol: \ 1; \ \frac{1}{2}+\frac{\sqrt{3}}{2}i; \ -\frac{1}{2}-\frac{\sqrt{3}}{2}i \right)$$

b) Find out the fourth roots of the unit in their standard form and draw them.

(Sol: ±1; ± i)

c) Find out the fifth roots of the unit in their standard form and draw them.

(Sol: $1_{0°}$; $1_{72°}$; $1_{144°}$; $1_{216°}$; $1_{288°}$)

d) Find out the sixth roots of the unit in their standard form and draw them.

$$\left(Sol: \ ±1; \ ±\frac{1}{2}±\frac{\sqrt{3}}{2}i \right)$$

59. Solve the following equations and draw the solutions:

a) $x^3 + 8 = 0$ b) $x^4 - 16 = 0$ c) $ix^4 + 16 = 0$ d) $x^4 + 1 = 0$

Review exercises

1. Divide and write the answers in standard form:

a) $\dfrac{3-i}{2+7i}$ b) $\dfrac{3}{9-i}$ c) $\dfrac{8i}{1+2i}$ d) $\dfrac{6-9i}{2i}$

(Sol.: a) $-\dfrac{1}{53}-\dfrac{23}{53}i$;   b) $\dfrac{27}{82}+\dfrac{3}{82}i$;   c) $\dfrac{16}{5}+\dfrac{8}{5}i$;   d) $-\dfrac{9}{2}-3i$)

2. Calculate:

a) $\dfrac{1-(2+3i)^2(1-2i)}{2\,i^{77}-i^{726}}$ $\left(Sol: \ : \ -\dfrac{62}{5}+\dfrac{14}{5}i\right)$

b) $\dfrac{(2+3i)(3-2i)-(2-3i)^2}{17\,(1-i^{13})}$ $(Sol: \ \ i)$

c) $-2-5i-\dfrac{10-10i-5(1+i)}{8+2i-(5+3i)}$ $(Sol: \ \ -5-i)$

d) $\dfrac{(2-3i)^2-(2+3i)(3-2i)}{3\,i^{17}-1}$ $\left(Sol: \ \cdot \ -\dfrac{17}{5}+\dfrac{34}{5}i\right)$

e) $\dfrac{(3+i)(3-2i)-(2i-3)^2}{2\,i^{20}-i^{13}}+\dfrac{4}{5i}$ $\left(Sol: \ \dfrac{3}{5}+4i\right)$

f) $\dfrac{(2-3i)^2-(2+3i)(3-2i)}{3\,i^{17}-1}-\dfrac{4}{5i^{-25}}$ $\left(Sol: \ -\dfrac{17}{5}+6i\right)$

3. Determine a so that complex number $z = (3-6i)\cdot(2-ai)$ is:

a) A real number. What number?   b) A purely imaginary. What number?

c) Such that its affix is at the bisectrix of the 1st and 3rd quadrants. What is the number?

(Sol: a) a=(-4); 30;   b) a=1; -15i;   c) a=6; -30 -30i)

4. Determine m so that complex number $z = \dfrac{2-mi}{8-6i}$ is:

a) A real number. What number?   (Sol: m = 3/2; 1/4)

b) A purely imaginary. What number?   (Sol: m = -8/3; i/3)

c) Such that its affix is at the bisectrix of the 2nd and 4th quadrants. What is the number?

(Sol: m = 14; 1 - i)

5. Determine a so that complex number $z = (2 + 3i)\cdot(-2 + 6i)$ is:

a) A real number.   (Sol: a = 3)

b) A purely imaginary.   (Sol: a = -4/3)

c) Such that its affix is at the bisectrix of the 1st and 3rd quadrants.   (Sol: a = - 10)

6. a) Given $z = 2_{45°}$, find out $\overline{z}$, in its polar form.   (Sol: $2_{315°}$)

b) Given $z = 1_{30°}$, find out $-z$.

c) If $z = 2_{30°}$, find out its conjugate and its opposite.

d) Find out a complex number and its opposite knowing its conjugate is $\overline{z} = 3_{70°}$.

- 147 -

7. Represent the following zones in the complex plane:

a) $\text{Im}(z) = -2$ b) $\text{Re}(z) = \text{Im}(z)$ c) $-1 < \text{Re}(z) \le 3$ d) $\text{Im}(z) < 2$

e) $|z| = 5$ f) $|z| < 3$ g) $-1 \le |z| < 3$ h) $\text{Arg}(z) = 30°$

i) $\text{Re}(z) = -3$ j) $|z| \ge 4$ k) $\text{Arg}(z) = 90°$

8. Check that $|\mathbf{z}| = \sqrt{z \cdot \overline{z}}$.

9. If $\mathbf{z} = r_\alpha$: a) what is the relationship between $r_{\alpha+180°}$ and r_α ?

b) And between $r_{360°-\alpha}$ and r_α ?

10. Can product of two purely imaginary complex numbers be a real? Give an example.

11. What is the relationship between argument of a complex number and the one of its opposite?

12. What condition must a complex number, **z**, satisfy so that $|\mathbf{z}| = \dfrac{1}{z}$?

13. Calculate in the most suitable way and express the results in their standard form:

a) $\dfrac{(\sqrt{2}-\sqrt{2}\,\text{i})^2\,(-1-\text{i})^4}{(-1+\text{i})^3\,\text{i}^7}$ $\left(\textit{Sol:}\ \ 4\sqrt{2}_{\,135°} = -4 + 4i\right)$

b) $\dfrac{(-2\sqrt{3}-2\,\text{i})^5}{(-4+4\sqrt{3}\,\text{i})^3\,2\text{i}}$ $\left(\textit{Sol:}\ \ 1_{\,240°} = -\dfrac{1}{2}-\dfrac{\sqrt{3}}{2}i\right)$

c) $\dfrac{(-2\sqrt{3}-2\text{i})^4}{(-1+\sqrt{3}\,\text{i})^3\,(2-2\text{i})^2}$ $\left(\textit{Sol:}\ \ 4_{\,210°} = -2\sqrt{3} - 2i\right)$

d) $\dfrac{\left(2-2\sqrt{3}\,\text{i}\right)^3}{\left(-\sqrt{3}+\text{i}\right)^4 \cdot \text{i}}$ $\left(\textit{Sol:}\ \ 2_{\,210°} = -\sqrt{3} - i\right)$

e) $\left[\dfrac{\left(-\sqrt{3}+\text{i}\right)\left(-\dfrac{3}{2}+\dfrac{3\sqrt{3}}{2}\text{i}\right)}{-6}\right]^3$ $(\textit{Sol:}\ \ -i)$

14. Given the complex numbers $z_1 = \sqrt{3} - i$, $z_2 = 3i$ and $z_3 = 1+i$, calculate the following expressions, giving the results in their standard form:

a) $\dfrac{z_1 + z_2}{z_3}$ b) $z_1 \cdot z_3$ c) $(z_1)^4$ d) $\overline{z_2}$

$\left(\textit{Sol}: a)\dfrac{2+\sqrt{3}}{2}+\dfrac{2-\sqrt{3}}{2}i;\ \ b)(\sqrt{3}+1)+(\sqrt{3}-)i;\ \ c)-8+8\sqrt{3}i;\ \ d)-3i\right)$

15. Given the complex number $\mathbf{z} = \sqrt{2} - \sqrt{2}i$, calculate $z^5 \cdot \overline{z}$. *(Sol: – 64)*

16. Work out: $\dfrac{(3-2i)^2 - (1-i)(2-i)}{-3+i}$ *(Sol: $-\dfrac{19}{10} + \dfrac{37}{10}i$)*

17. Calculate z and express the results in their standard form:

$$\sqrt[4]{z} = \dfrac{-\sqrt{3}+i}{\sqrt{2i}}$$

(Sol: $-2 - 2\sqrt{3}i$)

18. Find out a and b that satisfy the equality:

$$(a - 2i) = (3 + i)(b - i)$$

(Sol: b = (–7), a = (–4))

19. Solve the equation: $z^2 - 10z + 29 = 0$ *(Sol: $z_1 = 5 + 2i$, $z_2 = 5 - 2i$)*

20. Calculate the value of x so that norm of $\dfrac{x+2i}{1-i}$ is 2. *(Sol: x = 2 and x = (-2))*

21. Find out the side of the triangle whose vertices are the affixes of the cubic roots of

$$4\sqrt{3} - 4i$$

(Sol: $2\sqrt{3}$)

22. Represent the following zones:

    a) $1 \le \text{Im}(z) \le 5$ b) $|z| = 3$ c) $z + \overline{z} = -4$

23. Find out two complex numbers whose quotient is $2_{150°}$ and its product is $18_{90°}$.

(Sol: Two solutions: $6_{120°}$ and $3_{330°}$; $6_{300°}$ and $3_{150°}$)

24. Check that $\left| z \cdot \overline{z} \right| = \left| z \right|^2$

25. Calculate $\cos 120°$ and $\text{sen} 120°$ from the product $1_{90°} \cdot 1_{30°}$.

(Sol: $\cos 120° = -\dfrac{1}{2}$; $\text{sen } 120° = \dfrac{\sqrt{3}}{2}$)

26. Find out the complex number that is obtained when transforming $(2 + 3i)$ by a 30°

rotation about the origin. *(Sol: $\dfrac{2\sqrt{3}-3}{2} + \dfrac{2+3\sqrt{3}}{2}i$)*

Unit 6.- Vectors

1. Vectors and their operations

Definition, elements and operations with vectors

A *vector* is an orientated segment, determined by two points. Its elements are:

➤ *Application point:* where it is applied.
➤ *Magnitude:* its length (the higher the applied force is, the longer the vector is).
➤ *Direction:* straight line in which it is.

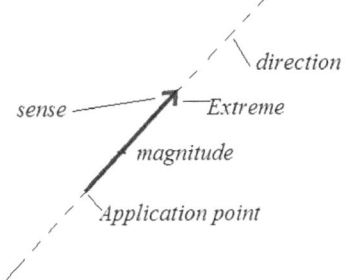

Notice that in quotidian life, we use the word *"direction"* in a different way. We say: "I am going in the direction to Madrid", meaning the *sense*.

In mathematical terms, *direction* would be the straight road joining Murcia and Madrid. And in that road, there are cars going in the *sense* Murcia → Madrid, and other ones in the *sense* Madrid → Murcia.

Notation of vectors

Vectors can be represented by a letter with an arrow on it, $\vec{u}$, $\vec{v}$, $\vec{w}$, or by their origin and extreme, $\overrightarrow{OA}$, $\overrightarrow{OB}$, $\overrightarrow{AB}$.

Product of a number by a vector

Product of a number, k, by a vector, $\vec{u}$, is another vector, $k \cdot \vec{u}$, whose:

> ➤ Magnitude: magnitude of $\vec{u}$, multiplied by absolute value of k.
> ➤ Direction: The same of $\vec{u}$.
> ➤ Sense:
>> ➤ The same sense than $\vec{u}$ if $k > 0$
>> ➤ Opposite to the sense of $\vec{u}$ if $k < 0$

Opposite of a vector

Opposite of a vector, $\vec{u}$, is the product of (-1) by it, and it is denoted as $-\vec{u}$. It is a new vector, with:

> ➤ Magnitude: The same of $\vec{u}$.
> ➤ Direction: The same of $\vec{u}$.
> ➤ Sense: Opposite to the sense of $\vec{u}$.

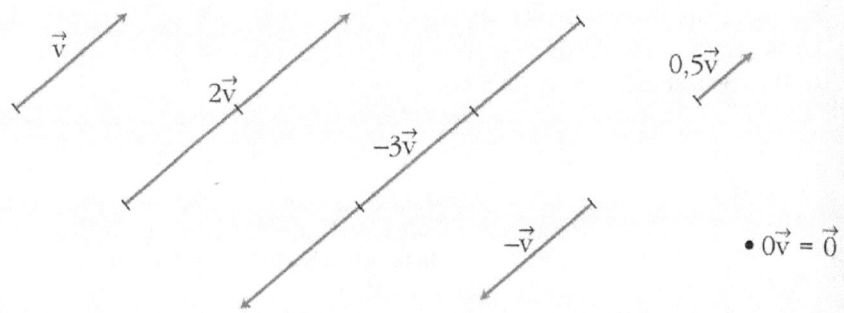

Addition of 2 vectors

2 vectors, $\vec{u}$ and $\vec{v}$, can be *added* in two ways:

> ➤ Locate the origin of $\vec{v}$ at the extreme of $\vec{u}$. The ***addition vector***, $\overrightarrow{u+v}$ is the vector with origin at the origin of $\vec{u}$ and extreme at the extreme of $\vec{v}$.

> ➤ Join both origins, draw a parallelogram. The ***addition vector***, $\overrightarrow{u+v}$ is the vector with origin at the origin of both vectors and extreme at the opposite vertex of the parallelogram.

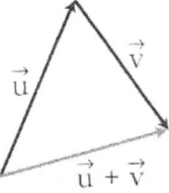
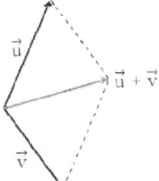

Linear combination of vectors

Given 2 vectors, $\vec{u}$ and $\vec{v}$, and 2 numbers, *a* and *b*, a ***linear combination*** of $\vec{u}$ and $\vec{v}$ is this operation:

$$a \cdot \vec{u} + b \cdot \vec{v}$$

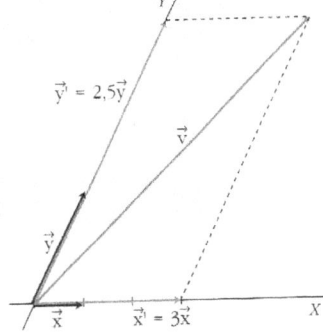

1. Given vectors $\vec{a}$ and $\vec{b}$, draw, on different axes:

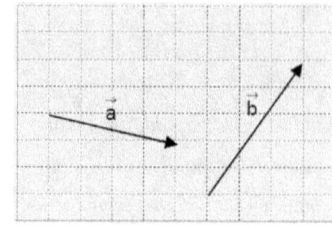

(Exercise 1)   *(Exercise 2)*

a) $\vec{a} + \vec{b}$ b) $\vec{a} - \vec{b}$ c) $3\,\vec{a}$ d) $3\,\vec{a} + 2\,\vec{b}$ e) $2\,\vec{a} - 3\,\vec{b}$

2. Copy the following vectors in your notebook and represent:

a) $\vec{a} + \vec{b}$ b) $\vec{a} - \vec{c}$ c) $3\,\vec{c} + 2\,\vec{a}$ d) $\dfrac{1}{3}\,\vec{a}$

2. Coordinates and magnitude of a vector

A vector, v, is numerically determined by its coordinates, a pair of numbers, $(v_x,\ v_y)$.

If vector v, is determined by points A and B, with coordinates $A(a_x,\ a_y)$ and $B(b_x,\ b_y)$, its coordinates can be determined as $v = \boldsymbol{AB} = (b_x - a_x,\ b_y - a_y)$

Example: Calculate the coordinates of vector **AB**, with A(2, 7) and B(6, 12).

Solution: $AB = (6 - 2,\ 12 - 7) = \boxed{(4,\ 5)}$

Example: Coordinates of vector **AB** are (2, 1). Knowing coordinates of point A(5, 2), calculate coordinates of point B.

Solution:

$$\left. \begin{array}{l} A = (5,2) \\ B = (b_x, b_y) \end{array} \right\} \rightarrow AB = (2,1) = (b_x - 5,\ b_y - 2) \rightarrow \left\{ \begin{array}{l} b_x - 5 = 2 \rightarrow b_x = 5 + 2 = 7 \\ b_y - 2 = 1 \rightarrow b_y = 2 + 1 = 3 \end{array} \right. \rightarrow \quad So,\ B(7,3)$$

Once we already know coordinates of a vector, we can know how long it. This is, its *magnitude*. We are going to do it by using Pythagoras' theorem. As you can see in the figure below, coordinates v_x and v_y are the legs of a right triangle, whose hypotenuse is *magnitude* of the vector. So,

$$v^2 = v_x^2 + v_y^2 \quad \rightarrow \quad v = \sqrt{v_x^2 + v_y^2}$$

Magnitude of a vector is calculated as $v = \sqrt{v_x^2 + v_y^2}$

Example: Points C(2, 5) and D(0, 4) determine vector CD. Calculate a) its coordinates and b) its magnitude.

Solution:

a) $\mathbf{CD} = (0 - 2, 4 - 5) = \boxed{(-2, -1)}$

b) $v = \sqrt{(-2)^2 + (-1)^2} = \sqrt{4+1} = \boxed{\sqrt{5}}$

3. Vector **AB** is applied on point A(1,2) and its head is on B(5, 8).

    a) Draw both points and the vector.

    b) Calculate the coordinates of the vector.

    c) Determine its magnitude.

4. Work out the radius of the circle with the center (8, -2) passing through the point (1, 4).

5. Calculate the magnitude (or length) of vectors: **AB** = (35, -21) and **CD** = (-10, 6)

6. Given point A = (1, -2), decide what is the nearest point from A, B or C, with B(4, 0) and C(0, 1).

7. Work out the coordinates of D, so that the quadrilateral of vertex: A(-1, -2), B(4, -1), C(5, 2) and D; is a parallelogram.

8. Given the vector **a** = (5, 3) and the point A = (4, -1), find the coordinates of the point B so that the vector **AB** is the same as the vector a.

Exercises

3. Operations with coordinates

Vectors can be added, subtracted, multiplied by a number, or a combination of these operations. Rules for these operations are shown in the following chart, for vectors $v(v_x, v_y)$ and $w(w_x, w_y)$.

> **Addition:** $v + w = (v_x+w_x, v_y+w_y)$.   **Subtraction:** $v - w = (v_x-w_x, v_y-w_y)$.
> **Product by a number:** $a \cdot v = (a \cdot v_x, a \cdot v_y)$.
> **Linear combination:** $a \cdot v + b \cdot w = (a \cdot v_x + b \cdot w_x, a \cdot v_y + b \cdot w_y)$.

Exercises

9. Given vectors **u**(3,5) and **v**(-2,0), calculate: a) **u+v**; b) **u-v**; c) **3·v-u**
10. Given vectors **a** = (9, 3) and **b** = (-5, 4), calculate the coordinates of 2**a** + 3**b**.
11. Complete:

a) (……, 5) + (2, …..) = (-3,0)   c) (-4, …..) – (……., 2) = (-4,5)

b) 2·(0, 3) + 3(….., ….) = (-3, 0)   d) (2, 3) – 2·(…., ….) = (-2, -7)

12. Given vectors **u** (2,1) and **v**(1,-2), calculate coordinates of vectors:

a) **u** + **v**   b) **u** – **v**   c) 2**u** + $\frac{1}{2}$**v**

13. Calculate m and n so that **w** = m·**u**+n·**v**, if **u** = (4,-8), **v** = (0,2), **w** = (2,-1).

Midpoint of a segment

Example: Segment AB is determined by points A(2, 5) and B(6, 1). Calculate the coordinates of the midpoint of this segment.

Solution: Midpoint can be named M, with unknown coordinates M(m_x, m_y). Because of being the midpoint, two vectors **AM** and **MB** are equal. So,

AM = (m_x – 2, m_y – 5)   **MB** = (6 - m_x, 1 - m_y)   As **AM** = **MB**, we have:

(m_x – 2, m_y – 5) = (6 - m_x, 1 - m_y) $\rightarrow$ $\begin{array}{l} m_x - 2 = 6 - m_x \rightarrow 2 \cdot m_x = 8 \rightarrow m_x = 4 \\ m_y - 5 = 1 - m_y \rightarrow 2 \cdot m_y = 6 \rightarrow m_y = 3 \end{array}$

So, midpoint of the segment is $\boxed{M(4, 3)}$

A different method: We can use a different equality to calculate midpoint of a segment. Notice that vector **AB** is twice **AM**, so, **AB** = 2 · **AM**

$$\textbf{AB} = (4, -4) \qquad \textbf{AM} = (m_x - 2, m_y - 5) \qquad \text{As} \quad 2 \cdot \textbf{AM} = \textbf{AB}, \text{ we have:}$$

$$2 \cdot (m_x - 2, m_y - 5) = (4, -4) \quad \rightarrow \quad \begin{array}{l} 2 \cdot m_x - 4 = 4 \ \rightarrow \ 2 \cdot m_x = 8 \ \rightarrow \ m_x = 4 \\ 2 \cdot m_y - 10 = -4 \ \rightarrow \ 2 \cdot m_y = 6 \ \rightarrow \ m_y = 3 \end{array}$$

So, midpoint of the segment is $\boxed{M(4,\ 3)}$

Exercises

14. Calculate central point of a segment determined by points (1,4) and (5,-8).

15. A town A is located at point (-2, 5) and another one, B, at point (4,10). They decide to build a shopping center in the central point of the road that joins them. Where will they build it?

16. Central point of a segment AB is located at point M(3,5). If A is (-4,2), where is B located?

17. Calculate the symmetric point of A(1,2) about B(-2,7). A scheme might be useful for you.

18. Given the segment determined by A(0,4) and B(9,-2), determine the location of two points that divide it into three similar portions.

19. A frog moves by jumping on a straight line. Its first jump starts at point A(-4, -4) and finishes at point B(2,1), where it starts a new jump. Determine the point in which it will finish its fifth jump from A.

20. Given points A(3,-2), B(1,3) and C(-6,0), find out point D so that ABCD is a parallelogram.

21. Find coordinates of points that divide segment with extremes A(-2,3) and B(6,2) into three equal parts.

4. Scalar product of two vectors

A new operation you can do with two vectors is its scalar product, denoted as $v \bullet w$

It is very important for you to remember that scalar product of two vectors is a ***number***.

Scalar product of two vectors, $v(v_x, v_y)$ and $w(w_x, w_y)$, is a ***number***, calculated by two different ways:

➤ $v \bullet w = \left|\overrightarrow{v}\right| \bullet \left|\overrightarrow{w}\right| \bullet \cos\left(\overrightarrow{v}, \overrightarrow{w}\right)$

➤ $v \bullet w = v_x \bullet w_x + v_y \bullet w_y$

You will use the best expression, depending on the information you are given.

Example: Given vectors **u** (2,3) and **v**(1,-2), calculate: a) $u \bullet v$ b) Angle $\left(\overrightarrow{v}, \overrightarrow{w}\right)$

Solution:

a) $v \bullet w = v_x \bullet w_x + v_y \bullet w_y = 2 \bullet 1 + 3 \bullet (-2) = 2 - 6 = \boxed{(-4)}$

b) Now, we use the first expression: $\left|\overrightarrow{v}\right| = \sqrt{2^2 + 3^2} = \sqrt{13}$ $\left|\overrightarrow{w}\right| = \sqrt{1^2 + (-2)^2} = \sqrt{5}$

$$v \bullet w = \left|\overrightarrow{v}\right| \bullet \left|\overrightarrow{w}\right| \bullet \cos\left(\overrightarrow{v}, \overrightarrow{w}\right)$$

$$-4 = \sqrt{13} \bullet \sqrt{5} \bullet \cos\left(\overrightarrow{v}, \overrightarrow{w}\right) \implies -4 = \sqrt{65} \bullet \cos\left(\overrightarrow{v}, \overrightarrow{w}\right)$$

$$\cos\left(\overrightarrow{v}, \overrightarrow{w}\right) = \frac{-4}{\sqrt{65}} = -0.496 \implies \left(\overrightarrow{v}, \overrightarrow{w}\right) = \boxed{119°\ 44''}$$

Properties of scalar product of two vectors

➤ *Commutative:* $v \bullet w = w \bullet v$

➤ *Associative:* $a(v \bullet w) = (av) \bullet w$

➤ *Distributive:* $u \bullet (v + w) = u \bullet v + u \bullet w$

Xercises

22. Given vectors **a**(-3,4) and **b**(5,-1), find out:

    a) a.b   b) |a| c) |b| d) angle between **a** and **b**.

23. Given vectors **a**(2, 3) and **b**(a, -1), find out the value of a so that **a** and **b** are perpendicular.

24. If $|v| = 3$, $|w| = 5$ and $v \bullet w = -2$, find out angle (**u,v**).

25. Calculate $v \bullet (w + v)$ and $w \bullet (w - v)$ knowing that $|v| = 3$, $|w| = 5$, (**v,w**) = 120°.

26. Given vectors v(3, -4) and w(-1, 3), calculate:

    a) $v \bullet w$ and $w \bullet v$ b) |v|, |w| and (**u,v**)

    c) Value of k so that vector (4, k) is perpendicular to **v**.

    d) A vector with magnitude 1 and perpendicular to **v**.

Review exercises

1. Observe the rhombus of the figure and calculate:

    a) $\overrightarrow{AB} + \overrightarrow{BC}$

    b) $\overrightarrow{OB} + \overrightarrow{OC}$

    c) $\overrightarrow{OA} + \overrightarrow{OD}$

    d) $\overrightarrow{AB} + \overrightarrow{CD}$

    e) $\overrightarrow{AB} + \overrightarrow{AD}$

    f) $\overrightarrow{DB} - \overrightarrow{CA}$

2. According to the figure, represent:

$$-\vec{u} + \vec{v}, \quad \vec{u} - \vec{v}, \quad \vec{u} + \vec{v},$$

$$-\vec{u} - \vec{v}, \quad -\vec{u} + 2\vec{v}, \quad \vec{u} - 2\vec{v}$$

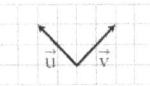

3. If coordinates of v and w are (3, -5) and (-2, 1), find out the coordinates of:

    a) $-2\vec{u} + \dfrac{1}{2}\vec{v}$ b) $-\vec{u} - \dfrac{3}{5}\vec{v}$ c) $\dfrac{1}{2}(\vec{u} + \vec{v}) - \dfrac{2}{3}(\vec{u} - \vec{v})$

4. Find out the vector **b** that verifies $c = 3a - \dfrac{1}{2}b$, where **a** = (-1, 3) and **c** = (7, -2).

5. Given **a**(3, -2), **b**(-2, 2) and **c**(0, -5), calculate m and n so that **c** = m**a** +n**b** .

6. Express vector **a**(-1, -8) as a linear combination of **b**(3, -2) and **c**(4, $-\dfrac{1}{2}$).

Clue: Calculate m and n so that $a = mb + nc$.

- 159 -

7. Given vectors **x**(5, –2), **y**(0, 3) and **z**(–1, 4), calculate:

    a) $x \cdot y$ a) $x \cdot z$ a) $z \cdot y$

8. Given vectors **u**(2, 3), **v**(–3, 1) y **w**(5, 2), calculate:

    a) $(3\vec{u} + 2\vec{v})\vec{w}$ b) $\vec{u} \cdot \vec{w} - \vec{v} \cdot \vec{w}$ c) $(\vec{u} \cdot \vec{v})\vec{w}$ d) $\vec{u}(\vec{v} \cdot \vec{v})$

9. Calculate the magnitudes of the following vectors:

    $\vec{u}(3, 2)$ $\vec{v}(-2, 3)$ $\vec{w}(-8, -6)$

    $\vec{z}\left(\dfrac{\sqrt{2}}{2}, \dfrac{\sqrt{2}}{2}\right)$ $\vec{t}(5, 0)$ $\vec{r}(1, 1)$

10. Find out the value of m so that the magnitude of **u**$(m, -\dfrac{2}{3})$ is 1.

11. Calculate the value of x so that scalar product of **v**(3, –2) and **w**(x, 1) is 7. What is the angle between **v** and **w**?

12. Find out the angle between the following pairs of vectors:

    a) (5, 2), (1, –2)   b) (2, 6), (5, –2)   b) $(-\dfrac{1}{2}, -2), (\dfrac{\sqrt{3}}{2}, -2)$

13. Given vector **w**(–5, k) calculate the value of k so that:

    a) it is perpendicular to (4, –2).   b) The magnitude of **w** is $\sqrt{34}$.

14. Find out the coordinates of a vector **v**(x, y), perpendicular to **u**(3, 4) whose length is double than **u**.

15. Given vectors **a**(2, 1) and **b**(6, 2), find a vector, **v**, perpendicular to **b** and so that $v \cdot a=1$.

16. If **v**(5, -b) and **w**(a, 2), find out the values of a and b, knowing that **u** and **w** are perpendicular and that $|w| = \sqrt{13}$.

17. Given vectors **a** = 2**u** – **y** and **b** = -3**u** + k**v**, where **u** = (2, 3) and **v** = (-3, 0), find out k so that (**a** + **b**) is perpendicular to (**a** – **b**).

18. Find out the value of k so that vectors **x** = k**a** + **b** and **y** = k**a** – **b** are perpendicular, where **a**(2,4) and **b**(1,5).

19. Given vectors **u**(m, –6), and **w**(2, n), calculate m and n so that $|u| = 10$ and **u** and **w** are perpendicular.

20. Find out the coordinates of a vector, **u**, so that $|u| = 1$ and **u·v** = 1, where **v**(2,3).

21. About vectors **a** and **b**, it is known that $|a| = 3$ and $|b| = 5$ and that they are separated by an angle of 120°. Calculate $|a - b|$.

22. If $|u| = 3$ and (**u** + **v**) · (**u** – **v**) =– 11, calculate $|v|$.

23. Knowing that $|u| = 3$, $|v| = 5$ and **u** ⊥ **v**, find out $|u+v|$ and $|u - v|$.

24. If $|u| = 4$, $|v| = 3$ and $|u+v| = 5$, what is the angle between **u** and **v**?

25. Calculate m so that vectors **a**(7, 1) and **b**(1, m) are separated by a 45° angle.

26. Calculate a so that vectors **a**(3, a) and **b**(5, 2) are separated by a 60° angle.

27. Find out a vector, **x**, knowing it is separated 60° from **a**(2,4) and that they have the same magnitude.

28. Find out a vector, **y**, knowing it is separated 30° from **a**(1,2) and that $|y| = \sqrt{3} \cdot |a|$.

Unit 7.- Analytical geometry

1. Equations of a straight line

Equation of a straight line is an algebraic relationship between coordinates x (abscise) and y, (ordinated) of all its points.

<table>
<tr>
<td rowspan="1">Exercises</td>
<td>

1. Check straight line with equation $2x - 3y = 5$ goes through point $(1, -1)$.
2. Find out value of a knowing straight line $ax + 2y = -3$ goes through point $(2, -1)$.
3. Given straight line with equation $x - 5y = 2$:

    a) indicate three points, A, B and C, by which it goes through.

    b) check that vectors AB and AC are proportional.
4. Find out the intersection point of straight lines $3x - 5y = 2$ and $5x + y = 22$.
5. Find out the intersection point of straight lines $x - 3y = 2$ and $-3x + 9y = -6$. What do you observe?
6. Find out the intersection point of straight lines $2x - y = 4$ and $3x - \frac{3}{2}y = 6$.

</td>
</tr>
</table>

A straight line is determined by two elements:

➢ A point by which it goes through, $A(a_x, a_y)$

➢ A direction vector, $\mathbf{u}(u_x, u_y)$

Vectorial equation of a straight line

Look at the graph at the right. We name as $X(x,y)$ a generic point, representing whatever point of the straight line.

Notice vector $\overrightarrow{x}$ is the sum of vector $\overrightarrow{a}$ plus $\overrightarrow{AX}$, which is a number of times (t) vector $\overrightarrow{u}$

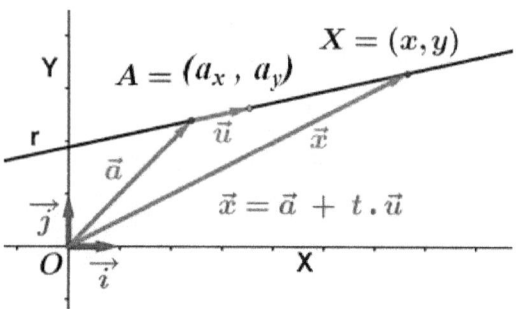

All the points of the straight line can be determined in this way, only changing the value of t.

Vectorial equation: $\quad \overrightarrow{x} = \overrightarrow{a} + t \cdot \overrightarrow{v}$

An example of vectorial equation: $(x,y) = (1, -2) + t \cdot (2, 5)$.

Parametrical equation of a straight line

Parametrical equation is obtained from *vectorial* equation: $\quad \overrightarrow{x} = \overrightarrow{a} + t \cdot \overrightarrow{v}$

$(x, y) = (a_x, a_y) + t \cdot (u_x, u_y)$

$(x, y) = (a_x, a_y) + (t \cdot u_x, t \cdot u_y)$

$(x, y) = (a_x + t \cdot u_x, a_y + t \cdot u_y) \quad \Rightarrow \quad \begin{cases} x = a_x + t \cdot u_x \\ y = a_y + t \cdot u_y \end{cases}$

Parametrical equation:	$\begin{cases} x = a_x + t \cdot u_x \\ y = a_y + t \cdot u_y \end{cases}$

Example: Give the vectorial and parametrical equations of the straight line that goes through point $(2, -1)$ and whose direction vector is $(-2, 3)$.

Solution: $A = (2, -1)$, $\boldsymbol{u} = (-2, 3)$.

$(x, y) = (2, -1) + t \cdot (-2, 3)$ *(Vectorial equation)*

$(x, y) = (2, -1) + (-2t, 3t)$

$(x, y) = (2 - 2t, -1 + 3t)$ $\Rightarrow$ $\begin{cases} x = 2 - 2t \\ y = -1 + 3t \end{cases}$ *(Parametrical equation)*

Continuous equation of a straight line

Continuous equation is obtained from *parametrical* one, only isolating parameter t in both equations and expressing the equality of both expressions:

$\begin{cases} x = a_x + t \cdot u_x & \rightarrow & t = \dfrac{x - a_x}{u_x} \\ \\ y = a_y + t \cdot u_y & \rightarrow & t = \dfrac{y - a_y}{u_y} \end{cases}$ $\Rightarrow$ $\dfrac{x - a_x}{u_x} = \dfrac{y - a_y}{u_y}$

Continuous equation:	$\dfrac{x - a_x}{u_x} = \dfrac{y - a_y}{u_y}$

Example: Give the continuous equation of the straight line that goes through point $(2, -1)$ and whose direction vector is $(-2, 3)$.

Solution: Only by substituing: $\dfrac{x - 2}{-2} = \dfrac{y + 1}{3}$

- 163 -

Example: Give the direction vector and a point of the straight line $x + 3 = \dfrac{y}{2}$.

<u>Solution:</u> It may be easier writing all the terms: $\dfrac{x+3}{1} = \dfrac{y-0}{2}$. So, $\boldsymbol{v} = (1,2)$ and $A = (3,0)$.

Example: Give the direction vector and a point of the straight line $\dfrac{-x+1}{5} = y - 2$.

<u>Solution:</u> It may be easier writing all the terms: $\dfrac{x-1}{-5} = \dfrac{y-2}{1}$. So, $\boldsymbol{v} = (-5,1)$ and $A = (1,2)$.

Implicit or general equation of a straight line

General equation is obtained from *continuous* one, by *cross-multiplying* and rearranging terms.

$$\frac{x-a_x}{u_x} = \frac{y-a_y}{u_y} \qquad \rightarrow \qquad u_y \cdot (x - a_x) = u_x \cdot (y - a_y)$$

$$u_y \cdot x - u_y \cdot a_x = u_x \cdot y - u_x \cdot a_y$$

$$u_y \cdot x - u_x \cdot y + (u_x \cdot a_y - u_y \cdot a_x) = 0$$

$$A \cdot x + B \cdot y + C = 0$$

Implicit or general equation: $\quad A \cdot x + B \cdot y + C = 0$

Example: Give the implicit equation of the straight line $\dfrac{x-2}{-2} = \dfrac{y+1}{3}$.

<u>Solution:</u> $\dfrac{x-2}{-2} = \dfrac{y+1}{3} \quad \rightarrow \quad 3x - 6 = -2y - 2 \quad \rightarrow \quad \boxed{3x + 2y - 4 = 0}$

Example: Given the straight line x – 2y = 5, obtain its direction vector and its continuous equation:

<u>Solution:</u> Although you can solve this question by comparing the given equation and expression $u_y \cdot x - u_x \cdot y + (u_x \cdot a_y - u_y \cdot a_x) = 0$, there is an easier option: obtain 2 points of the line. You can freely choose 2 *x-coordinates* and calculate their corresponding *y-coordinates*.

For example: $\begin{array}{lllll} x_1 = 1 & \rightarrow & 1 - 2y_1 = 5 & \rightarrow & y_1 = -2 & \rightarrow & A_1 = (1, -2) \\ x_2 = -3 & \rightarrow & -3 - 2y_2 = 5 & \rightarrow & y_2 = -4 & \rightarrow & A_2 = (-3, -4) \end{array}$

Direction vector is $\overrightarrow{A_1 A_2} = (-3 - 1, -4 - 2) = \boxed{(-4, -6)}$

For continuous equation, we will use this direction vector and point $A_1(1, -2)$: $\boxed{\dfrac{x-1}{-4} = \dfrac{y+2}{-6}}$

7. Given vector $u(-1, 7)$ and point $A(0,3)$, give vectorial, parametrical, continuous and general equations of the straight line.

8. Given straight line with continuous equation $\dfrac{-x-1}{2} = -y$, give the other equations.

9. Given straight line with implicit equation $x + 3y = -2$, give the other equations.

10. Given straight line with general equation $5x - 2y = 0$, give the other equations.

Point-slope equation of a straight line

Slope or *gradient, m,* of a straight line is the *tangent* of the angle between the straight line and the positive part of the x-axis.

Slope of a straight line passing by points $A(a_x, a_y)$ and $B(b_x, b_y)$ is calculated as

$$m = \frac{b_y - a_y}{b_x - a_x}$$

Given two points with coordinates $A(x_1, y_1)$ and $B(x_2, y_2)$, we can calculate the slope of the straight line that joins them. We have only to calculate variation of *y-variable* (increasing or decreasing) when *x- variable* varies, passing from point A to B.

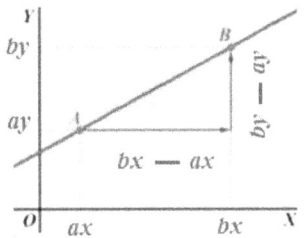

Example: Calculate the slope of the straight line going through points with coordinates $A(-1, 1)$ and $B(1, 5)$.

Solution: $\quad m = \dfrac{y_2 - y_1}{x_2 - x_1} = \dfrac{5-1}{1-(-1)} = \dfrac{4}{2} = 2$

Now, look at the right triangle between straight line and positive part of the x-axis:

Opposite side: 5; Adjacent side: 2 $\rightarrow$ $tg\alpha = \dfrac{5}{2.5} = 2$

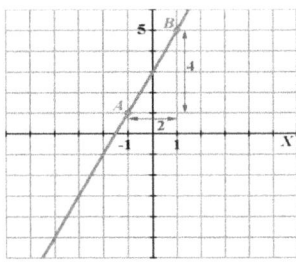

Gradient of a perpendicular straight line

Given a straight line, with slope m, the slope, **m´**, of a perpendicular direction is:

$$m' = -\frac{1}{m}$$

Example:

a) Plot the points A (1, 2) and B (4, 11), join them and calculate the gradient of AB.

b) On the same set of axes, plot the points P (5, 4) and Q (8, 3), join them and calculate the gradient of PQ.

c) With your set of rules, check these segment-lines are perpendicular and, then, check their gradients fulfil the relationship: $m' = -\dfrac{1}{m}$

Solution:

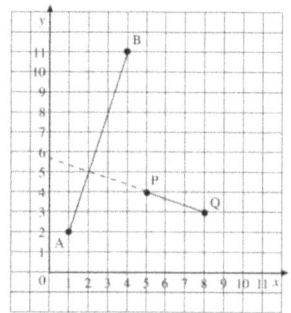

a) $m = \dfrac{11-2}{4-1} = \dfrac{9}{3} = 3$

b) $m = \dfrac{3-4}{8-5} = \dfrac{-1}{3} = -\dfrac{1}{3}$

c) Notice that $\dfrac{-1}{3} = -\dfrac{1}{3} \;\rightarrow\; m' = -\dfrac{1}{m}$

11. In each case, decide whether the lines AB and PQ are parallel, perpendicular or neither.

a) A(4, 3) B(8, 4)   P(7, 1) Q(6, 5)

b) A(−2, 0) B(1, 9)   P(2, 5) Q(6, 17)

c) A(8, −5) B(11, −3)   P(1, 1) Q(−3, 7)

d) A(3, 1) B(7, 3)   P(−3, 2) Q(1, 0)

12. A triangle has vertices A (3, 1), B (7, 5) and C (1, 3). Show that the triangle is a right-angled triangle.

13. The coordinates of the points A, B, C and D are A(3, 0), B(0, 1), C(1, 4) and D(4, 3). Show that ABCD is a square.

14. The points A(3, 2), B(6, 0), C(5, 4) and D(2, 6) are the vertices of a quadrilateral.
a) Show that this is *not* a rectangle.   b) Show that this is a parallelogram.

15. The lines AB and PQ are perpendicular. The coordinates of the points are
A(3, 2)   B(7, 4)   P(3, 7) and Q(6, q)   Determine the value of q.

It is also possible to calculate the *slope* of a vector from its coordinates:

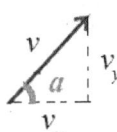

$$m = tg\,\alpha = \frac{v_y}{v_x}$$

Point-slope equation is obtained from *continuous* one, *cross-multiplying*:

$$\frac{x-a_x}{u_x} = \frac{y-a_y}{u_y} \qquad \rightarrow \qquad \frac{u_y}{u_x}\cdot(x-a_x) = y-a_y$$

$$m\cdot(x-a_x) = y-a_y \quad or:$$

$$y-a_y = m\cdot(x-a_x)$$

Slope-point equation: $y-a_y = m\cdot(x-a_x)$

Explicit equation of a straight line

Explicit equation is obtained from *slope-point* one, by only *rearranging:*

$y-a_y = m\cdot(x-a_x)$ *(slope-point equation)*

$y-a_y = m\cdot x - m\cdot a_x$

$y = m\cdot x - m\cdot a_x + a_y$

$y = m\cdot x + n$

where *m* is the *slope* of the line and *n* is named *y-intersect*, because when x = 0, y = n.

Explicit equation: $y = m \cdot x + n$

Important: Notice when writen in its *explicit* form, *slope* of the straight line is the *coefficient* of "*x*".

Example: Write the equation of the straight line going through points $A(-1, 1)$ and $B(1, 5)$.

Solution: Direction vector, **AB**, can be calculated as: $(1 - (-1), 5 - 1) = (2, 4)$

$$m = \frac{5-1}{1-(-1)} = \frac{4}{2} = 2, \text{ so, equation is: } y = 2x + n.$$

To calculate n, you can substitute coordinates of A or B. For example, using B:

$B(1, 5) \rightarrow 5 = 2 \cdot 1 + n \rightarrow n = 3.$ So, equation is: $\boxed{y = 2x + 3}$.

Example: Indicate the slope of the following straight lines:

a) $y = 5x + 2 \rightarrow m = 5.$

b) $y = 3 - x \rightarrow m = (-1).$

c) $2x + 3y = 3 \rightarrow 3y = -2x + 3 \rightarrow y = -\frac{2}{3}x + 1 \rightarrow m = -\frac{2}{3}.$

Example: Find the equation of a straight line, parallel to the one with equation $y = 4x - 2$, knowing it goes through the point $(1, 9)$.

Solution: The straight line we are looking for is parallel to $y = 4x - 2$. So, its slope is $m = 4$, and its equation will be $y = 4x + n$.

As it passes by point $(1, 9)$, this coordinates must verify $9 = 4 \cdot 1 + n$, so $n = 9 - 4 = 5$.

Our straight line is $\boxed{y = 4x + 5}$.

<table>
<tr><td rowspan="2">Exercises</td><td>16. Write the equation of the straight line going through point A(2,7) and whose gradient is m = -3.</td></tr>
<tr><td>17. Write the equation of the straight line going through point A(0,0) and whose gradient is m = -7.</td></tr>
</table>

Exercises

18. a) Calculate the slope of the straight line whose direction vector is (2,5).

b) Write the equation of the straight line going through point A(1,2) and whose direction vector is (2,5).

19. Write the equation of the straight lines going through point A(2,3) and B(-1,7).

20. Write the equation of the straight line going through point A(2,-2) and whose gradient is m = -1.

21. Write the equation of the sides of the triangle whose vertices are A(2,3), B(-1,6), C(0,-2).

Mixed exercises

Exercises

22. Given point A(5,3) and direction vector $v(1, -2)$, calculate:

a) The equation of the straight line, r, determined by them, in all its forms.

b) Indicate another three different points of this straight line.

c) Check, in a numerical way, if points P(2,-1) and Q(3,7) $\in$ r.

d) Draw this straight line and check, graphically, previous question.

23. Given points A(1,3) and B(-1,6), calculate:

a) The equation of the straight line, s, determined by them, in all its forms.

b) Indicate another three different points of this straight line.

c) Check, in a numerical way, if points P(7,-6) and Q(2,2) $\in$ s.

d) Draw this straight line and check, graphically, previous question.

24. a) Given point A(3,5) and direction vector $v(2, -4)$, find out the equation of the straight line determined by them, in all its forms and represent it. *(Sol: 2x+y-11=0)*

b) Idem for A(3,1) and $v(4, -2)$. *(Sol: x+2y-5=0)*

c) Idem for A(3,1) and $v(0, 2)$. *(Sol: x=3)*

d) Idem for A(3,-1) and $v(5, 0)$. *(Sol: y=-1)*

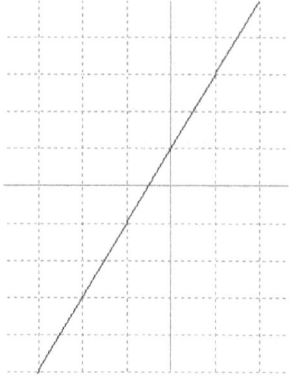

25. Given the straight line in the figure, find out its equation:

   a) Directly, in its continuous form.

   b) General form, from the continuous one.

   c) Directly, in its slope-point form.

   d) Directly, in its explicit form.

26. Find out the equation of the straight line going through points A(3,2) and B(1,-4) in all its forms and represent it.   *(Sol: 3x-y-7=0)*

27. Represent the following straight lines:

   a) 2x+3y-7=0   **b)** x=3   **c)** y=2   **d)** $\begin{cases} x = 3 - \lambda \\ y = -5 + 2\lambda \end{cases}$ **e)** $\dfrac{x-1}{2} = \dfrac{y+3}{-1}$

28. Write in its explicit form and calculate their slopes:

   a) $\dfrac{x-3}{2} = \dfrac{y+5}{-1}$ **b)** 5x+3y+6=0   **c)** $\begin{cases} x = 2 + t \\ y = 5 - 3t \end{cases}$

   (Sol: a) – ½ ; b) – 5/3;   c) –3)

29. Decide if straight line 3x-2y+5=0 goes through point P(2,-1).

   And through point (1,4)?   *(Sol: NO; YES)*

30. Given straight line, *r*, ax+5y+4=0, find out value of *a* so that *r* goes through point P(2,-2).   *(Sol: a=3)*

31. Decide, numerically, if points A(3,1), B(5,2) and C(1,0) are aligned.   *(Sol: YES)*

32. Ídem for A(1,1), B(3,4) and C(4,6).   *(Sol: NO)*

33. Find out the value of k so that points A(1,7), B(-3,4) and C(k,5) are aligned.

   (Sol: k=-5/3)

34. Calculate the equation of the straight line going through point A(-2, 1/3) and is parallel to straight line going through points P(2,1) and Q(3,4). *(Sol: y = 3x +19/3)*

35. Given straight line determined by points A(1,0) and B(3,4), calculate:

   a) its parametrical, continuous, slope-point and explicit form.

   b) its slope   c) Does it go through point (2,2)?   *(Sol: m = 2; YES)*

36. Calculate the equation of the straight line going through point A(2,1) and forming an angle of 120° with the positive part of the X-axis. *(Sol: $y-1=-\sqrt{3}\cdot(x-2)$)*

37. What is the angle that straight line x+y+5=0 forms with the positive part of the X-axis? *(Sol: 135°)*

38. Given straight line 5x-3y+7=0, find out the length of the segments that it determines on the axes. *(Sol: 7/5; 7/3)*

39. Find out the area determined by the straight line 5x+y-5=0 and the abscise and ordinated axes. *(Sol: 5/2)*

40. Calculate the equation of the straight line going through point P(3,1) and forming an angle of 45° with axis OX⁺. *(Sol: $y = x + 2$)*

41. a) What is the angle between the straight line 3x-2y+6=0 and abscise axis?

b) What is the angle between the straight line 2x-y+5=0 and ordinated axis?

c) Calculate n so that straight line 3x+ny-2=0 forms an angle of 60° with OX⁺.

(Sol: a) 56° 18' 36''; b) 26° 33' 54''; c) n = $-\sqrt{3}$)

42. Solve the following systems in order to find out the intersection point of the given straight lines:

a) $\left.\begin{array}{l} 2x+3y=11 \\ 3x-2y=-3 \end{array}\right\}$
b) $\left.\begin{array}{l} 2x+3y=11 \\ 6x+9y=33 \end{array}\right\}$
c) $\left.\begin{array}{l} 2x+3y=11 \\ 6x+9y=3 \end{array}\right\}$

(Sol: a) (1,3)   b) ∞ solutions   c) No solution)

43. Find out the equation of the straight line going through intersection point of the straight lines 2x+3y-4=0 and x-y=0 and through point A(2,1). *(Sol: x-6y+4=0)*

44. Straight line y+2=m(x+3) goes through intersection point of the straight lines 2x+3y+5=0 and 5x-2y-16=0. Calculate the value of *m*. *(Sol: m=-1/5)*

2. Relative position of straight lines

Two straight lines can be: the same straight line, parallel and secant lines.

There are different methods to determine relative positions of 2 lines:

2.1. Graphically

Example: Represent the linear plots corresponding to both straight line equations.

a) $\begin{cases} 2x + y = 7 \\ 4x + 2y = 14 \end{cases}$
b) $\begin{cases} 2x + y = 7 \\ 2x + y = 0 \end{cases}$
c) $\begin{cases} 2x + y = 7 \\ -2x + 5y = 10 \end{cases}$

<u>Solution:</u> Feel free to give values to "x", calculate corresponding "y-values", and you must obtain plots like following:

a) $\begin{cases} 2x + y = 7 \\ 4x + 2y = 14 \end{cases}$
b) $\begin{cases} 2x + y = 7 \\ 2x + y = 0 \end{cases}$
c) $\begin{cases} 2x + y = 7 \\ -2x + 5y = 10 \end{cases}$

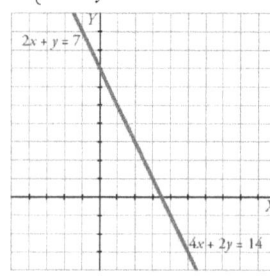 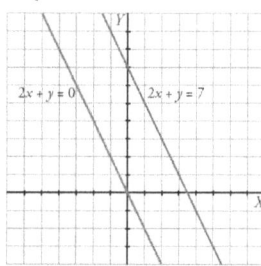 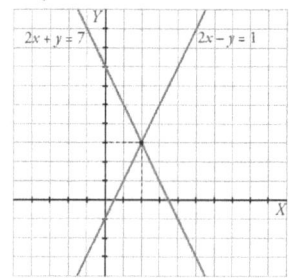

As you can see, both linear plots are *coincident*, so, they have infinite common points.

As you can see, both linear plots are *paralel*, so, they do not have any common point.

As you can see, both linear plots are *secant*, so, they have one common point.

2.2. By solving the system

Remember solution of a system are the coordinates of the intersection point of both straight lines. So:

Number of solutions	1	Infinite	None
Number of common points	1	Infinite	None
Position of straight lines	Secant	Coincident	parallel

2.3. By comparing coefficients of the general equation

Consider above systems and calculate quotients of corresponding coefficients:

a) $\begin{cases} 2x+y=7 \\ 4x+2y=14 \end{cases} \rightarrow \dfrac{2}{4}=\dfrac{1}{2}=\dfrac{7}{14}$ b) $\begin{cases} 2x+y=7 \\ 2x+y=0 \end{cases} \rightarrow \dfrac{2}{2}=\dfrac{1}{1}\neq\dfrac{7}{0}$ c) $\begin{cases} 2x+y=7 \\ -2x+5y=10 \end{cases} \rightarrow \dfrac{2}{-2}\neq\dfrac{1}{5}$

As a general rule: Given two straight lines with equations $\begin{cases} Ax+By=C \\ A'x+B'y=C' \end{cases}$

$\dfrac{A}{A'}=\dfrac{B}{B'}=\dfrac{C}{C'}$	$\dfrac{A}{A'}=\dfrac{B}{B'}\neq\dfrac{C}{C'}$	$\dfrac{A}{A'}\neq\dfrac{B}{B'}$
Coincident	***Parallel***	***Secant***

---

45. Only by looking at them, indicate the relative position of the following pairs of lines:

a) $\begin{cases} 2x-y=1 \\ 4x-2y=8 \end{cases}$ b) $\begin{cases} x-2y=5 \\ 2x-4y=10 \end{cases}$ c) $\begin{cases} 5x+2y=-1 \\ 4x-y=7 \end{cases}$ d) $\begin{cases} x-2y=5 \\ 2x-4y=-3 \end{cases}$

46. Complete the following pair of straight lines so that first pair intersect at $(x=3, y=-2)$, second are parallel and third and fourth are coincident.

a) $\begin{cases} 3x+2y=\ldots \\ \ldots-y=8 \end{cases}$ b) $\begin{cases} x+y=5 \\ 2x+2y=\ldots \end{cases}$ c) $\begin{cases} 3x-2y=4 \\ 6x-4y=\ldots \end{cases}$ d) $\begin{cases} -x+2y=7 \\ \ldots-4y=\ldots \end{cases}$

47. Decide the relative position of the following pair of straight lines:

a) $\begin{cases} x-y+3=0 \\ 5x+2y-4=0 \end{cases}$ b) $\begin{cases} x-y+3=0 \\ 2x-2y+6=0 \end{cases}$ c) $\begin{cases} x-y+3=0 \\ x-y-2=0 \end{cases}$

48. Decide the relative position of the following pair of straight lines:

a) $\begin{aligned} r &\equiv x-y+3=0 \\ r &\equiv 2'x-2y+6=0 \end{aligned}$ b) $\begin{aligned} s &\equiv x-y+3=0 \\ s' &\equiv 5x+2y-4=0 \end{aligned}$ c) $\begin{aligned} t &\equiv x-y+3=0 \\ t' &\equiv x-y-2=0 \end{aligned}$

49. Given the following straight lines, decide which are parallel and the same.

r: 2x+3y-4=0   u: 4x+6y-8=0
s: x-2y+1=0   v: 2x-4y-6=0
t: 3x-2y-9=0   w: 2x+3y+9=0

(Sol: r=u; s//v; r//w)

Exercises

50. Check, by two methods, if the following straight lines are secant, parallel or coincident:

a) $\left.\begin{array}{l} 3x + 2y - 5 = 0 \\ 3x + 2y + 7 = 0 \end{array}\right\}$
b) $\left.\begin{array}{l} x + 3y - 4 = 0 \\ x + 2y - 5 = 0 \end{array}\right\}$
c) $\left.\begin{array}{l} x + y - 3 = 0 \\ 2x + 2y - 6 = 0 \end{array}\right\}$

(Sol: a) parallel; b) secant; c) coincident)

51. Find out values of m and p so that straight lines $mx+3y+5=0$ and $2x+6y-p=0$ are coincident.

(Sol: m=1; p=-10)

52. Given the following straight lines, calculate m so that they are parallel.

a) $\begin{array}{l} 3x-4y+1=0 \\ mx+8y-14=0 \end{array}$
b) $\begin{array}{l} 4x-3y+1=0 \\ mx+6y+4=0 \end{array}$

(Sol: a) $m = -6$; b) $m = -8$)

2.4. From their direction vectors

If you consider two direction vectors, they can be:

➢ **Proportional:** So, they are parallel vectors: Lines can be parallel or coincident.
➢ **Non-proportional:** So, they are non-parallel vectors: Lines are secant.

Notice this method allows us to identify secant or non-secant lines, but if they are not secant, we cannot determine if they are parallel or coincident. In that case, we will have to use one more method.

Special case: Perpendicular lines

Remember scalar prodct of 2 vectors is:

$$v \bullet w = \left|\overrightarrow{v}\right| \bullet \left|\overrightarrow{w}\right| \bullet \cos\left(\overrightarrow{v}, \overrightarrow{w}\right)$$

$$v \bullet w = v_x \bullet w_x + v_y \bullet w_y$$

Notice if vectors v and w are perpendicular, $\left(\overrightarrow{v}, \overrightarrow{w}\right) = 90°$ and $\cos\left(\overrightarrow{v}, \overrightarrow{w}\right) = 0$

This means you can easily determine if 2 vectors are perpendicular: they are so if $v \bullet w = 0$.

> **Perpendicular straight lines:** If straight lines, r and s, are perpendicular, their direction vectors, v_r and v_s satisfy that

$$v \bullet w = 0$$

2.5. From their slope and y-intercept

Consider straight lines $\begin{array}{l} r: \quad y = mx + n \\ s: \quad y = m^*x + n^* \end{array}$

> If m = m*, then they are parallel or coincident $\begin{cases} - If \ n = n^* & \rightarrow & Coincdent \\ - If \ n \neq n^* & \rightarrow & Parallel \end{cases}$

> If m ≠ m*, then they are secant.

Exercises

53. For each of the following straight lines, calculate its slope and obtain the equation of the perpendicular straight line that goes through given point:

a) x-2y+3=0; P(3,-1) *(Sol: 2x+y-5=0)*   b) 3x+2y+1=0; P(1,-1) *(Sol: 2x-3y-5=0)*

c) y-4=2(x-1); P(1,1) *(Sol: x+2y-3=0)*   d) y=2x-5; P(-2,3)   *(Sol: x+2y-4=0)*

e) y-3=2(x+1); 0(0,0) *(Sol: x+2y=0)*   f) $\left. \begin{array}{l} x = 1 + \lambda \\ y = 2 - 3\lambda \end{array} \right\}$ *(Sol: x-3y+8=0)*

54. Find out the equation of the straight line perpendicular to the segment of extremes A(5,6) and B(1,8) at its midpoint. What is the name of this line? Draw it.

(Sol: 2x-y+1=0; mediatrix)

55. Straight line 3x-2y-6=0 intersects both axes at points A and B. Find them and calculate the mediatrix of AB.   *(Sol: 4x+6y+5=0)*

56. Given the straight line x+2y+1=0, find the symmetric point of P(2,-3) about this line.   *(Sol: P'(16/5,-3/5))*

57. Straight line 2x-y+1=0 is the mediatrix of segment AB and A = (2,-3). Calculate B.

(Sol: P'(-22/5, 1/5))

3. Angles between straight lines

Angle between two straight lines can be calculated by two different methods, depending on the information we have.

3.1. From direction vectors

We will use their scalar product:

$$\boldsymbol{v} \bullet \boldsymbol{w} = \left| \vec{v} \right| \cdot \left| \vec{w} \right| \cdot \cos\left(\vec{v}, \vec{w} \right)$$

$$\boldsymbol{v} \bullet \boldsymbol{w} = v_x \bullet w_x + v_y \bullet w_y$$

Notice: $v_x \bullet w_x + v_y \bullet w_y = \left| \vec{v} \right| \bullet \left| \vec{w} \right| \bullet \cos\left(\vec{v}, \vec{w} \right) \implies \left(\vec{v}, \vec{w} \right) = \arccos\left\{ \dfrac{\vec{v} \bullet \vec{w}}{\left| \vec{v} \right| \bullet \left| \vec{w} \right|} \right\}$

$$\left(\vec{v}, \vec{w} \right) = \arccos\left\{ \dfrac{\vec{v} \bullet \vec{w}}{\left| \vec{v} \right| \bullet \left| \vec{w} \right|} \right\}$$

Example: Given straight lines, r, $\dfrac{x-1}{2} = \dfrac{y}{3}$, and s, $\begin{cases} x = 5+t \\ y = -8 - 2t \end{cases}$, calculate the angle between them.

<u>Solution:</u> Notice their direction vectors are: $\mathbf{v_r} = (2,3)$ and $\mathbf{v_s} = (1,-2)$.

We are calculating their scalar product by using both expressions:

$$\mathbf{v_r} \bullet \mathbf{v_s} = 2 \cdot 1 + 3 \cdot (-2) = 2 - 6 = \boxed{(-4)}$$

$$\left| \vec{v} \right| = \sqrt{2^2 + 3^2} = \sqrt{13} \qquad \left| \vec{w} \right| = \sqrt{1^2 + (-2)^2} = \sqrt{5} \qquad \boldsymbol{v} \bullet \boldsymbol{w} = \boxed{\sqrt{13} \cdot \sqrt{5} \cdot \cos\left(\vec{v_r}, \vec{v_s} \right)}$$

Both quantities are equal:

$$-4 = \sqrt{13} \bullet \sqrt{5} \bullet \cos\left(\vec{v_r}, \vec{v_s} \right) \implies -4 = \sqrt{65} \bullet \cos\left(\vec{v_r}, \vec{v_s} \right)$$

$$\cos\left(\vec{v_r}, \vec{v_s} \right) = \dfrac{-4}{\sqrt{65}} = -0.496 \implies \left(\vec{v_r}, \vec{v_s} \right) = \boxed{119° \ 44''}$$

3.2. From slopes

Slope of a straight line is the tangent of the angle if forms with OX^+ axis.

Remember the expression for the tangent of the difference of two angles:

$$tg(A-B) = \frac{tgA - tgB}{1 + tgA \cdot tgB}$$

If we have two straight lines r_A and r_B, forming angles of A° and B° with OX^+ axis, their slopes are $m_A = tgA$ and $m_B = tgB$.

So, the angle between them, A – B, verifies:

$$tg(A-B) = \frac{m_A - m_B}{1 + m_A \cdot m_B} \qquad \rightarrow \qquad (A-B) = arctg\left(\frac{m_A - m_B}{1 + m_A \cdot m_B}\right)$$

$$(A-B) = arctg\left(\frac{m_A - m_B}{1 + m_A \cdot m_B}\right)$$

Example: Given straight lines, r, y = 3x – 1, and s, 4x + 2y + 1 = 0, calculate the angle between them.

Solution: $m_A = 3$ and $m_B = -2$ $\rightarrow$ $(A-B) = arctg\left(\frac{3-(-2)}{1+3\cdot(-2)}\right) = arctg\left(\frac{5}{-5}\right) = arctg(-1) = \boxed{135°}$

58. Calculate the angle between the following pairs of straight lines:

a) 2x-3y+4=0 5x-2y-3=0 *(Sol: 34° 31')*

b) 2x+3y-5=0 x-y+7=0 *(Sol: 78° 41')*

c) x-2y+4=0 3x-y-1=0 *(Sol: 45°)*

d) y=2x-3 y=-2x+1 *(Sol: 53° 8')*

e) y=3x-5 y=3x+2 *(Sol: 0°)*

f) -x+2y+1=0 3x+y+5=0 *(Sol: 81° 52')*

g) -x+2y+5=0 2x-3y+4=0 *(Sol: 7° 8')*

Exercises

59. Calculate the angle between the following pairs of straight lines:

a) $\dfrac{x-1}{2} = \dfrac{y}{3}$ $-2x+3y-5=0$

b) $\dfrac{x+2}{3} = \dfrac{y-1}{4}$ $3x+4y=0$

c) $3x+4y-12=0$ $5x-12y+8=0$

d) $\left.\begin{array}{l} x = 3 + t \\ y = 5 - 2t \end{array}\right\}$ $\left.\begin{array}{l} x = -3 + 4\lambda \\ y = -1 + 3\lambda \end{array}\right\}$

e) $y=7x+54$ $3x-4y+128=0$

(Sol: a) 22° 37'   b)90°   c) 59° 30' d) 79° 42'   e) 45°)

60. Without operating, decide if the following pairs of straight lines are perpendicular:

a) $2x+3y-4=0$ $4x+6y-8=0$

b) $2x+3y-4=0$ $6x-4y+5=0$

c) $3x-2y+7=0$ $4x+6y-3=0$

d) $x+y-8=0$ $2x+3y+6=0$

(Sol: NO; YES; YES; NO)

61. Is straight line $2x+3y+4=0$ perpendicular to another one having slope 3/2?

(Sol: YES)

62. Calculate parameter m so that straight lines $2x-4y+12=0$ and $mx+8y-15=0$ are perpendicular.

(Sol: m=16)

63. Calculate the value of a so that straight lines $ax + (a-1)y - 2(a+2) = 0$ and $3ax - (3a+1)y - (5a+4) = 0$ are:

a) Parallel   b) Perpendicular.

(Sol: a=0 or a=1/3; a=-1/2)

64. Calculate coefficients m and n of straight lines $mx - 2y + 5 = 0$ and $nx + 6y – 8 = 0$ if they are perpendicular and first one goes through point (1,4).

(Sol: m=3; n=4)

65. Given the straight line $ax + by = 1$, calculate a and b if this line is perpendicular to $2x+4y=11$ and it goes through point (1,3/2).

(Sol: a=4; b=-2)

66. Find out the value of a so that straight lines $\begin{cases} x = 2 - t \\ y = 2t \end{cases}$ and $\begin{cases} x = 1 + 2t \\ y = 2 + at \end{cases}$ are separated an angle of 45°.

(Sol: a1=6, a2=-2/3)

67. Give the equation of the straight line that goes through A(5,-2) and is separated an angle of 45° of line with equation $3x + 7y – 12 = 0$.

(Sol: $y + 2 = \dfrac{2}{5}(x - 5)$; $y + 2 = -\dfrac{5}{2}(x - 5)$)

4. Distances

4.1. Distance between two points

Remember the ***magnitude*** of a vector is its length. We are going to calculate it by using Pythagoras' theorem. As you can see in the figure below, coordinates v_x and v_y are the legs of a right triangle, whose hypotenuse is *magnitude* of the vector. So,

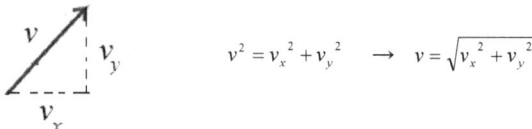

$$v^2 = v_x^{\,2} + v_y^{\,2} \quad \rightarrow \quad v = \sqrt{v_x^{\,2} + v_y^{\,2}}$$

So, you can calculate the distance between two points A and B:

> ➤ **Step 1:** Calculate the coordinates of the vector **AB**: $(x_B - x_A, y_B - y_A)$
> ➤ **Step 2:** The distance between A and B is the magnitude of vector **AB**:
>
> $$d(A,B) = \sqrt{(x_B - x_A)^2 + (y_B - y_A)^2}$$

Example: Calculate the distance between the points C(2, 5) and D(0, 4).

Solution:

a) **CD** $= (0 - 2, 4 - 5) = \boxed{(-2, -1)}$

b) $d = \sqrt{(-2)^2 + (-1)^2} = \sqrt{4+1} = \boxed{\sqrt{5}}$

4.2. Distance between a point and a straight line

Given a point $A(x_0, y_0)$ and the straight line with equation $Ax + By + C = 0$, the ***perpendicular distance*** between them is:

$$d(A,r) = \frac{|A \cdot x_0 + B \cdot y_0 + C|}{\sqrt{A^2 + B^2}}$$

Example: Calculate the distance between the point P(0,6) and the straight line $\dfrac{x-2}{3}=\dfrac{y+1}{5}$

Solution: First of all, you must rewrite the equation:

$$\frac{x-2}{3}=\frac{y+1}{5} \rightarrow 5\cdot(x-2)=3\cdot(y+1) \rightarrow 5x-10=3y+3 \rightarrow 5x-3y-13=0$$

Now, you can use the previous expression: $d(A,r)=\dfrac{5\cdot 0-3\cdot 6-13}{\sqrt{5^2+(-3)^2}}=\dfrac{-31}{\sqrt{34}}$

As you know, a distance cannot be a negative number. So, we must calculate its absolute value, as well as rationalize it:

$$d(A,r)=\left|\frac{-31}{\sqrt{34}}\right|=\boxed{\frac{31\sqrt{34}}{34}}$$

68. Calculate the distance between P and Q:

    a) P(3, 5), Q(3, –7)   b) P(–8, 3), Q(–6, 1)

    c) P(0, –3), Q(–5, 1)   d) P(–3, 0), Q(15, 0)

69. a) Find out the mid-point of the segment-line with extremes A(–2, 0) and B(6, 4).

    b) Check that the distances from the mid-point to both extremes are equal.

70. Check that the triangle with vertices A(–1, 0), B(3, 2), C(7, 4) is an isosceles one. Which are the equal sides?

71. By using the Pythagoras' Theorem, check that the triangle with vertices A(–2, –1), B(3, 1) and C(1, 6) is a right one.

72. Calculate the distance between point P(1,2) and line 3x-4y+1=0.   *(Sol: 4/5)*

73. Idem for P(2,-1) and 3x+4y=0.   *(Sol: 2/5)*

74. Idem for origin and $\begin{cases} x=1+2t \\ y=-2-t \end{cases}$.   *(Sol: $\dfrac{3\sqrt{5}}{5}$)*

75. Idem for origin and y = 4.   *(Sol:4)*

76. Idem for P(1, – 3) and $\dfrac{x-1}{2}=y+5$.   *(Sol: $\dfrac{4\sqrt{5}}{5}$)*

77. Idem for P(2,4) and y=-2x+3.   *(Sol: $\sqrt{5}$)*

78. Idem for P(-1,7) and y-3=2(x+3).   *(Sol: 0)*

79. Calculate the distance from the origin to the straight line going through A(-2,1) and B(3,-2). *(Sol: $\sqrt{34}/34$)*

80. Calculate the distance from point (-1,1) to the straight line that intersects axes OX^+ and OY^+ at distances 3 and 4 from origin. *(Sol: 13/5)*

81. Calculate the length of the segment determined by straight line x-2y+5=0 when intersecting coordinated axes. *(Sol: $2\sqrt{5}/5$)*

82. Calculate the distance from origin to the straight line going through point (0,2) and whose slope is -1. *(Sol: $\sqrt{2}$)*

83. Find out the value of c so that distance from straight line x-3y+c=0 to the point (6,2) is $\sqrt{10}$. *(Sol: c= 10)*

84. Find out the value of a so that distance from straight line ax+2y-2=0 to the point (1,2) is $\sqrt{2}$. *(Sol: a= 2)*

85. Calculate the equation of the two straight lines that go through point A(1,-2) and are at a distance of 2 units from point B(3,1). A draw may be useful.

$$(Sol: y+2=\frac{5}{12}(x-1); \ x=1)$$

86. Find out the equation of the two parallel straight lines with slope 3/4 that are at a distance of 2 units from point (2,3). Try to solve it by mean of explicit equation. A draw may be useful. *(Sol: $y=\frac{3}{4}x+4; \ y=\frac{3}{4}-1$)*

.

4.3. Distance between two parallel straight lines

Given two parallel straight r and s, lines with equations $Ax + By + C = 0$ and $A'x + B'y + C' = 0$, the **perpendicular distance** between them is:

$$d(A,r) = \frac{|C - C'|}{\sqrt{A^2 + B^2}}$$

Exercises

87. Calculate the distance between straight lines 2x+3y-6=0 and 2x+3y+7=0.

(Sol: $\sqrt{13}$)

88. Calculate the distance between straight lines $\left.\begin{array}{l}x=2-3t\\y=1+t\end{array}\right\}$ and $\dfrac{x+3}{-3}=y+5$.

(Sol: $23\sqrt{10}/10$)

89. Calculate the distance between straight lines 3x-4y+16=0 y 2x-5y+2=0.

(Sol:0)

90. Calculate the distance between straight lines 3x-4y+16=0 and $y=\dfrac{3x}{4}-1$.

(Sol: 4)

91. Given the straight line 3x-4y+19=0:

a) Find out the equation of the straight line parallel to it and going through P(5,6).

b) Calculate the distance between them.

c) Find out the angle than these lines form with straight line 7x-y+3=0.

(Sol: a) 3x-4y+9=0;   b) 2 u;   c) 45°)

92. a) Find out the equation of the straight line that goes through P(– 1; 2) and is parallel to y = 3x – 1.

b) Calculate the distance between them.

c) Find out the angle between r and t: x-2y+4=0.

(Sol: a) 3x-y+5=0;   b) $3\sqrt{10}/5$ u;   c) 45°)

93. Given the following pairs of straight lines, calculate m so that they are parallel and calculate the distance between them:

a) 3x-4y+1=0   and   mx+8y-14=0   b) mx+y=12   and   4x-3y=m+1

c) 4x-3y+1=0   and   mx+6y+4=0

(Sol: a) m=-6; d=6/5; b) m=-4/3; d=107/15;   c) m=-8; d=3/5)

94. Calculate the value of c so that distance between straight lines 4x+3y-6=0 and 4x+3y+c=0 is 3.

(Sol: c_1=9, c_2=-21)

95. Given straight line r: x+y-3=0 and point P(-1,2):

a) Write all the forms of the equation of straight line perpendicular to r and going through P.

(Sol: x-y+3=0)

b) Find the intersection point, M, of both lines.

(Sol: (0,3))

c) Find the symmetric point of P about r. Draw an scheme.

(Sol: (1,4))

Review exercises

1. a) Write parametrical equations of the straight line that goes through point P(3,2) and has the same direction than vector **v**(1,-2)

   b) Give three points of r

   c) Check if points A(7,-6) and B(-3,7) are in the straight line r. *(Sol: YES; NO)*

2. Write parametrical equations of the straight line that goes through points A(-1,3) and B(5,-1).

3. Given straight line $\begin{cases} x = 7 - 2t \\ y = -4 + 3t \end{cases}$, find out a parallel and a perpendicular straight lines that go through point M(1,-2).

4. Given following straight lines,

   $r_1: \begin{cases} x = 5 - 2k \\ y = 2 + 4k \end{cases}$ $r_2: \begin{cases} x = -4k \\ y = 1 - 2k \end{cases}$ $r_3: \begin{cases} x = 3 + k \\ y = 6 - 2k \end{cases}$

   study the relative position and find out the intersection point, if possible, of:

   a) r_1 and r_2 b) r_1 and r_3 *(Sol: a) Secant (22/5, 16/5);   b) Coincident)*

5. Given straight line r: $3x - 2y + 6 = 0$ and point P(5,-1), find out the equations of the straight lines s and p that go through P and are:

   a) s parallel to r

   b) p perpendicular to r   *(Sol: a) 3x – 2y – 17 = 0 ;   b) 2x + 3y – 7 = 0)*

6. Given straight lines r and s, calculate the angle between them.

   r: $\begin{cases} x = t \\ y = 4 - 2t \end{cases}$ s: $x - y = 0$

   (Sol: 71° 33′ 54″)

7. Find out the equation of the straight line that goes through point P(3,5) and is separated an angle of 45° from straight line r: $2x + 3y - 6 = 0$.

   (Sol: x – 5y + 22 = 0 and 5x+ y - 20 = 0)

8. Find out the equation of the straight lines parallel to r: $2x - y + 3 = 0$ that are at a distance of $\sqrt{5}$ units from r.   *(Sol: r_1: 2x – y – 2 = 0; r_2: 2x – y + 8 = 0)*

9. In the triangle of vertices A(0,-1), B(8,3) and C(6,-1), calculate the length of the height that starts at B.

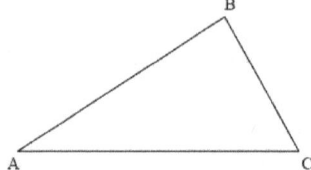

(Sol: 4)

10. Find out the coordinates of point P', symmetric of P(3,-2) about line r: 2x–y+8 = 0.

(Sol: (– 49/5, 22/5))

11. Find out the coordinates of point of straight line r:y=-3x+2 that is at the same distance from points A(5,1) and B(3,-2). *(Sol: (–1/14, 31/14))*

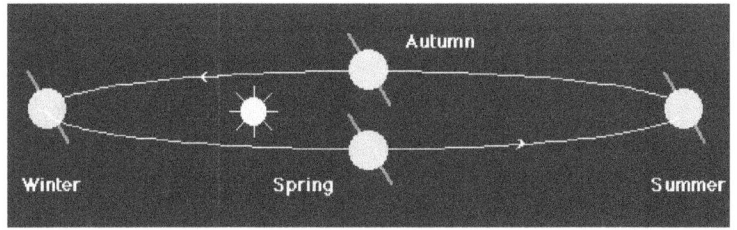

Autumn

Winter   Spring   Summer

Unit 8.- Conic sections

A *conic section* is the intersection of a plane and a double right circular cone. By changing the angle and location of the intersection, we can produce different types of conics. There are four basic types:

- Circles
- Ellipses
- Hyperbolas
- Parabolas.

* None of the intersections will pass through the vertices of the cone.

Parabola

Circle

Ellipse

Hyperbola

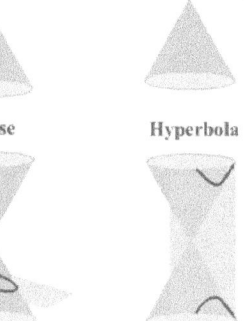

1. Circumference

A *circumference* is formed by all the points that are at the same distance from a common point. This point is called *centre* and the distance from the centre to all the points of the circumference is called *radius*.

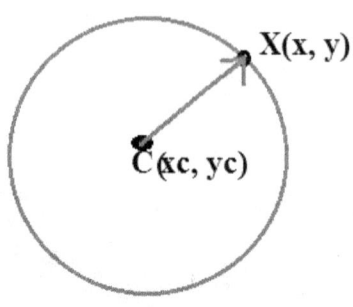

Vector **CX**: **CX** = $((x - x_c, y - y_c))$

Magnitude of vector **CX**, |CX|:

$$|CX| = \sqrt{(x - x_C)^2 + (y - y_C)^2}$$

Notice magnitude of vector **CX** is exactly the radius of the circumference:

$$|CX| = \sqrt{(x - x_C)^2 + (y - y_C)^2} = R$$

$$(x - x_C)^2 + (y - y_C)^2 = R^2$$

The equation of a circumference with centre $C(x_c, y_c)$ and radius R, is:

$$(x - x_C)^2 + (y - y_C)^2 = R^2$$

Example: Write the equation of the circumference with centre at (3, -2) and radius 5.

Solution:

$$(x - 3)^2 + (y + 2)^2 = 5^2 \rightarrow x^2 - 6x + 9 + y^2 + 4y + 4 = 25 \rightarrow \boxed{x^2 + y^2 - 6x + 4y - 12 = 0}$$

Example: Find out centre and radius of circumference: $x^2 + y^2 + 8x - 2y - 1 = 0$.

Solution: You can solve this example by comparing the given equation and the development of the general equation:

$$(x - x_C)^2 + (y - y_C)^2 = R^2 \rightarrow x^2 + x_C^2 - 2 \cdot x_C \cdot x + y^2 + y_C^2 - 2 \cdot y_C \cdot y = R^2$$

$$x^2 + y^2 - 2 \cdot x_C \cdot x - 2 \cdot y_C \cdot y + (x_C^2 + y_C^2 - R^2) = 0$$

$$x^2 + y^2 \quad + 8x \quad - 2y \quad - 1 \quad = 0.$$

By comparison: $\begin{cases} -2 \cdot x_C = 8 \rightarrow x_C = -4 \\ -2 \cdot y_C = -2 \rightarrow y_C = 1 \\ (x_C^2 + y_C^2 - R^2) = -1 \rightarrow (-4)^2 + 1^2 - R^2 = -1 \rightarrow 16 + 1 - R^2 = -1 \rightarrow R = \sqrt{16} = 4 \end{cases}$

So, we have: $\boxed{C(-4, 1) \text{ and } R = 4}$.

1. Write the equation of the circumference of centre C and radius r:

   a) C(4, −3), r = 3   b)C(0, 5), r = 6   c) C(6, 0), r = 2

   d)C(0, 0), r = 5

2. Find out the centre and the radius of the circumferences with equations:

   a) $(x-2)^2 + (y+3)^2 = 16$ b) $(x+1)^2 + y^2 = 81$ c) $x^2 + y^2 = 10$

3. Write the equation of the circumference of centre C(2, -1) and radius 5.

4. Find out the equation of the circumference whose centre is the point P(1, 3), and that is tangent to straight line r : 4x + 3y − 1 = 0.

   (Sol: $25x^2+25y^2-50x-150y-106 = 0$)

5. a) Calculate centre and radius of circumference: $2x^2 + 2y^2 - 8x - 12y + 24 = 0$

   b) Write equation of concentric circumference to previous one, with radius 3.

   (Sol: (2, 3); 1; $x^2 + y^2 - 4x - 6y + 4 = 0$)

Relative position of a straight line and a circumference

A straight line can be exterior, tangent or secant to a circumference. You are able to determine it only by solving the system:

Exterior	Tangent	Secant
No intersection point	1 intersection point	2 intersection points
No solution	1 solution	2 solutions

6. Study relative position of straight line r: 2x - 3y + 5 = 0 and circumference

C: $x^2 + y^2$ - 6x - 2y + 6 = 0.   *(Sol: exterior)*

7. a) Find out centre and radius of circumference: $x^2 + y^2$ - 2x - 3 = 0.

8. b) Study relative position of straight line r: 2x - y = 0 and previous circumference.

(Sol: (1, 0); 2; secant)

9. Find out the value of *k* so that straight line 3x + 4y + k = 0 is tangent to

circumference $x^2 + y^2$ + 4y - 5 = 0.   *(Sol: k = 23 and (-7))*

Power of a point to a circumference

Power of a point, P, to a circumference is defined as:

$$P = d^2 - r^2$$

where *d* is the distance between point P and the centre of the circumference, and *r* is radius of the circumference.

Notice:

➢   If P > 0 → Exterior

➢   If P < 0 → Interior

➢   If P = 0 → Point is on the circumference

2. Ellipse

An *ellipse* is formed by all the points that verify that the sum of their distances to two points, called, *foci*, is constant.

Note: *foci* is the plural of *focus.*

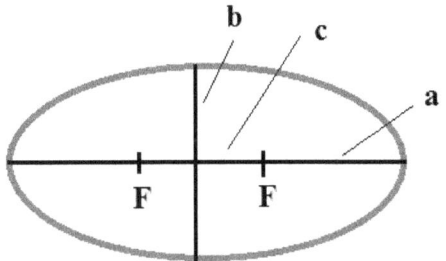

Its main elements are:

- Bigger semi-axis, **a**.
- Lower semi-axis, **b**.
- Focal semi-distance, **c**.

Reduced equation of an ellipse

The reduced equation of an ellipse is:

$$\frac{x^2}{a^2} + \frac{y^2}{b^2} = 1 \qquad \text{with} \qquad a^2 = b^2 + c^2$$

Important: Previous equation is valid not only for horizontal ellipses, but also for vertical ones. You must only pay attention to write parameter "a" below coordinate "x" and parameter "b" below coordinate "y".

Important: Given equation is only for ellipses with centre at (0, 0). General equation for an ellipse with centre at point $C(x_c, y_c)$ is:

$$\frac{(x - x_c)^2}{a^2} + \frac{(y - y_c)^2}{b^2} = 1$$

Example: Give the equation for ellipse with centre at (0, 0) that intersects coordinated axes at points (3, 0) and (0, 2).

Solution: Notice a = 3 and b = 2. So:

$$\frac{x^2}{9} + \frac{y^2}{4} = 1 \qquad \text{or} \qquad 4x^2 + 9y^2 = 36$$

Eccentricity of an ellipse

Eccentricity of an ellipse is defined as: $e = c/a$

$$0 < e < 1$$

If you think of an ellipse as a 'squashed' circle, the eccentricity of the ellipse gives a measure of just how 'squashed' it is. The bigger e is, the more squashed the ellipse is.

Exercises

10. Write the equation of the following ellipses and give their centres, semi-axes, foci and eccentricities.

a)
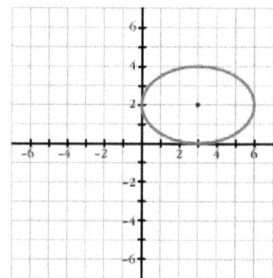

b)
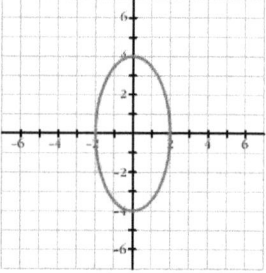

3. Hyperbola

A *hyperbola* is formed by all the points that verify that the differences of their distances to two points, called, *foci*, is constant.

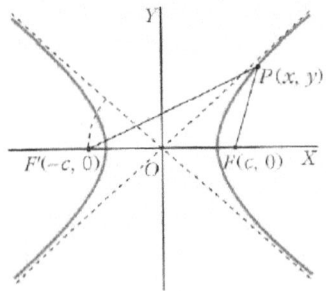

Its main elements are:

- **O**: Centre of the hyperbola.

- Vertices: closest poits to the centre.

- Semi-axis, **a**.

- Focal semi-distance, **c**.

Reduced equation of a hyperbola

Main axis: OX;   Centre: C(0, 0)

$$\frac{x^2}{a^2} - \frac{y^2}{b^2} = 1 \qquad \text{with} \qquad c^2 = a^2 + b^2$$

Main axis: OY;   Centre: C(0, 0)

$$-\frac{x^2}{b^2} + \frac{y^2}{a^2} = 1$$

Main axis: OX;   Centre: C(x_c, y_c)

$$\frac{(x - x_c)^2}{a^2} - \frac{(y - y_c)^2}{b^2} = 1$$

Main axis: OX;   Centre: C(x_c, y_c)

$$-\frac{(x - x_c)^2}{b^2} + \frac{(y - y_c)^2}{a^2} = 1$$

Eccentricity of a hyperbola

Eccentricity of an ellipse is defined as: $e = c/a$ $\qquad\qquad$ e < 1

The bigger e is, the more squashed the hyperbola is.

Exercises

11. Write the equation of the following hyperbolas and give their semi-axes, foci and eccentricities.

a)

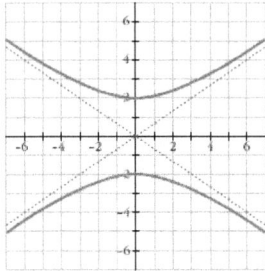

b)

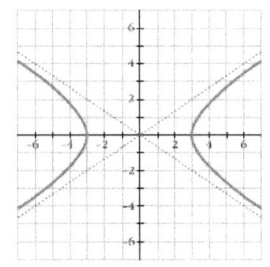

4. Parabola

A *parabola* is formed by all the points that are at the same distance from a point, called *focus,* and a straight line, called *directrix* straight line.

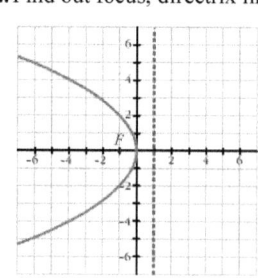

Its main elements are:

- Focus, F.

- Vertex, V.

- Directrix straight line.

- p: distance from focus to directrix straight line.

Reduced equation of a hyperbola

Vertex in (0, 0)	Vertex in (a, b)
- Vertical, open up: $y = 2px^2$	- Vertical, open up: $(y-b) = 2p(x-a)^2$
- Vertical, open down: $y = -2px^2$	- Vertical, open down: $(y-b) = -2p(x-a)^2$
- Horizontal, open right: $y^2 = 2px$	- Horizontal, open right: $(y-b)^2 = 2p(x-a)$
- Horizontal, open left: $y^2 = -2px$	- Horizontal, open left: $(y-b)^2 = -2p(x-a)$

Xercises

12. Find out focus, directrix line and equation of the following parabola:

5. Locus

In Maths, *locus* is a set of points that verify a condition.

For example, circumference is defined as *"the locus of points on a plane that are at a certain distance from a central point"* .

Locus is usually given by an expression of coordinates x and y that their points must verify.

Example: Give the equation of the locus whose distances to points A(0, 1) and B(0, -1) are similar. Is it a known geometric element?

<u>Solution:</u> A generic point of this locus is X(x, y).

Distance from X to A: d(X,A) $= d(X,A) = \sqrt{(x-0)^2 + (y-1)^2} = \sqrt{x^2 + y^2 - 2y + 1}$

Distance from X to B: d(X,B) $= d(X,B) = \sqrt{(x-0)^2 + (y+1)^2} = \sqrt{x^2 + y^2 + 2y + 1}$

As $\quad d(X,A) = d(X,B) \quad \Rightarrow \quad \sqrt{x^2 + y^2 - 2y + 1} = \sqrt{x^2 + y^2 + 2y + 1}$

$$x^2 + y^2 - 2y + 1 = x^2 + y^2 + 2y + 1$$

$$\boxed{y = 0}$$

This is OX-axis. Notice it is the mediatrix of segments with extremes at A and B.

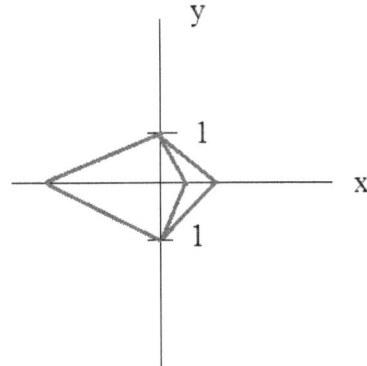

13. Find out the equation of the locus whose sum of distances to points A(0, 1) and B(0, -1) is 8. Is it a known figure? $(Sol: \frac{x^2}{3} + \frac{y^2}{4} = 1;$ *ellipse)*

14. Find out the equation of the locus of the points whose distance to Q(2, 4) is 3. Is it a known figure? $(Sol: x^2 + y^2 - 4x - 8y + 11 = 0;$ *circumference)*

15. Identify and find out the equation of the locus of the points whose distance to straight line r_1: x + y + 1 = 0 is the same than to straight line r_2: 2x + 2y + 4 = 0.
$(Sol: 2x + 2y + 3 = 0)$

16. Find out the locus of the points, P(x, y), that verify that triangle ABP is a right one in P, where A(2, 1) and B(-6, 1). $(Sol: (x+2)^2 + (y-1)^2 = 16)$

17. Find out the equation of the locus of the points whose squared of the distances to points A(-4, 0) and B(4, 0) is 40. Identify the figure you obtain. $(Sol: x^2 + y^2 = 4)$

18. Find out the equation of the locus of the points whose distance to point A(1, 0) is three times their distance to straight line x = 2. Identify the figure you obtain.
$(Sol: 8x^2 - y^2 - 34x + 35 = 0;$ *hyperbola)*

19. Find out the equation of the locus of the points, P, that verify that $\frac{d(P,A)}{d(p,r)} = 2$, where A(1, 0) and r: y = 4. What is the figure you obtain?
$(Sol: x^2 - 3y^2 - 2x + 32y - 63 = 0;$ *hyperbola)*

20. Find out the equation of the locus of the points, P, whose distance to A(2, 0) is twice the distance to B(-1, 0). Identify the figure you obtain.
$(Sol: (x+2)^2 + y^2 = 4;$ *circumference)*

Review exercises

1. Find out the equation of the circumference tangent to straight line 3x - 4y + 5 = 0 and whose centre is point C (2, 1). $(Sol: 25x^2 + 25y^2 - 100x - 50y + 76 = 0)$

2. Write the equation of the circumference that goes through points P (2, -1), Q (3, 0) and R (0, -2). $(Sol: x^2 + y^2 + 3x - 7y - 18 = 0)$

3. Find centre and radius of circumference whose centre is in the straight line y = 3x and that goes through points (3, 2) and (1, 4).　　*(Sol: C(1/2, 3/2); R = $\sqrt{26}$/2)*

4. Find out the equation of the circumference that goes through points A(-1,2) and B(1,4) and has its centre in the straight line y = 2x.　　*(Sol: $x^2 + y^2$ -2x - 4y + 1 = 0)*

5. Write the equation of the circumference with centre at point (2, -3) and that is tangent to straight line 3x - 4y + 5 = 0.　　*(Sol: $25x^2 + 25y^2$ -100x + 150y - 204 = 0)*

6. Write the equation of the circumference with radius 2 and that goes through points (1, 0) and (3, 2).　　*(Sol: $x^2 + y^2$ - 6x + 5 = 0 and $x^2 + y^2$ - 2x - 4y + 1 = 0)*

7. a) Find out centre and radius of circumference $2x^2 + 2y^2$ - 8x - 12y + 8 = 0

   b) Write the equation of the circumference with radius 5, being concentric to previous one.　　*(Sol: C(2, 3); R = 3; $x^2 + y^2$ -4x - 6y - 12 = 0)*

8. Find out the equation of the circumference tangent to straight line 4x + 3y - 25 = 0 and whose centre is the intersection point of straight lines 3x - y - 7 = 0 and 2x + 3y - 1 = 0.　　*(Sol: $x^2 + y^2$ -4x + 2y - 11 = 0)*

9. Study relative position of straight line r: 2x+y=1 and circumference x^2+y^2-4x-2y-4=0.
   (Sol: secant)

10. Study relative position of straight line r: 3x+4y-25 = 0 and circumference x^2+y^2-25 = 0. If they intersect at any point, find out its coordinates.　　*(Sol: tangent (3,4))*

11. Find the value of k so that straight line s: x + y + k = 0 is tangent to circumference $x^2 + y^2 + 6x + 2y + 6 = 0$.　　*(Sol: k = 4+2$\sqrt{2}$ and k= 4-2$\sqrt{2}$)*

12. Study relative position of straight line r: $y = \dfrac{8 - 4x}{3}$ and circumference

   x^2+y^2-12x-6y+20=0.　　*(Sol: tangent)*

12. Study relative position of straight line r: x+y=2 and circumference x^2+y^2+2x+4y+1=0.
   (Sol: exterior)

13. a) Describe the following conic section: $16x^2 + y^2 = 16$.

   b) Locate their foci.　　*(Ellipse; F(0, $\sqrt{15}$), F'(0, $-\sqrt{15}$))*

14. a) Describe the following conic section: $4y^2 - x^2 = 4$.

   b) Locate their foci.　　*(Hyperbola; F(0, $\sqrt{5}$), F'(0, $-\sqrt{5}$))*

15. Identify the following conic section and represent it: $9x^2 + 25y^2 = 225$.

16. Obtain the equation of the conic section whose graph is the following:

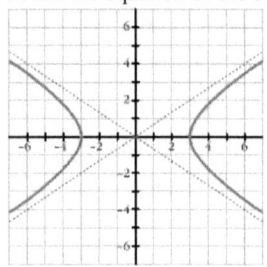

$$(Sol: \frac{x^2}{9} - \frac{y^2}{4} = 1)$$

17. Describe the following conic section and represent it: $36x^2 + 4y^2 = 144$.

18. Write the equation of the following conic section:

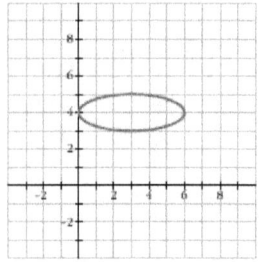

$$(Sol: \frac{(x-3)^2}{9} + \frac{(y+4)^2}{1} = 1)$$

19. Identify the following conic section and represent it: $4y^2 - 9x^2 = 36$.

20. Describe the following conic sections and represent them:

    a) $4x^2 + 25y^2 = 100$ b) $4y^2 - x^2 = 4$

$$Sol:\ a)\ Ellipse: \begin{cases} a = 5 \\ b = 2 \\ F(\sqrt{21},0);\ \ F'(-\sqrt{21},0) \\ e = \sqrt{21}/5 \end{cases} \qquad b)\ Hyperbola: \begin{cases} a = 1 \\ F(0,\sqrt{5});\ \ F'(0,-\sqrt{5}) \\ e = \sqrt{5} \end{cases}$$

21. Identify the following conic sections and find out their main elements:

    a) $\frac{(x-2)^2}{16} - \frac{y^2}{4} = 1$ b) $\frac{x^2}{25} + \frac{y^2}{49} = 1$

$$Sol:\ a)\ Hyperbola: \begin{cases} Centre = (2,0) \\ F(2+2\sqrt{5},0);\ \ F'(2-2\sqrt{5},0) \\ e = \sqrt{5}/2 \end{cases} b)\ Ellipse: \begin{cases} C(0,0) \\ a = 7 \\ b = 5 \\ F(0,\sqrt{24});\ \ F'(0,-\sqrt{24}) \\ e = \sqrt{24}/7 \end{cases}$$

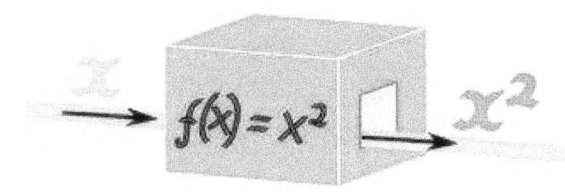

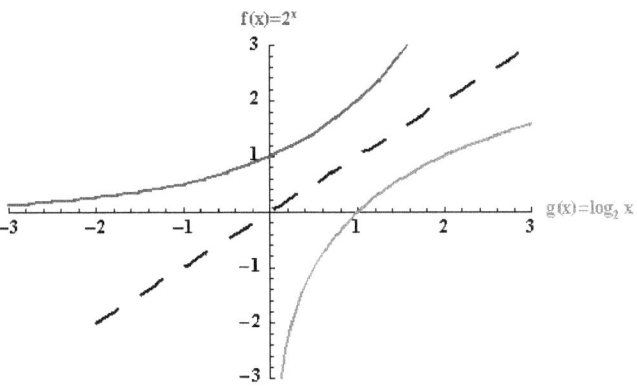

Unit 9.- Functions

1. Concept of function. Domain and range of a function

A **function of R in R** is a relationship where each input has a ***single*** output

It is often written as f(x), where x is the input value.

Example: f(x) = 3x -1 *(f of x is three times x minus one)* is a function of x because each input x has a single output:

f(2) = 3·2 – 1 = 5
f(3) = 3·3 – 1 = 8
f(10) = 3·10 – 1 = 29.

Domain and range of a function

> ➤ **_Domain_** of a function is the set of values that independent variable **_x_** is able to take. It is denoted **_Df_** or **_Domf_**.
>
> ➤ **_Range_** of a function is the set of values that dependent variable **_y_** takes. It is denoted **_Rf_** or **_Rgf_**.

For example, for $f(x) = 3x - 1$, as x is allowed to have whatever real value, it is said: **_Domf = R_**. In this case, also **_Rgf = R_**.

Restrictions to domain of a function

Imagine $f(x) = \dfrac{x+5}{x}$. Notice x is allowed to have whatever value, except x = 0, as $f(0) = \dfrac{5}{0}$ and this does not exist.

Another example, imagine $f(x) = \sqrt{x}$. Notice x is not allowed to have any negative value, as square roots of negative numbers do not exist.

As you can see, there are restrictions to domain of a function.

> ➤ **Polynomials:** Domf = R
> ➤ **Quotients:** $f(x) = \dfrac{n(x)}{d(x)}$ Domf = R − {x / d(x) = 0}
> ➤ **Odd index roots:** Domf = R
> ➤ **Even index roots:** $f(x) = \sqrt[n]{r(x)}$ Domf = {x / r(x) ≥ 0}
> ➤ **Logarithms:** $f(x) = \log_a^{r(x)}$ Domf = {x / r(x) > 0}

Exercises	**1.** Indicate which of the following graphs are a function. Justify your answer: a) 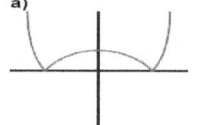 b) 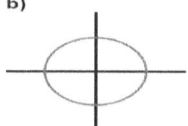

2. This graph represents function y = f(x):

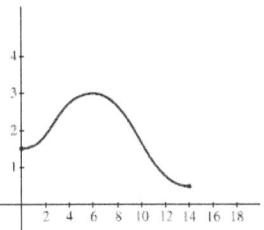

    a) What is its domain?

    b) Indicate intervals in which it increases and decreases.

    c) At what point does it have a maximum?

3. Given the following functions:

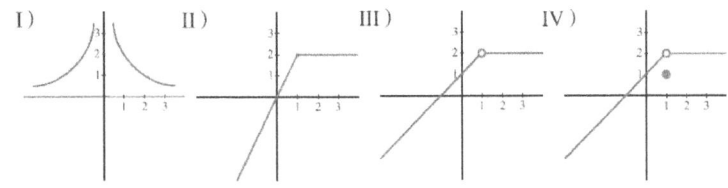

    a) Indicate which of them are continuous.

    b) Find out the image of x = 1, this is, f(1).

4. Given the graph:

    a) Indicate if f(x) is a continuous function. Justify your answer.

    b) Find out f (-1), f (0), f (2) y f (3).

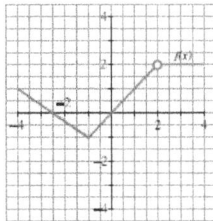

5. Find out f(−1), f(0) and f(2), with: $f(x) = \begin{cases} 3x^2 - 1 & if \ x \le -1 \\ x+1 & if \ -1 < x \le 2 \\ x^2 & if \ x > 2 \end{cases}$

6. Indicate domain and range of the following functions:

a)

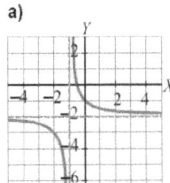

b)

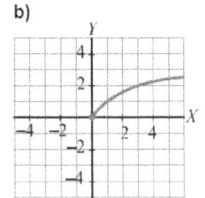

c)

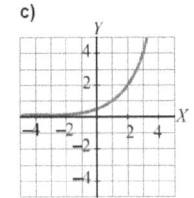

7. This graph represent function $y = f(x)$:

a) What is its domain?

b) Indicate intervals in which it increases and decreases.

c) At what point does it have a maximum?

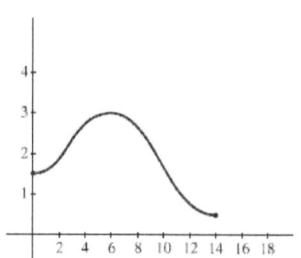

8. Find out domain of following functions:

a) $y = \dfrac{1}{x^2 - 16}$ b) $y = \sqrt{1 + 2x}$ c) $y = \dfrac{x}{x^2 - 4}$ d) $y = \sqrt{2x}$ e) $y = \dfrac{1}{x^2 + 4}$

f) $y = \dfrac{1}{\sqrt{x - 2}}$ g) $y = \dfrac{1}{x^2 - 2x}$ h) $y = \sqrt{6 + 3x}$ i) $y = \dfrac{3}{(x - 5)^2}$ j) $y = \sqrt{2x - 4}$

k) $y = \dfrac{1}{x^2 - 9}$ l) $y = \sqrt{x - 2}$ m) $y = \dfrac{2 + x}{x^2}$ n) $y = \sqrt{3x - 1}$ ñ) $y = \dfrac{x + 1}{\sqrt{x}}$

o) $y = \dfrac{1}{3x - x^2}$ p) $y = \sqrt{x^2 - 1}$ q) $y = \dfrac{2x}{(x - 3)^2}$

9. Consider all rectangles with perimeter 30 cm. If "x" is the length of the base, area will be: $A = x(15 - x)$. What is domain of this function?

$15{-}x\ \boxed{}$
x

10. Indicate the domain of the following functions:

1) $f(x) = x^2 - 2x + 6$ 2) $g(x) = \dfrac{x - 1}{x^2 + 3x + 2}$ 3) $h(x) = \dfrac{x - 2}{x^2 + 4}$

4) $i(x) = \dfrac{4 - x}{x^3 - 5x^2 + 6x}$ 5) $j(x) = \dfrac{8}{2x^2 + 3x - 2}$ 6) $k(x) = \sqrt{x - 1}$

7) $l(x) = \sqrt{2x - 7}$ 8) $m(x) = \sqrt{x^2 - 9x + 20}$ 9) $n(x) = \sqrt{x^3 - 4x}$

11. Indicate the domain of the following functions:

a) $f(x) = x^2 - 4x + 3$ b) $f(x) = \dfrac{2x + 3}{x^2 - 4x + 3}$ c) $f(x) = \sqrt[3]{x^2 - 4x + 3}$

d) $f(x) = \sqrt{x^2 - 4x + 3}$ e) $f(x) = \dfrac{\sqrt{x^2 - 4x + 3}}{x + 1}$ f) $f(x) = \dfrac{x + 1}{\sqrt{x^2 - 4x + 3}}$

g) $f(x) = \sqrt{\dfrac{x + 1}{x^2 - 4x + 3}}$

2. Intersection points with the axes

When studying functions, it is interesting knowing if they intersect axes and where they do it.

> ➢ **Intersection points with ordinated axis**
> Intersection point of a function $y = f(x)$ with the Y-axis has abscise = 0. So, its coordinates are $(0, f(0))$.
>
> ➢ **Intersection points with abscise axis**
> Intersection point of a function $y = f(x)$ with the X-axis has ordinated 0. So, they are the points whose x coordinate are solution of equation $f(x) = 0$.

Example: Find out the intersection points of function $y = \dfrac{3}{5}x - 3$ with the coordinated axes.

Solution: For $x = 0$ → $y = -3$ → $Q(0, -3)$

For $y = 0$ → $\dfrac{3}{5}x - 3 = 0$ → $x = 5$ → $P(5, 0)$

Look at the plot of the function:

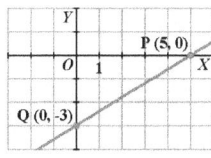

---

12. Write the coordinates of five points located on the abscise axis. What do they have in common?
13. Write the coordinates of five points located on the ordinated axis. What do they have in common?
14. Find out the intersections with the axes for these functions:

   a) $y = 2x - 1$ b) $y = 3x + 2$ c) $y = -3x + 2$ d) $y = 2$ e) $y = x^2 - 2x - 3$

   f) $y = x^2 - 2x + 1$ g) $y = 4 - x^2$

15. Find out the intersections with the axes for these linear functions and represent them:
   a) $2y = x - 4$ b) $3y - 6 = 2x$

Exercises

3. Symmetries. Even and odd functions

Plots of functions are geometric figures, so, they can be symmetric about an axis or a point.

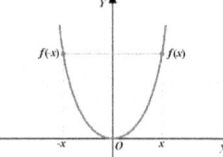

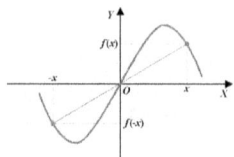

Symmetry about the ordinated axis

Symmetry about center of coordinated axes.

Symmetry of function $y = x^2$

x	$y = x^2$
− 3	9
− 2	4
− 1	1
0	0
1	1
2	4
3	9

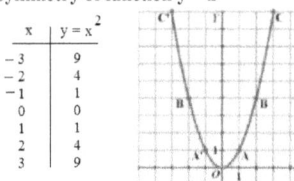

Symmetry of function $y = x^3$

x	$y = x^3$
− 3	− 27
− 2	− 8
− 1	− 1
0	0
1	1
2	8
3	27

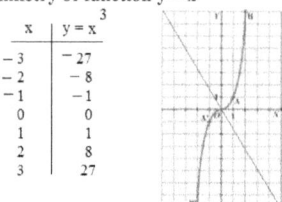

Notice this function is symmetric about *y-axis*.

These functions are named *EVEN functions*.

We could have a mirror along the *y-axis*.

As you can see in the values table, in even functions,

$$f(-a) = f(a)$$

Notice this function is symmetric about origin.

These functions are named *ODD functions*.

We could locate a mirror where drawn a blue line.

As you can see in the values table, in odd functions,

$$f(-a) = - f(a)$$

Example: Study the symmetry of function $y = x^4 - 6x^2$.

Solution: Substitute x by $-x$: $\quad f(-x) = (-x)^4 - 6(-x)^2 = x^4 - 6x^2 = f(x) \quad \rightarrow \quad f(-x) = f(x)$
So, $f(x)$ is an even function, and its plot is symmetric about the *y-axis*.

16. Complete:

					In general
a)	$f(x) = x^2$	$f(1) =$ $f(-1) =$	$f(2) =$ $f(-2) =$	$f(3) =$ $f(-3) =$	$f(x) =$ $f(-x) =$
b)	$f(x) = 2x$	$f(1) =$ $f(-1) =$	$f(2) =$ $f(-2) =$	$f(3) =$ $f(-3) =$	$f(x) =$ $f(-x) =$
c)	$f(x) = x^2-5$	$f(1) =$ $f(-1) =$	$f(2) =$ $f(-2) =$	$f(3) =$ $f(-3) =$	$f(x) =$ $f(-x) =$
d)	$f(x) = x^3-2x$	$f(1) =$ $f(-1) =$	$f(2) =$ $f(-2) =$	$f(3) =$ $f(-3) =$	$f(x) =$ $f(-x) =$
e)	$f(x) = 5x^4 - 2x^2$	$f(1) =$ $f(-1) =$	$f(2) =$ $f(-2) =$	$f(3) =$ $f(-3) =$	$f(x) =$ $f(-x) =$
f)	$f(x) = x^5-4x$	$f(1) =$ $f(-1) =$	$f(2) =$ $f(-2) =$	$f(3) =$ $f(-3) =$	$f(x) =$ $f(-x) =$

17. Indicate if these functions are symmetric and the type of symmetry they have got:
a) $f(x) = x^3 - 5x$ b) $g(x) = \dfrac{x^2}{x^4 - 1}$ c) $h(x) = \dfrac{2x}{x+1}$ d) $i(x) = 4x^2 - 5$

Exercises

4. Continuity of a function

Continuity of a function is a very intuitive idea. A function will be said to be a continuous one if it can be drawn by an only stroke. In the same way, one function will be said to be a discontinuous one if it has some discontinuity.

Definition of continuity

A function is said to be a **continuous** one if it does not have any kind of discontinuity. A function can be **continuous in an interval** if it has discontinuities only outside that interval.

Functions are continuous into their domains.

Types of discontinuities

There are several reasons by which a function can be discontinuous at one point:

a) Asymptotic discontinuity: It has infinite vertical asymptotes at that point. This means y values of the function increase or drop infinitely when x coordinate gets closer and closer to that point.

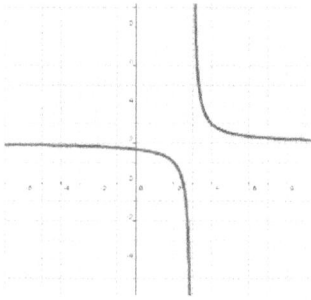

b) Jump discontinuity: When moving along the curve of the function, we come across a jump.

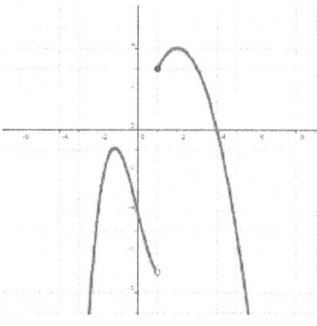

c) Point discontinuity: At one point, the function is not defined or it is defined at a point outside the curve.

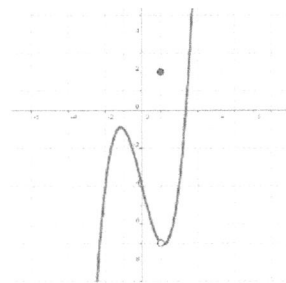

Notice a point discontinuity can be easily "repaired" by only adding the correct point in its definition:

For example, observe the curve of the function defined as $f(x) = \begin{cases} x \text{ if } x < 1 \\ 3 \text{ if } x = 1 \\ x \text{ if } x > 1 \end{cases}$

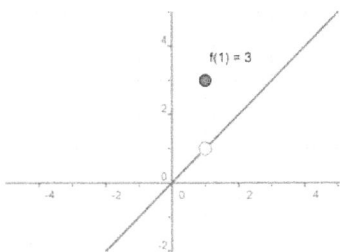

This function has a point discontinuity at x = 1, but it can be "repaired" if we change the 3 by a 1 in its definition: $f(x) = \begin{cases} x \text{ if } x < 1 \\ 1 \text{ if } x = 1 \\ x \text{ if } x > 1 \end{cases}$ Now, it is a continuous function.

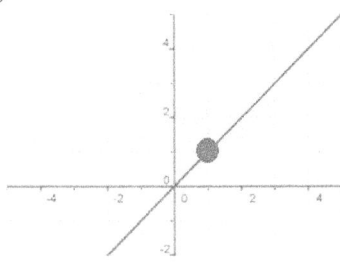

That is why point discontinuities are named ***avoidable discontinuities***.

$$\text{Types of discontinuities} \begin{cases} \text{Unavoidable discontinuities} \begin{cases} \textit{First kind} \rightarrow \textit{Asymptotic discontinuities} \\ \textit{Second kind} \rightarrow \textit{Jump discontinuities} \end{cases} \\ \textit{Avoidable discontinities} \rightarrow \textit{Point discontinuities} \end{cases}$$

5. Affine function y = mx + n

As this part of the unit has already been studied by you in a previous unit, *Analytical geometry*, you will only be proposed some questions to be answered. You may need to review your notes.

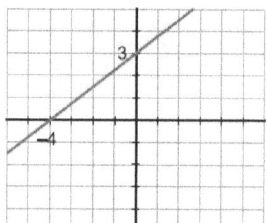

Affine functions are those ones whose graph is a straight line that does not need to go through the coordinate's origin.

Its algebraic expression is $y = mx + n$.

Algebraic expression for an ***affine function*** is $y = mx + n$, where:

➤ m is the ***slope*** or *gradient* of the function.

➤ n is the *y-intercept:* straight line intersects ordinated axis at point **(0, n).**

18. Represent:

    a) $y = \dfrac{3}{2}x - 2$ b) $y = -0.5x + 3.5$ c) $y = -\dfrac{3}{5}x + 1$ d) $f(x) = \dfrac{4 - 2x}{5}$

19. Write the equation of the straight line that goes through points (3, - 4) and (-2, 3).

20. Write the equation of the straight lines that have the following graphs:

a)   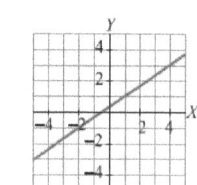   **b)**   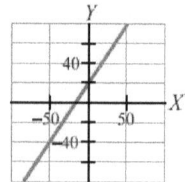

21. Write the equation of the straight line that goes through point (1, - 4) and whose slope is – 2.

22. Represent the following functions:

a) $y = \begin{cases} 2x & \text{if } x \le -1 \\ -2 & \text{if } -1 < x \le 3 \\ x - 5 & \text{if } x > 3 \end{cases}$ b) $y = \begin{cases} -3 & \text{if } x < 0 \\ 2x + 1 & \text{if } x \ge 0 \end{cases}$ c) $y = \begin{cases} -x + 3 & \text{if } x < 1 \\ 2 & \text{if } 1 \le x < 2 \\ x & \text{if } x \ge 2 \end{cases}$

23. Decide which of the following functions corresponds to the given graph:

a) $f(x) = \begin{cases} 2x + 5 & \text{if } -3 \le x \le -1 \\ x + 5 & \text{if } 0 \le x < 3 \\ 2x & \text{if } 3 \le x \le 8 \end{cases}$ b) $g(x) = \begin{cases} 2x + 5 & \text{if } -3 \le x < 0 \\ 5 - x & \text{if } 0 \le x < 3 \\ 2 & \text{if } 3 \le x \le 8 \end{cases}$

c) $h(x) = \begin{cases} 2 & \text{if } -3 < x < 0 \\ -1 & \text{if } 0 < x < 3 \\ 0 & \text{if } 3 < x < 8 \end{cases}$

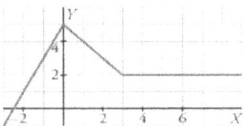

24. Define the function whose graph is given below:

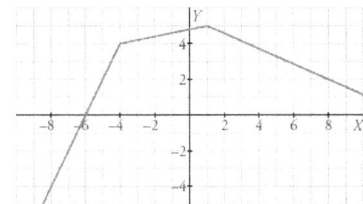

---

6. Quadratic functions $y = ax^2 + bx + c$

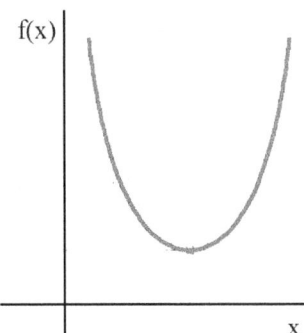

f(x)

x

> *Quadratic functions* are those whose algebraic expression is $y = ax^2 + bx + c$.
> Its graph is a parable, whose vertex is located at the point: $\left(x_v = \dfrac{-b}{2a}, y_v = f(x_v) \right)$
> IMPORTANT: The parable will be **open up** if a > 0 and **open down** if a < 0.

25. Find out the vertex of the following parables:

    a) $y = 3x^2 - 12x + 1$ b) $2y - 3x + x^2 = 5 - x - x^2$

26. Join each quadratic function to its curve:

    a) $y = x^2$

    b) $y = (x - 3)^2$

    c) $y = x^2 - 3$

    d) $y = x^2 - 6x + 6$

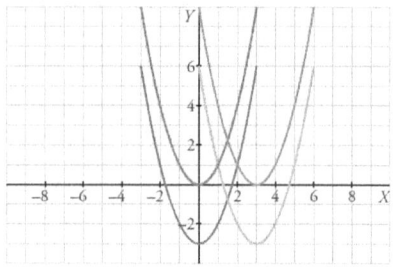

27. Represent the curve of the following parables, finding out he vertex, some close points and the intersection points with the axes:

    a) $y = (x + 4)^2$ b) $y = x^2 + 2x$ c) $y = -3x^2 + 6x - 3$ d) $y = -x^2 + 5$

28. Locate the vertex (abscise and ordinated) of the following parables and indicate if it is a maximum or a minimum:

    a) $y = x^2 - 5$ b) $y = 3 - x^2$ c) $y = -2x^2 - 4x + 3$

    d) $y = 3x^2 - 6x$ e) $y = 5x^2 + 20x + 20$ f) $y = -x^2 + 5x - \dfrac{3}{2}$

29. Represent the graph of the following functions:

    a) $y = -x^2 + 4x - 1$ b) $y = (x + 1)^2 - 3$ c) $y = -x^2 + 4$ d) $f(x) = -2x^2 + 4x$

Example: Solve, graphically and algebraically, the following system: $\begin{cases} y = x^2 + 2x - 3 \\ y = 1 - x \end{cases}$

Solution:

Graphically:

We represent the curve of the parable $y = x^2 + 2x - 3$

    - Vertex: $x_v = \dfrac{-b}{2a} = \dfrac{-2}{2} = -1 \;\rightarrow\; y_v = 1 - 2 - 3 = -4 \;\Rightarrow\; V(-1, -4)$

    - Intersections with the axes:

    Y-axis $\rightarrow$ x = 0 $\rightarrow$ y = -3 $\rightarrow$ (0, -3)

    X-axis $\rightarrow$ y = 0 $\rightarrow$ $x^2 + 2x - 3 = 0 \;\rightarrow\; x = \dfrac{-2 \pm \sqrt{4 + 12}}{2} = \dfrac{-2 \pm 4}{2} \begin{cases} x_1 = 1 & (1,0) \\ x_2 = -3 & (-3,0) \end{cases}$

- Points close to the vertex:

X	-4	-2	-1	0	2
Y	5	-3	-4	-3	5

We represent the straight line $y = 1 - x$

x	1	0
y	0	1

Notice both graphs intersect at the points (-4, 5) and (1, 0).

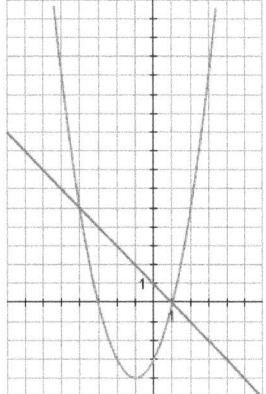

Algebraically: As y is already isolated at both expressions, we can equalize them:

$$\begin{cases} y = x^2 + 2x - 3 \\ y = 1 - x \end{cases}$$

$$x^2 + 2x - 3 = 1 - x \rightarrow x^2 + 3x - 4 = 0 \Rightarrow x = \frac{-3 \pm \sqrt{9+16}}{2} = \frac{-3 \pm 5}{2} \begin{cases} x_1 = -4 \\ x_2 = 1 \end{cases}$$

$$\begin{cases} x_1 = -4 \rightarrow y = 1 + 4 = 5 \\ x_2 = 1 \rightarrow y = 0 \end{cases} \rightarrow \text{So, solutions are } \boxed{(-4, 5) \text{ and } (1, 0)}.$$

Exercises

30. Solve, graphically and algebraically, the following systems:

a) $\begin{cases} y = x^2 - 4x + 5 \\ x - y - 3 = 0 \end{cases}$ b) $\begin{cases} y = 2x^2 + 8x - 11 \\ y + 3 = 0 \end{cases}$

31. Check, graphically and algebraically, that the following systems do not have solution:

a) $\begin{cases} y = \dfrac{1}{2}x^2 - x - \dfrac{3}{2} \\ y = \dfrac{x}{2} - 3 \end{cases}$ b) $\begin{cases} y = \dfrac{1}{x-1} \\ y = -x + 1 \end{cases}$

32. Represent the following functions:

a) $f(x) = \begin{cases} -1-x & if \ x < -1 \\ 1-x^2 & if \ -1 \le x \le 1 \\ x-1 & if \ x > 1 \end{cases}$ b) $f(x) = \begin{cases} x^2 & if \ x < 0 \\ -x^2 & if \ x \ge 0 \end{cases}$

33. a) Find out the equation of the straight line going through points (-2, -1) and (1, 3), and represent it.

b) Find out the intersection points of previous straight line and function $y = -x^2+4x$.

34. a) Find out the equation of the straight line going through points (-1, 3) and whose slope is − 3.

b) Find out the intersection points of previous straight line and function $y = -x^2+4$.

7. Inverse proportionality functions $y = K / x;$ $y = (ax + b) / (cx + d)$

f(x)

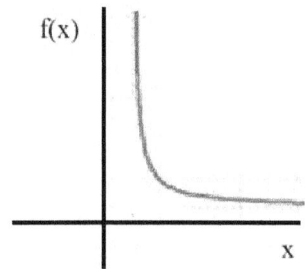

X

> Function $f(x) = \dfrac{k}{x}$, is named *inverse proportionality function*.
> Its curve is an equilateral hyperbole and it depends on the sign of k.
> Its curve approaches to both coordinated axes.

35. Represent the curve of the function $y = \dfrac{-2}{x}$ by completing the following values chart:

x	-4	-3	-2	-1	0	1	2	3	4
f(x)									

The curve of the function $f(x) = \dfrac{a\,x + b}{c\,x + d}$ is similar to $f(x) = \dfrac{K}{x}$. The only difference is that:

- ➢ It approaches to the vertical line $x = \dfrac{-d}{c}$
- ➢ It approaches to the horizontal line $y = \dfrac{a}{c}$

Exercises

36. Represent the curve of the function $y = \dfrac{2x+1}{x-3}$ by completing the following values chart:

x	4	5	6	7	8	9	10	11	3
f(x)									

x	2	1	0	-1	-2	-3	-5	-7	-9
f(x)									

37. Do the same for the function $y = \dfrac{x-1}{x-2}$.

38. Represent the curve of the following functions, studying, first of all, the straight lines to which they approach:

a) $y = \dfrac{3x-1}{x+2}$ b) $y = \dfrac{x+4}{2x+7}$ c) $y = \dfrac{3x-7}{2x+5}$ d) $y = \dfrac{x}{x+3}$

8. Radical functions $\quad y = \sqrt{f(x)}$

Exercises

39. Represent the curve of the following functions:

a) $y = 1 - \sqrt{-3x}$ b) $y = \sqrt{3x-1}$ c) $y = \sqrt{2x+3} - 1$

9. Exponential and logarithmical functions $y = e^{f(x)}$; $y = \log_a[f(x)]$

40. Represent the curve of the following functions:

a) $y = 2^{1-x}$ b) $y = \log_{1/4} x$ c) $y = 1 - \log_2 x$ d) $y = \left(\dfrac{1}{4}\right)^{x+2}$ e) $y = 3^{x+1}$

41. Consider the following functions:

a) Write their analytical expressions.

b) Indicate their domains.

c) Indicate intervals in which they increase and decrease.

a) 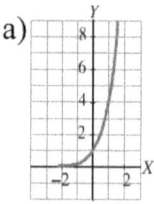   b)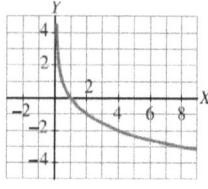

c) 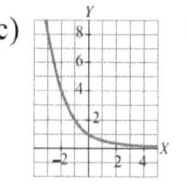   d)

10. Functions defined as intervals

42. Represent the curve of the following functions:

a) $y = \begin{cases} 2x^2 & si \quad x < -1 \\ 2x + 4 & si \quad x \geq -1 \end{cases}$ b) $y = \begin{cases} x^2 - 1 & si \quad x \leq 2 \\ 3 & si \quad x > 2 \end{cases}$ c) $y = \begin{cases} (-x + 1)/2 & si \quad x \leq -1 \\ -x^2 & si \quad x > -1 \end{cases}$

11. Functions with absolute value

43. Given representation of functions y = f(x), represent y = |f(x)|:

a)

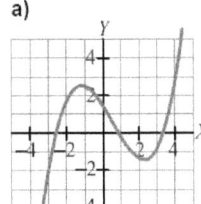

b)

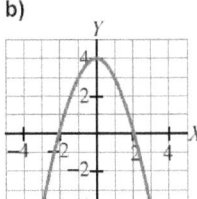

c)

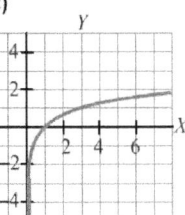

d)

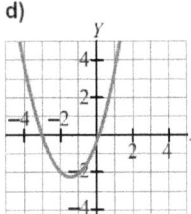

e)

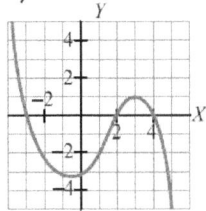

44. Define as intervals the following absolute value functions:

a) $y = |2x + 4|$

b) $y = |-x + 3|$

c) $y = \left|\dfrac{x+1}{2}\right|$

d) $y = |3x - 2|$

e) $y = \left|\dfrac{3x+1}{2}\right|$

12. Transformation of functions

Known the graph of a function, y = f(x), we can draw some others, only by transforming it:

> ➤ y = f(x) + k → Move the graph, vertically, k units up.
> ➤ y = f(x) – k → Move the graph, vertically, k units down.

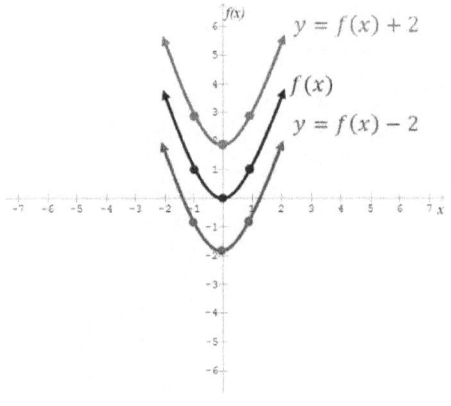

> ➤ y = f(x+k) → Move the graph, horizontally, k units to the left.
> ➤ y = f(x – k) → Move the graph, horizontally, k units to the right.

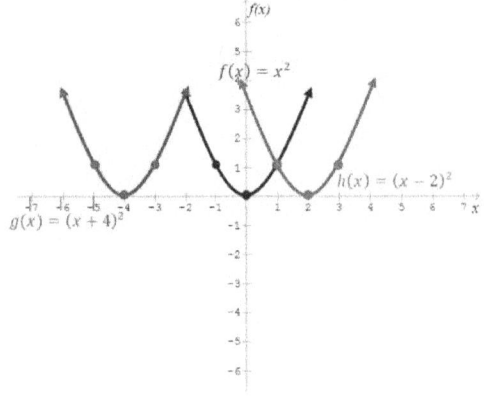

> ➢ y = – f(x)   →   Reflect the graph, about OX axis.

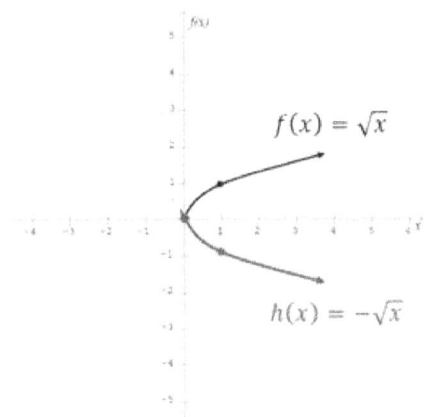

$f(x) = \sqrt{x}$

$h(x) = -\sqrt{x}$

Exercises

45. Given representation of function y = f(x), draw a) y = f(x) – 3 and b) y = f(x+2).

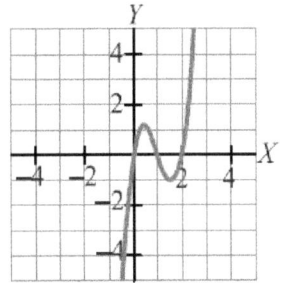

46. Given representation of function y = f(x), draw a) y = f(x) + 2 and b) y = – f(x).

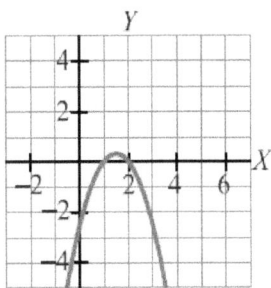

47. Given representation of function y = f(x), draw a) y = f(x − 1) and b) y = f(x) − 1.

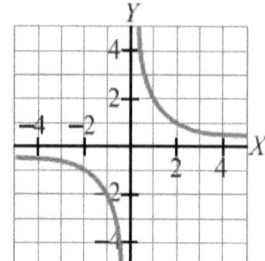

48. Given representation of function y = f(x), draw a) y = f(x − 2) and b) y = − f(x).

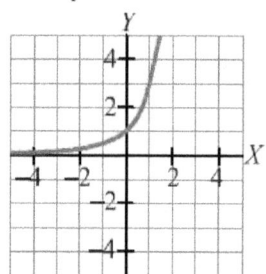

13. Composition of functions   (g o f)(x)

Composition is a new operation we can do with functions.

For example, given $f(x) = x^2$ and $g(x) = x + 1$, composition (g o f)(x) is calculated as g[f(x)]:

(g o f)(x) = g[f(x)] = f(x) + 1 = x^2 + 1.

(f o g)(x) = f[g(x)] = $(x+1)^2$ = x^2 + +2x + 1.

Notice composition of functions is **NOT** a commutative operation, this is, (g o f)(x) ≠ (f o g)(x).

Example: Given $f(x) = x^2 + 1$ and $g(x) = \sqrt{x}$, work out: a) (g o f)(x)   b) (f o g)(x).

Solution:

(g o f)(x) = g[f(x)] = $\sqrt{f(x)}$ = $\boxed{\sqrt{x^2 +1}}$.

(f o g)(x) = f[g(x)] = f($\sqrt{x}$) = $(\sqrt{x})^2 +1$ = $\boxed{x + 1}$.

49. Given functions $f(x) = \dfrac{-3x+2}{4}$ and $g(x) = x^2 + 1$, find out:

   a) $(f \circ g)(x)$ b) $(g \circ f)(x)$

50. Given functions $f(x) = \dfrac{x^2}{3}$ and $g(x) = x+1$, find out:

   a) $(f \circ g)(x)$ b) $(g \circ f)(x)$

51. Given functions $f(x) = 3x^2$ and $g(x) = \dfrac{1}{x+2}$, explain how you can obtain, by composition, functions:

   a) $p(x) = \dfrac{3}{(x+2)^2}$ b) $q(x) = \dfrac{1}{3x^2+2}$

52. Explain how you can obtain, by composition, functions p(x) and q(x), from f(x) and g(x), where: $f(x) = 2x-3$, $g(x) = \sqrt{x-2}$, $p(x) = 2\sqrt{x-2}-3$ and $q(x) = \sqrt{2x-5}$.

53. Functions f and g are defined as: $f(x) = \dfrac{x-1}{3}$ and $g(x) = \sqrt{x}$. Explain how you can obtain, from them, functions: $p(x) = \sqrt{\dfrac{x-1}{3}}$ and $q(x) = \dfrac{\sqrt{x}-1}{3}$.

Inverse of a function

Given a function, *f(x)*, its inverse, $\boldsymbol{f^{-1}(x)}$, is the function that satisfy: $\boldsymbol{(f^{-1} \circ f)(x) = x}$

A practical way to obtain inverse of a function, y = f(x), is permuting *"x"* and *"y"* in its analytical expression and, after that, isolating *"y"*:

Example: Given $f(x) = \sqrt{\dfrac{x-1}{3}}$, calculate its inverse, $f^{-1}(x)$.

Solution: $y = \sqrt{\dfrac{x-1}{3}}$ $\rightarrow$ permute *"x"* and *"y"* $\rightarrow$ $x = \sqrt{\dfrac{y-1}{3}}$

$x^2 = \dfrac{y-1}{3}$ $\rightarrow$ $3x^2 = y-1$ $\rightarrow$ $y = 3x^2 + 1$ So, $\boxed{f^{-1}(x) = 3x^2 + 1}$

CHECK: $(f^{-1} \circ f)(x) = f^{-1}(f(x)) = 3(fx)^2 + 1 = 3\left(\sqrt{\dfrac{x-1}{3}}\right)^2 + 1 = 3 \cdot \dfrac{x-1}{3} + 1 = x - 1 + 1 = x$

54. Find out the inverse of the following functions:

a) $y = \dfrac{2x-1}{3}$ b) $y = \dfrac{2-3x}{4}$ c) $y = \dfrac{-x+3}{2}$ d) $y = \dfrac{-2x-1}{5}$ e) $y = \dfrac{-2+7x}{3}$

(Sol: a) $y = \dfrac{3x+1}{2}$ b) $y = \dfrac{2-4x}{3}$ c) $y = 3-2x$ d) $y = \dfrac{-5x-1}{2}$ e) $y = \dfrac{3x+2}{7}$)

Review exercises

1. Given $f(x) = \sqrt{x}$:

   a) Justify that it is a function   b) Calculate f(4), f(1), f(0), f(-9), f(1/4), f(2) and f(− 2)

   d) Indicate domain and range of f.

2. Which of the following graphs are functions?

a) 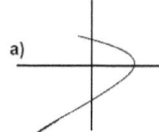   b) 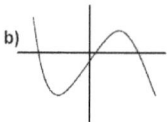   c) 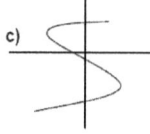   d)

3. Indicate domain and range of the following functions:

a) 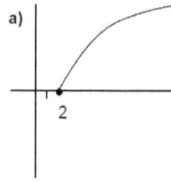   b) 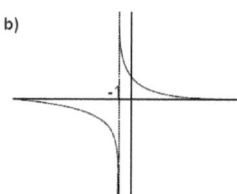   c)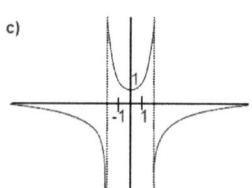

4. In a square of side 4 cm, we cut in the corners isosceles triangles whose equal sides measure x cm.

   a) Write area of octagon as a function of x.

   b) Which is domain and range of that function?

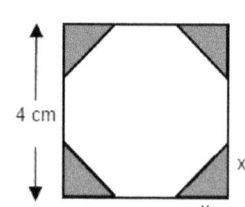

4 cm

x

x

5. Indicate the domain of the following functions:

a) $f(x) = x^3 + x^2 - 3x + 1$

b) $f(x) = \dfrac{8x}{x+5}$

c) $f(x) = \dfrac{1}{x^2 - 2x - 8}$

d) $f(x) = \dfrac{2}{4x - x^2}$

e) $f(x) = \dfrac{2x}{x^2 - 16}$

f) $f(x) = \dfrac{2x}{x^2 + 16}$

g) $f(x) = \sqrt{x+5}$

h) $f(x) = \dfrac{1}{\sqrt{x+5}}$

i) $f(x) = \sqrt{2x - 5}$

j) $f(x) = \sqrt{4 - x}$

k) $f(x) = \sqrt{x^2 - 9}$

l) $f(x) = \sqrt{x^2 + 2x - 8}$

m) $f(x) = \sqrt{x^2 + 5x + 4}$

n) $f(x) = \sqrt{\dfrac{x}{x^2 - 16}}$

o) $f(x) = \dfrac{x+1}{(2x-3)^2}$

p) $f(x) = \sqrt{\dfrac{x+3}{x^2 - x - 6}}$

q) $f(x) = \dfrac{1}{\sqrt{3x - 12}}$

r) $f(x) = \dfrac{3x}{x^2 + 4}$

s) $f(x) = \dfrac{1}{\sqrt{x^2 - 5x + 6}}$

t) $f(x) = \dfrac{14}{x^2 + 2x + 1}$

u) $f(x) = \sqrt[3]{x^2 + 5x + 4}$

v) $f(x) = \sqrt{x^2 + 2x + 1}$

(Soluc: a) IR; b) IR-[-5}; c) IR-{-2,4}; d) IR-{0,4}; e) IR-{±4}; f) IR; g) [-5,∞); h) (-5,∞); I) [5/2,∞); j) (-∞,4]; k) (-∞,-3]U[3,∞); I) (-∞,-4]U[2,∞); m) (-∞,-4]U[-1,∞); n) (-4,0]U(4,∞); o) IR-{3/2}; p) [-3,-2)U(3,∞); q) (4,∞); r) IR; s) (-∞,2)U(3,∞); t) IR-{-1}; u) IR; v) IR)

6. Calculate the intersection points of the following functions with the coordinated axes. In a, b, c and d, represent them, only with this information:

a) $y = 2x - 6$

b) $f(x) = x^2 + 2x - 3$

c) $f(x) = x^2 + x + 1$

d) $f(x) = x^3 - x^2$

m) $y = \dfrac{x^2 + 4}{x + 2}$

e) $y = \dfrac{x^2 - 4}{x + 2}$

f) $f(x) = \sqrt{2x + 4}$

g) $f(x) = \sqrt{2x} + 4$

n) $f(x) = \dfrac{4}{x - 4}$

h) $y = \dfrac{x + 4}{2x + 2}$

i) $y = \dfrac{x^2 - 3}{x^2 - 1}$

o) $f(x) = x^4 - 1$

j) $f(x) = \sqrt{x^2 + x - 2}$

k) $y = \sqrt{x^2 + 9}$

l) $f(x) = x^3 - 6x^2 + 11x - 6$

(Soluc: a) (3,0),(0,-6); b) (-3,0),(1,0),(0,-3); c) (0,1); d) (0,0),(1,0); e) (-2,0),(2,0),(0,-2); f) (-2,0),(0,2); g) (0,4); h) (-4,0),(0,2); i) (√3,0),(-√3,0),(0,3); j) (-2,0),(1,0); k) (0,3); I) (1,0),(2,0),(3,0),(0,-6); m) (0,2); n) (0,-1); o) (-1,0),(1,0),(0,-1))

7. Indicate the symmetry of the following functions:

a) $f(x) = x^4$

b) $f(x) = x^3$

c) $f(x) = x^4 - x^2$

d) $f(x) = x^2 - x^3$

e) $f(x) = 2x - 3$

f) $f(x) = x^5 - x^3$

g) $f(x) = \dfrac{2x^2 + 1}{x^2 - 1}$

h) $y = \dfrac{x^2 + 1}{x}$

i) $y = \dfrac{2x^3}{x^2 + 1}$

j) $f(x) = \dfrac{x^2}{x^2 + 6}$

k) $y = \dfrac{3x}{2x^2 - 1}$

l) $f(x) = \dfrac{x}{x - 5}$

m) $y = \dfrac{5x^2}{x - 1}$

n) $f(x) = x + \dfrac{x^2 + 1}{x^2 + 3}$

o) $y = \dfrac{\sqrt{x - 2}}{x^2 + 3}$

(Sol: a) even; b) odd; c) even; d) none; e) none; f) odd; g) even; h) odd; i) odd; j) even; k) odd; l) none; m) none; n) none; o) none)

8. Calculate the intersection points of the following functions with the coordinated axes and study the symmetry of the following functions:

a) $f(x) = \dfrac{4}{x^2 + 1}$

b) $y = \dfrac{x + 3}{x^2 + 1}$

c) $y = \dfrac{14}{x^3}$

d) $y = \dfrac{x^2 - 9}{x^2 + 1}$

e) $f(x) = \dfrac{4x + 12}{3x + 6}$

9. Represent the following functions: a) $y = |3x - 6|$ b) $y = 2^{x-1}$

10. Represent the following functions: a) $y = |x + 3|$ b) $y = -\dfrac{x^2}{2} + 2x - 2$

11. Represent the following functions: a) $y = 1 - \dfrac{x^2}{4}$ b) $y = \left(\dfrac{1}{3}\right)^{x+1}$

12. Represent the following functions: a) $y = \left|\dfrac{x - 1}{3}\right|$ b) $y = 2^{-x}$

13. Represent the following functions: a) $f(x) = (x - 1)^2 + 2$ b) $f(x) = 3^{1-x}$

14. Join each of the following graphs with their function:

a) $y = \dfrac{2}{3}x$ **b)** $y = 2x^2 - 3$ **c)** $y = 3,5x - 0,75$ **d)** $y = -x^2 + 4$

I)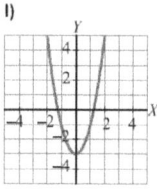

II)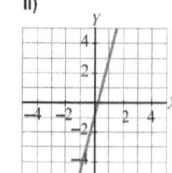

III)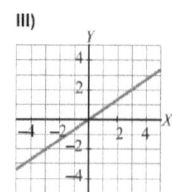

IV)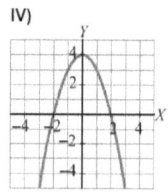

15. Join each of the following graphs with their function:

a) $y = \dfrac{-3x^2}{4}$ b) $y = \dfrac{-3x}{4}$ c) $y = 2x^2 - 2$ d) $y = 2x - 2$

I) 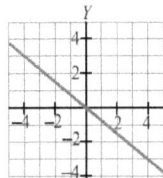 II) 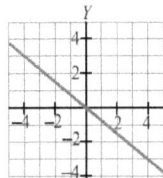 III) 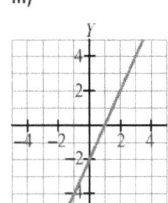 IV)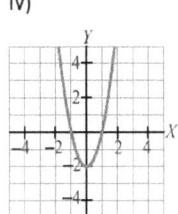

16. Join each of the following graphs with their function:

a) $y = \dfrac{1}{x-4}$ b) $y = \sqrt{2x}$ c) $y = \dfrac{1}{x} + 2$ d) $y = -\sqrt{x+1}$

I) 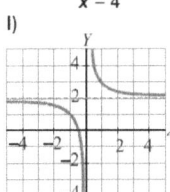 II) 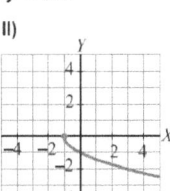 III) 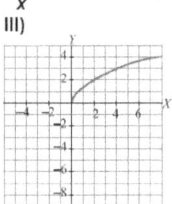 IV)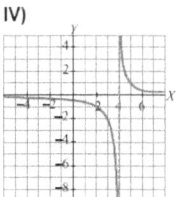

17. Join each of the following graphs with their function:

a) $y = \dfrac{1}{x} - 3$ b) $y = \sqrt{x-3}$ c) $y = \dfrac{1}{x-3} + 2$ d) $y = \sqrt{x+3}$

I)   II) 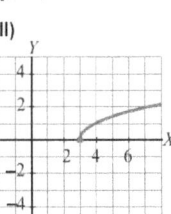 III) 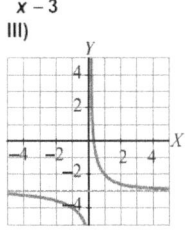 IV)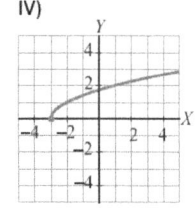

18. Join each of the following graphs with their function:

a) $y = 3^x$ b) $y = \left(\dfrac{1}{3}\right)^x$ c) $y = \log_3 x$ d) $y = \log_{1/3} x$

I) 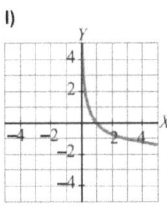 II) 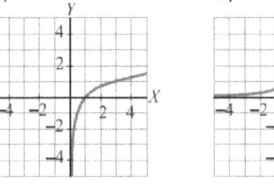 III) 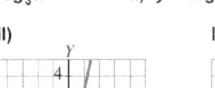 IV)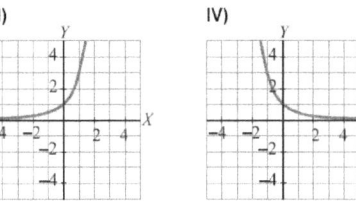

19. Join each of the following graphs with their function:

a) $y = \left(\dfrac{2}{3}\right)^x$　　　b) $y = \left(\dfrac{3}{2}\right)^x$　　　c) $y = log_2 x$　　　d) $y = log_{1/2} x$

I) 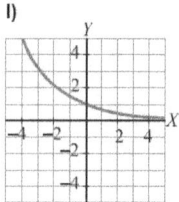　II) 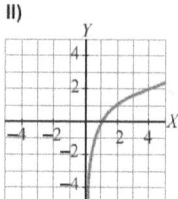　III) 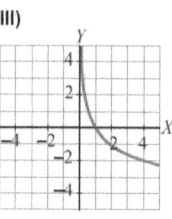　IV)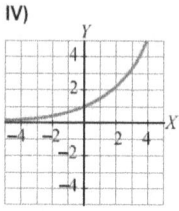

20. In some countries they use a different system to measure temperatures, called Farenheit degrees. Knowing that 10 °C = 50 °F and that 60 °C = 140 °F, give the equation that allows the transformation of temperatures from °C to °F.

21. With a 200 m metallic wall we want to build a rectangular fence, where one of its sides will be an existing wall.

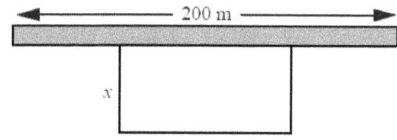

　a) If x is one of the sides of the fence, what is the measure of the other sides?

　b) Build a function for the area of the rectangle.

22. An iron bar is 30 cm long at 0 °C. When it is heated, its dilation is given by a linear expression, $L = a + bt$, where L is the length (in cm) and t is the temperature (in °C).

　a) Find out analytical expression for L, if L(1) = 30.0005 cm and L(3) = 30.0015 cm.

　b) Represent this function.

23. In a square with side x, we want to know the area of the shaded zone. Build the function that gives this area as a function of the side, x, and represent it.

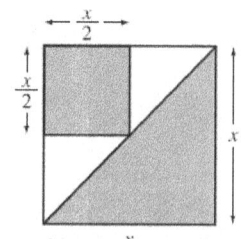

24. In a fruit shop they have 20 kg of apples that today are being sold at a price of 40 c/kg. Each day, 1 kg will be wasted and price will increase in 10 cent/kg.

  a) Write the equation that gives the quantity of money earned as a function of the days.

  b) Represent it.

25. Given representation of function y = f(x), draw a) y = f(x) + 1 and b) y = f(x + 1).

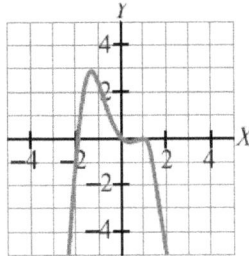

26. From graph of the function $y = \sqrt[3]{x^2}$, represent, on different axes:

  a) $y = \sqrt[3]{x^2} + 3$ b) $y = \sqrt[3]{(x+3)^2}$ c) $y = \sqrt[3]{x^2} - 2$ d) $y = \sqrt[3]{(x-2)^2}$

27. From graph of the function $y = \dfrac{1}{x^2 + 1}$, represent, on different axes:

  a) $y = \dfrac{2}{x^2 + 1}$ b) $y = \dfrac{1}{3(x^2 + 1)}$ c) $y = \dfrac{-1}{x^2 + 1}$

28. From graph of the function $y = \sqrt[3]{x}$, represent, on different axes: a) $y = \sqrt[3]{-x}$ b) $y = -\sqrt[3]{x}$

29. From graph of the function $y = \dfrac{1}{x}$, represent, successively and on different axes:

  a) $y = \dfrac{1}{x+2}$ b) $y = \dfrac{x+3}{x+2} = 1 + \dfrac{1}{x+2}$.

30. a) Represent hyperbola $y = -\dfrac{1}{x-2} + 3$ from graph of function $y = \dfrac{1}{x}$.

  b) Idem with $y = \sqrt[3]{-x+2}$ from $y = \sqrt[3]{x}$.

31. Write the analytic expression of the functions given as intervals whose graphs are given:

a)
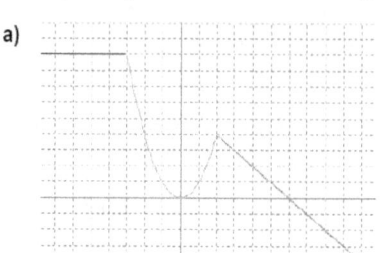

b)
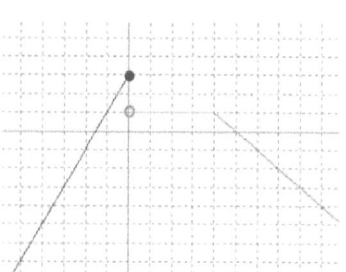

32. Given the following absolute value functions, indicate:

    i) Write them as intervals   ii) Represent them

    iii) Dom(f)   iv) Intersection points with axes   v) Continuity.

a) $f(x) = |x - 1|$ **b)** $f(x) = |-3x + 3|$ **c)** $f(x) = |3x + 6|$ **d)** $f(x) = |x^2 - 5x + 6|$

e) $f(x) = |x^2 - 4x + 3|$ **f)** $f(x) = |-x^2 - 4x - 5|$ **g)** $f(x) = \left|\frac{1}{2}x^2 - x - 4\right|$ **h)** $f(x) = |x^2 - 4x + 5|$

i) $f(x) = |-x^2 + x - 1|$ **j)** $f(x) = |x^2 - 4x|$ **k)** $f(x) = \frac{|x|}{x}$ **l)** $f(x) = |9 - x^2|$

m) $f(x) = |x| + x$ **n)** $f(x) = |x + 1| + |x - 1|$ **o)** $f(x) = |3x + 6| - |2x - 2|$ **p)** $f(x) = \sqrt{\frac{|x|}{x}}$

q) $f(x) = \begin{cases} \frac{x}{2} + 1 & \text{si } x < 2 \\ |2x-6| & \text{si } x \geq 2 \end{cases}$ **r)** $f(x) = \begin{cases} |x^2 + 4x + 3| & \text{si } x < -1 \\ (x-1)^2 - 2 & \text{si } x \geq -1 \end{cases}$

33. From graph of the function $y = |x|$, represent, successively and on different axes:

    a) $y = |x| + 3$ b) $y = |x - 2|$ c) $y = |2x|$ d) $y = \left|\frac{x}{3}\right|$

34. Calculate $(f \circ g)(x)$ and $(g \circ f)(x)$:

    a) $f(x) = 3x - 5$ $g(x) = 1/x$ | **c)** $f(x) = 1/x$ $g(x) = \sqrt{x}$ | **e)** $f(x) = \sqrt{x}$ $g(x) = \frac{1}{x^2 - 4}$

    b) $f(x) = x^2$ $g(x) = 3x - 5$ | **d)** $f(x) = x^2$ $g(x) = \sqrt{x}$

35. Check if the following pairs of functions are inverse and represent them on the same axes:

a) $f(x) = 2x$ $g(x) = \dfrac{x}{2}$ b) $f(x) = x^3$ $g(x) = \sqrt[3]{x}$ c) $f(x) = x + 1$ $g(x) = x - 1$

What do you observe?

36. Find out the inverse functions of:

a) $y = \sqrt{x + 4}$

b) $y = 3 + 2(x - 1)$

c) $y = 2\sqrt{x + 1}$

d) $y = 3/x$

e) $y = x^2 - 1$

f) $f(x) = \dfrac{1}{2x + 3}$

g) $y = 2 + \sqrt{x}$

h) $y = \dfrac{x + 1}{3x - 1}$

37. A manufacturer has calculated that production cost (in €) of a product are given by the following expression:

$$C(x) = x^2 + 20x + 40000$$

where x represents number of sold units. Moreover, each unit is sold at a price of 520 €.

a) Express, as a function of the number of sold products, x, benefit that is obtained and repreesent it.

b) How many units must be sold to get the maximum benefit? Which is the maximum benefit that can be obtained?

38. The dose of a drug starts at 10 mg and each day it must be increased in 2 mg till reach 20 mg. It must continue 15 days with this quantity and from that moment it must be decreased 4 mg per day.

a) Represent the function that describes this situation and write its analytical expression, as intervals.

b) Indicate its domain and range.

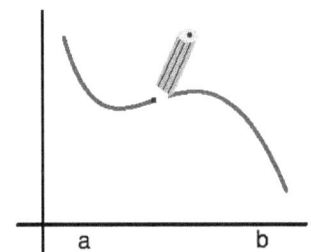

When does a limit exist?

Unit 10.- Limits and continuity

1. Introduction

When we talked about functions before, we paid attention at the values of functions at specific points. For example, the value of f(x) at x = 1. The idea behind limits is to analyse what the

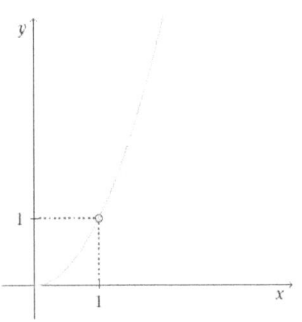

function is "approaching" when x "approaches" a specific value.

For example, this is the function $f(x)=x^2$. Let's focus on the point (1,1). We can see from the graph that when x approaches 1, the function f(x) approaches 1. When this happens, we say that: $\lim_{x \to 1}(x^2) = 1$

This is read *"the limit as x approaches 1 of x squared equals 1"*.

Why Limits are Useful

You might ask what this is useful for. You already know the function equals 1 when x equals 1, right? Well, the point is that sometimes we don't care what the function is at x=1. As an example of this, let's consider the following function:

$$f(x) = \begin{cases} x^2 & \text{if } x \neq 1 \\ 0 & \text{if } x = 1 \end{cases}$$

which means that this function equals x^2 when x is anything other than 1, and equals 0 when x equals 1. This function is the same as the one we saw before, but in this case it has a "hole" at x=1. Let's see what the graph looks like:

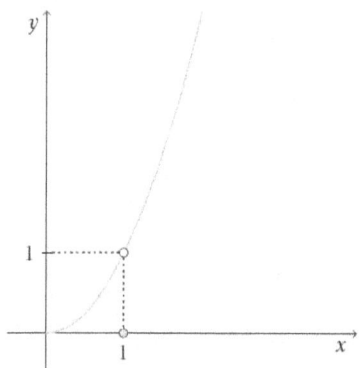

What does the function approach when x approaches 1? It also approaches 1, right? It doesn't matter that the function is other than 1 at that point! So,

$$\lim_{x \to 1} f(x) = 1$$

The Idea of Continuous Functions

Limits and continuity are often covered in the same chapter because they are very related. The basic idea of continuity is very simple, and the "formal" definition uses limits.

Basically, we say a function is **continuous** when you can *graph it without lifting your pencil from the paper.* Here's an example of what a continuous function looks like:

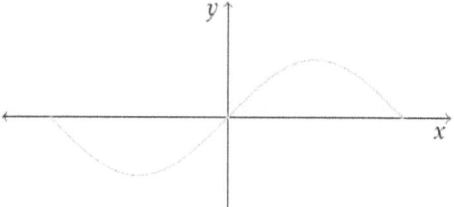

In this chapter, you will learn to recognize a continuous function when you see its analytical expression.

Now, what would a discontinuous function look like? A function essentially is discontinuous when it has any "gap". For example:

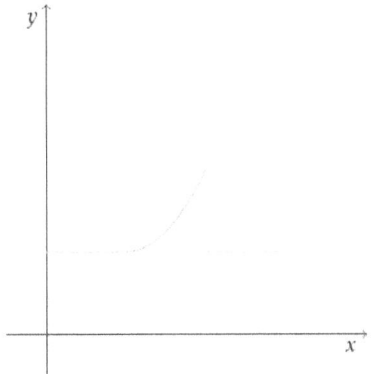

2. Concept of a limit

We say $\lim_{x \to a} f(x) = L$, and it is read "*the limit as x approaches a of f of x equals L*" if f(x) takes values closer and closer to L as x is getting closer and closer values to a.

$$\lim_{x \to c} f(x) = l$$

Example: Let $f(x) = \dfrac{x+7}{2+x}$, calculate $\lim_{x \to 3} f(x)$:

<u>Solution:</u> We will consider a value of "x" very close to 3, for example, x = 2.9999:

$$\lim_{x \to 3}\left(\frac{x+7}{2+x}\right) = \frac{2.9999+7}{2+2.9999} = \frac{9.9999}{4.9999} \approx \boxed{2}$$

2. Vertical asymptotes

Sometimes,

$$\lim_{x \to a} f(x) = +\infty \ (or \ -\infty).$$

In that case, vertical straight line $x = a$ is said to be a *vertical asymptote*.

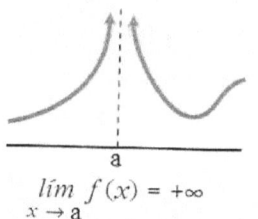

$$\lim_{x \to a} f(x) = +\infty$$

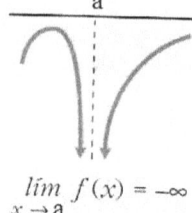

$$\lim_{x \to a} f(x) = -\infty$$

Example: Let $f(x) = \dfrac{2}{x^2 - 1}$, calculate $\lim_{x \to 1} f(x)$:

Solution: We will consider a value of "x" very close to 1, for example, x = 0.9999:

$$\lim_{x \to 1}\left(\frac{2}{x^2 - 1}\right) = \frac{2}{0.99980001 - 1} = \frac{2}{-0.00019999} \approx \boxed{-\infty}$$

When this happen we must calculate *lateral limits*:

> **Left hand limit** $\displaystyle\lim_{x \to 1^-}$: x is getting values very close to 1, and lower than 1.

For example, x = 0.9999 or x = 0.9999999999

> **Right hand limit** $\displaystyle\lim_{x \to 1^+}$: x is getting values very close to 1, and bigger than 1.

For example, x = 1.0001 or 1.00000000000001

> **Existence of the limit:** We say limit exists if both lateral limits are the *equal*.

In our example:

$$\begin{cases} \lim_{x \to 1^-}\left(\dfrac{2}{x^2 - 1}\right) = \dfrac{2}{0.99980001 - 1} = \dfrac{2}{-0.00019999} \approx -\infty \\[4mm] \lim_{x \to 1^+}\left(\dfrac{2}{x^2 - 1}\right) = \dfrac{2}{1.00020001 - 1} = \dfrac{2}{0.00020001} \approx +\infty \end{cases}$$ There is **NOT** limit as x → 1.

In this case, *f(x)* is said to have a vertical asymptote at *x = 1*. Graphically:

f(x)

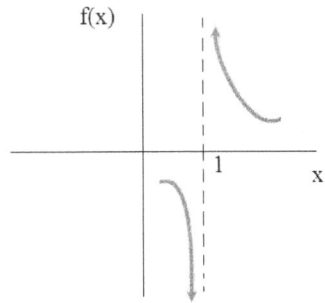

1

x

Exercises

1. Calculate the following limits:

   a) $\lim_{x \to 3}(x^2 - 5x + 7)$ b) $\lim_{x \to 3}(x-2)^7$ c) $\lim_{x \to 3}(x^2 + 5)$ d) $\lim_{x \to 1}\dfrac{x^2 + 1}{x + 1}$

   e) $\lim_{x \to 1}\dfrac{x-1}{x+1}$ f) $\lim_{x \to 0}\dfrac{x^4 + x^2}{3x + 5}$ g) $\lim_{x \to 3}\dfrac{x^3 - 1}{x^2 - 3}$ h) $\lim_{x \to 4}\dfrac{x^2 + 1}{x + 1}$

   i) $\lim_{x \to 6}\dfrac{x-1}{x+1}$ j) $\lim_{x \to 2}\dfrac{x^2 + 3}{3 - x}$ k) $\lim_{x \to 1}\dfrac{x+1}{x^2 + 1}$ l) $\lim_{x \to 1}\dfrac{2x+1}{x+1}$

   m) $\lim_{x \to 0}\dfrac{3x^3 + 2}{6x^5 - 1}$ n) $\lim_{x \to \infty}\dfrac{x^5 + 8x^3}{3x^5 + 1}$

   (Sol: 1, 1, 14, 1, 0, 0, 13/3, 17/3, 5/6, 7, 1, 3/2, (-2), 1/3)

2. From the following graph for f(x), calculate its limits:

   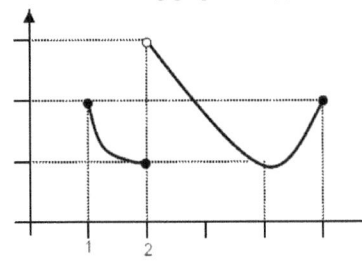

   a) as x approaches to 1

   b) as x approaches to 2

   c) as x approaches to 4

   d) as x approaches to 5

3. Infinite limits

Infinite limits, $\lim_{x \to +\infty} f(x)$ or $\lim_{x \to -\infty} f(x)$, are those ones calculated as x is taking values closer and closer to $+\infty$ or $-\infty$, respectively.

There are 2 possibilities:

a) f(x) goes to infinity:

$\lim_{x \to +\infty} f(x) = +\infty$	$\lim_{x \to +\infty} f(x) = -\infty$
f(x)	f(x)
X	X
$\lim_{x \to -\infty} f(x) = +\infty$	$\lim_{x \to -\infty} f(x) = -\infty$
f(x)	X
	f(x)

Example: Calculate: a) $\lim_{x \to +\infty} \left(x^2 - 5\right)$ b) $\lim_{x \to +\infty} \left(2 - x\right)$ c) $\lim_{x \to -\infty} \left(x^2 - 1\right)$

Solution: a) $+\infty$ b) $-\infty$ c) $+\infty$

b) f(x) approaches to a horizontal straight line. Horizontal asymptote

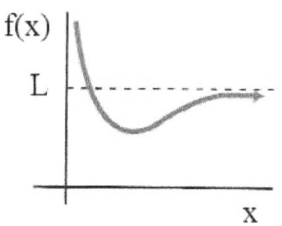

When:

$$\lim_{x \to \pm\infty} f(x) = L$$

then, horizontal straight line $y = L$ is said to be a *horizontal asymptote*.

4. Indeterminate forms of a limit

Imagine you are calculating $\lim_{x \to +\infty} (x^3 - x)$. You would say it is: $\infty - \infty$ *What is the limit?*

Or imagine you are calculating $\lim_{x \to +\infty} \left(\dfrac{x^2 + 2}{x^5 + 1} \right)$. You would say it is $\dfrac{\infty}{\infty}$ *What is the limit?*

This type of limits are named ***indeterminate form of a limit***, but we can solve them by using the following rules.

4.1. Indeterminate form $\dfrac{\infty}{\infty}$

We name $\dfrac{P(x)}{Q(x)}$

> ➤ If degree of P(x) > degree of Q(x), this limit is ∞ *(+∞ or -∞*, depending on the signs).
> ➤ If degree of Q(x) > degree of P(x), this limit is 0.
> ➤ If degree of P(x) = degree of Q(x), this limit is the quotient of the coefficients of the monomials of higher degrees in numerator and denominator.

Example: Calculate the limits $(x \rightarrow +\infty)$ of these functions and draw a scheme:

a) $f(x) = \dfrac{x^2 + 2}{x^5 + 1}$
b) $f(x) = \dfrac{x^2 - 4}{x + 1}$
c) $f(x) = \dfrac{3x^2 + 5x - 2}{2x^2 - 4x + 1}$

Solution:

a) As degree of numerator < degree of denominator, the limit of f(x) is $\boxed{0}$.

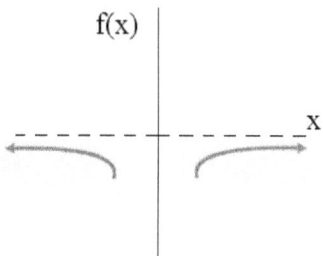

Notice there is a horizontal asymptote at *y = 0*.

b) As degree of numerator > degree of denominator, the limit of f(x) is $\boxed{+\infty}$.

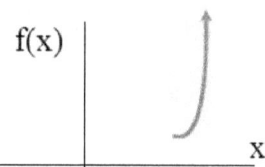

c) As degree of numerator = degree of denominator, the limit of f(x) is $\boxed{\dfrac{3}{2}}$.

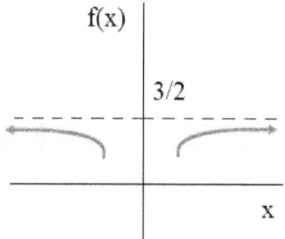

Notice there is a horizontal asymptote at *y = 3/2*.

3. Calculate the following limits:

a) $\lim\limits_{x \to \infty} \dfrac{\sqrt{x}}{x}$ b) $\lim\limits_{x \to +\infty} \dfrac{x^2 - 6x + 8}{x^2 - 2}$ c) $\lim\limits_{x \to -\infty} \dfrac{x^2 - 6x + 8}{x^2 - 2}$ d) $\lim\limits_{x \to +\infty} \dfrac{x^4 - 1}{x^2 - 1}$

e) $\lim\limits_{x \to -\infty} \dfrac{x^4 - 1}{x^2 - 1}$ f) $\lim\limits_{x \to +\infty} \dfrac{x^5 - 1}{x^7 - 1}$ g) $\lim\limits_{x \to -\infty} \dfrac{x^5 - 1}{x^7 - 1}$ h) $\lim\limits_{x \to \infty} \dfrac{3x^2 + x - 1}{2x^2 - x}$

i) $\lim\limits_{x \to \infty} \dfrac{x^3 - x^2 + 1}{4x^3 + x^2 - x}$ *(Sol: 0, 1, 1, $+\infty$, $+\infty$, 0, 0, 3/2, 1/4)*

4.2. Indeterminate form $\infty - \infty$ We name $P(x) - Q(x)$

➤ If degree of P(x) > degree of Q(x), this limit is $+\infty$ *(The sign is determined by P(x))*.
➤ If degree of Q(x) > degree of P(x), this limit is $-\infty$ *(Sign is determined by Q(x))*.
➤ If degree of P(x) = degree of Q(x), we will operate in different ways, depending if there are radicals or not:
a) If there are not radicals, we will operate the difference of fractions.
b) If there are radicals, we must multiply and divide the expression by its *conjugated* expression. The conjugated expression of *(a + b)* is *(a – b)*.

Example: Calculate the limits ($x \to +\infty$) of these functions:

a) $f(x) = x^2 - \sqrt{x^2 + 1}$ Solution: Degree$(x^2) = 2$ and degree $\left(\sqrt{x^2 + 1}\right) = 1$. So, $\lim f(x) = \boxed{+\infty}$.

b) $f(x) = x^2 - \sqrt{x^7 - 2}$ Solution: Degree$(x^2) = 2$ and degree $\left(\sqrt{x^7 - 2}\right) = \dfrac{7}{2}$. So, $\lim f(x) = \boxed{-\infty}$

c) $f(x) = \dfrac{x^2 + 1}{x} - \dfrac{x^3 - 2}{x^2 - 1}$ Solution: Degree $\left(\dfrac{x^2 + 1}{x}\right) = 1$ and degree $\left(\dfrac{x^3 - 2}{x^2 - 1}\right) = 1$.

As there are not radicals, we must operate the difference of fractions:

$$lim \dfrac{x^2 + 1}{x} - \dfrac{x^3 - 2}{x^2 - 1} = lim \dfrac{x^4 - 1 - x^4 + 2x}{x \cdot (x^2 - 1)} = lim \dfrac{-1 + 2x}{x \cdot (x^2 - 1)} = \boxed{0}.$$

d) $f(x) = x - \sqrt{x^2 + 1}$

Solution: Degree(x) = 1 and degree($\sqrt{x^2+1}$) = 1. So, we must multiply

and divide the expression by its conjugated expression, $x + \sqrt{x^2+1}$:

$$\lim \left(x - \sqrt{x^2+1}\right) = \lim \frac{\left(x - \sqrt{x^2+1}\right)\left(x + \sqrt{x^2+1}\right)}{x + \sqrt{x^2+1}}$$

Remember this identity: $(a + b) \cdot (a - b) = a^2 - b^2$

$$\lim \frac{\left(x - \sqrt{x^2+1}\right)\left(x + \sqrt{x^2+1}\right)}{x + \sqrt{x^2+1}} = \lim \frac{x^2 - \left(\sqrt{x^2+1}\right)^2}{x + \sqrt{x^2+1}} = \lim \frac{x^2 - \left(x^2+1\right)}{x + \sqrt{x^2+1}} = \lim \frac{x^2 - x^2 - 1}{x + \sqrt{x^2+1}} =$$

$$= \lim \frac{-1}{x + \sqrt{x^2+1}} = \frac{-1}{\infty} = \boxed{0}$$

4. Calculate the following limits:

a) $\lim\limits_{x \to \infty} \left(\frac{x^2 - 3}{x + 4} - x \right)$

b) $\lim\limits_{x \to \infty} \left(\sqrt{x^2 + x} - \sqrt{x^2 - 1} \right)$

c) $\lim\limits_{x \to -\infty} \frac{\sqrt{x-5}}{\sqrt{x-4}}$

d) $\lim\limits_{x \to -\infty} \left(-3x^3 + 4x^2 + 2x + 5 \right)$

e) $\lim\limits_{x \to \infty} \left(x^2 - 5x + 7 \right)$

f) $\lim\limits_{x \to +\infty} \left(-x^2 + 7x + 5 \right)$

g) $\lim\limits_{x \to +\infty} \left(x^3 + 9x^2 + 6 \right)$

h) $\lim\limits_{x \to -\infty} \left(x^2 - 2x + 1 \right)$

i) $\lim\limits_{x \to -\infty} \left(-x^2 + 5x + 7 \right)$

j) $\lim\limits_{x \to -\infty} \left(x^3 + 7x^2 + 1 \right)$

k) $\lim\limits_{x \to -\infty} \left(4x^2 - 3x^3 + 5 \right)$

l) $\lim\limits_{x \to +\infty} \sqrt{x^2 + x} - x$

m) $\lim\limits_{x \to \infty} \left(\sqrt{x+2} - \sqrt{x} \right)$

n) $\lim\limits_{x \to +\infty} \sqrt{x^2 + 1} - \sqrt{x^2 - 1}$

ñ) $\lim\limits_{x \to 0} \left(\frac{1}{x^2} - \frac{1}{x^4} \right)$

o) $\lim\limits_{x \to +\infty} \sqrt{(x+2) \cdot (x-3)} - x$

(Sol: (-4), ½, 1, +∞, +∞, −∞, +∞, +∞, −∞, −∞, +∞, ½, 0, 0, −∞, (-1/2))

4.3. Indeterminate form $\dfrac{0}{0}$

> ➤ If there are only polynomials, it is solved by factorizing numerator and denominator and, after that, simplifying.
> ➤ If there are radicals, we must multiply and divide the expression by *conjugate* of expression containing radicals. The conjugated expression of *(a + b)* is *(a – b)*.

Example: Calculate: a) $\lim\limits_{x \to 2}\left(\dfrac{x^4 - 4x^2}{x^2 - 2x}\right)$ b) $\lim\limits_{x \to 0}\left(\dfrac{3x}{1 - \sqrt{x+1}}\right)$

Solution:

a) First of all, substitute: $\lim\limits_{x \to 2}\left(\dfrac{x^4 - 4x^2}{x^2 - 2x}\right) = \dfrac{16 - 16}{4 - 4} = \dfrac{0}{0}$ We must factorize:

$$\lim_{x \to 2}\left(\frac{x^2 \cdot (x+2) \cdot (x-2)}{x \cdot (x-2)}\right) = \lim_{x \to 2}\left(\frac{x \cdot (x+2)}{1}\right) = \boxed{8}$$

b) First of all, substitute: $\lim\limits_{x \to 0}\left(\dfrac{3x}{1 - \sqrt{x+1}}\right) = \dfrac{0}{0}$ As there are radicals, we must multiply and divide by conjugate of expression with radical: $\left(1 + \sqrt{x+1}\right)$

$$\lim_{x \to 0}\left(\frac{3x \cdot \left(1 + \sqrt{x+1}\right)}{\left(1 - \sqrt{x+1}\cdot\left(1 + \sqrt{x+1}\right)\right)}\right) = \lim_{x \to 0}\left(\frac{3x \cdot \left(1 + \sqrt{x+1}\right)}{1 - (x+1)}\right) = \lim_{x \to 0}\left(\frac{3x \cdot \left(1 + \sqrt{x+1}\right)}{-x}\right) =$$

$$= \lim_{x \to 0}\left(\frac{3 \cdot \left(1 + \sqrt{x+1}\right)}{-1}\right) = \boxed{(-6)}$$

Exercises

5. Calculate the following limits:

a) $\lim\limits_{x \to 4} \dfrac{x^2 - 4x}{x - 4}$

b) $\lim\limits_{x \to -1} \dfrac{x^4 - 5x^2 + 4}{x^4 + 5x^3 + 5x^2 - 5x - 6}$

c) $\lim\limits_{x \to 2} \dfrac{x^3 - 2x^2}{x - 2}$

d) $\lim\limits_{x \to 1} \dfrac{2x^3 - 14x^2 + 12}{x^3 - 10x^2 + 27x - 18}$

e) $\lim\limits_{x \to 1} \dfrac{x^4 - 1}{x - 1}$

f) $\lim\limits_{x \to 0} \dfrac{(1+x)^2 - 1}{x}$

g) $\lim\limits_{x \to 0} \dfrac{x}{1 - \sqrt{x+1}}$

h) $\lim\limits_{x \to 3} \dfrac{\sqrt{x+1} - 2}{x - 3}$

i) $\lim\limits_{x \to 0} \dfrac{\sqrt{x+9} - 3}{\sqrt{x+16} - 4}$

j) $\lim\limits_{x \to 1} \dfrac{\sqrt{x-1} + \sqrt{x+1}}{\sqrt{x+1} - \sqrt{x-1}}$

k) $\lim\limits_{x \to -1} \dfrac{x^2 + 2x + 1}{x^3 + 3x^2 + 3x + 1}$

l) $\lim\limits_{x \to 3} \dfrac{\sqrt{x+1} - 2}{x - 3}$

(Sol: 4, (-3/2), 4, (-11/5), 4, 2, 2, ¼, 8/9, 1, ∃, 1/4)

4.4. Indeterminate form $\dfrac{k}{0}$

This indeterminate form is solved by calculating lateral limits.

➤ If both lateral limits are equal, this is the limit.
➤ If both lateral limits are different, this limit DOES NOT EXIST.

Example: Calculate: a) $\lim\limits_{x \to 1} \left(\dfrac{2}{x^2 - 1} \right)$

Solution: First of all, substitute: $\lim\limits_{x \to 1} \left(\dfrac{2}{x^2 - 1} \right) = \dfrac{2}{1 - 1} = \dfrac{2}{0}$ So, lateral limits:

$$\begin{cases} \lim\limits_{x \to 1^-} \left(\dfrac{2}{x^2 - 1} \right) = \dfrac{2}{0.99980001 - 1} = \dfrac{2}{-0.00019999} \approx -\infty \\[4mm] \lim\limits_{x \to 1^+} \left(\dfrac{2}{x^2 - 1} \right) = \dfrac{2}{1.00020001 - 1} = \dfrac{2}{0.00020001} \approx +\infty \end{cases}$$

As they are different, this limit **DOES NOT EXIST**.

6. Calculate the following limits:

a) $\lim\limits_{x \to 0} \dfrac{1}{x}$

b) $\lim\limits_{x \to 0} \dfrac{-1}{x^2}$

c) $\lim\limits_{x \to 1} \dfrac{1}{x-1}$

d) $\lim\limits_{x \to 3} \dfrac{3}{x-3}$

e) $\lim\limits_{x \to 2} \dfrac{x^2 - x - 2}{x^2 - 4x + 4}$

f) $\lim\limits_{x \to -1} \dfrac{x^2 + 2x + 1}{x^3 + 3x^2 + 3x + 1}$

g) $\lim\limits_{x \to -2} \dfrac{-5x}{(x+2)^2}$

(Sol: ∃, −∞, ∃, ∃, ∃, ∃, +∞)

4.5. Indeterminate form 1^{∞} or *"e" limit*

When calculating this limit: $\lim\limits_{x \to \infty} \left(\dfrac{x+4}{x} \right)^{x+2}$, we obtain its value is 1^{∞}. This kind of indeterminate form is solved by applying the following rule:

If $\lim [f(x)]^{g(x)} = 1^{\infty}$, its limit is e^{L}, where $L = \lim \left[f(x) - 1 \right] g(x)$

Example: Calculate: $\lim\limits_{x \to \infty} \left(\dfrac{x+4}{x} \right)^{x+2}$

Solution: $\lim \left(\dfrac{x+4}{x} \right) = 1$ and $\lim (x+2) = \infty$. So, this limit is 1^{∞} and it is an *"e" limit*.

We must apply the previous expression:

$$L = \lim \left(\dfrac{x+4}{x} - 1 \right) \bullet (x+2) = \lim \left(\dfrac{x+4-x}{x} \right) \bullet (x+2) = \lim \left(\dfrac{4}{x} \right) \bullet (x+2) = \lim \left(\dfrac{4x+8}{x} \right) = 4$$

As these limits are e^{L}, the limit of our progression is $\boxed{e^{4}}$.

7. Calculate the following limits:

a) $\lim\limits_{x\to 1}\left(\dfrac{3x+4}{2x+5}\right)^{3x-1}$ b) $\lim\limits_{x\to 2}\left(\dfrac{x^2+1}{2x+1}\right)^{\frac{1}{(x-2)^2}}$ c) $\lim\limits_{x\to\infty}\left(\dfrac{x+2}{x-1}\right)^{x^2}$

(Sol: 1, $+\infty$, $+\infty$)

Important: Limits of geometrical progressions a^n (with $a \neq 1$)

For example, when calculating $\lim\limits_{x\to\infty}\left(\dfrac{6x^2-x+5}{2x^2+2x+1}\right)^{x-3}$, we have:

$lim\left(\dfrac{6x^2-x+5}{2x^2+2x+1}\right)=3$ and $lim\ (x-3)=\infty$. So, this limit is 3^{∞} and it is **NOT** an *"e" limit*.

Notice if you multiply $3\cdot3\cdot3\cdot3\cdot\ldots\ldots$ a lot of times, this product has no limit, its limit is ∞.

Another example, when calculating $\lim\limits_{x\to\infty}\left(\dfrac{x+5}{2x+1}\right)^{x+1}$, we have:

$lim\left(\dfrac{x+5}{2x+1}\right)=\dfrac{1}{2}$ and $lim\ (x+1)=\infty$. So, this limit is $\left(\dfrac{1}{2}\right)^{\infty}$ and it is **NOT** an *"e" limit*.

Notice if you multiply $\dfrac{1}{2}\cdot\dfrac{1}{2}\cdot\dfrac{1}{2}\cdot\dfrac{1}{2}\cdot$ ..... a lot of times, this product decreases to 0, its limit is 0.

If $lim\ f(x) = a^{\infty}$, its limit is $\begin{cases}\infty & \text{if } a>1 \\ 0 & \text{if } a <1\end{cases}$

If $lim\ f(x) = a^{-\infty}$, its limit is $\begin{cases}0 & \text{if } a>1 \\ \infty & \text{if } a <1\end{cases}$ *(Remember $a^{-\infty}=1/a^{\infty}=1/\infty=0$)*

Be careful, if $a = 1$, this is a *"e" limit*.

5. Continuity of a function

Continuity of a function is a very intuitive idea. A function will be said to be a continuous one if it can be drawn by an only stroke. In the same way, one function will be said to be a discontinuous one if it has some discontinuity.

Definition of continuity

A function is said to be a *continuous* one if it does not have any kind of discontinuity. A function can be a *continuous in an interval* if it has discontinuities only outside that interval.

Conditions for continuity

For example, the function represented below is, obviously, not continuous at x = a.

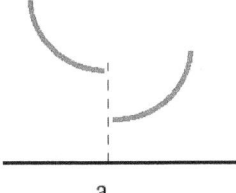

First condition for a function to be continuous is: $\lim_{x \to a^-} f(x) = \lim_{x \to a^+} f(x)$.

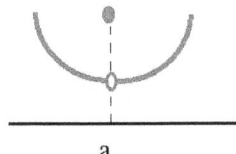

But this is not enough. Look at the function above. Both lateral limits are equal but, anyway, f(x) is not continuous. We need f(a) to have the same value than both lateral limits.

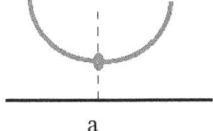

Now, this function is continuous. So, conditions for continuity of f(x) at x = a, are:

Conditions for continuity:

$$\lim_{x \to a^-} f(x) = \lim_{x \to a^+} f(x) = f(a)$$

Example: Decide if following functions are continuous:

a) $f(x) = \begin{cases} x+1 & \text{if } x < 5 \\ 6 & \text{if } x = 5 \\ 8 - 2e^{(x-5)} & \text{if } x > 5 \end{cases}$ b) $f(x) = \begin{cases} x^2 + 1 & \text{if } x < 1 \\ 5 & \text{if } x = 1 \\ 8 - 6x & \text{if } x > 1 \end{cases}$

Solution:

a) $\lim_{x \to 5^-} f(x) = 5 + 1 = 6$ b) $\lim_{x \to 1^-} f(x) = 1 + 1 = 2$

$\lim_{x \to 5^+} f(x) = 8 - 2 \cdot e^0 = 8 - 2 \cdot 1 = 6$ $\lim_{x \to 1^+} f(x) = 8 - 6 = 2$

$f(5) = 6 \quad \Rightarrow \quad Continuous \ at \ x = 5$ $f(1) = 5 \quad \Rightarrow \quad Discontinuous \ at \ x = 1$

Exercises

8. Check if the following functions are continuous:

a) $f(x) = \begin{cases} x^2 - 1 & \text{if } x < 0 \\ x - 1 & \text{if } x \geq 0 \end{cases}$ b) $f(x) = \begin{cases} (3 - x)/2 & \text{if } x < -1 \\ 2x + 4 & \text{if } x > -1 \end{cases}$

c) $f(x) = \begin{cases} (2 - x^2) & \text{if } x < 2 \\ (x/2) - 3 & \text{if } x \geq 2 \end{cases}$ d) $f(x) = \begin{cases} 3x & \text{if } x \leq 1 \\ x + 3 & \text{if } x > 1 \end{cases}$

(Sol: YES, NO, YES, NO)

Types of discontinuities

There are several reasons by which a function can be discontinuous at one point:

➢ **Asymptotic discontinuity:** It has infinite vertical asymptotes at that point. This means y values of the function increase or drop infinitely when x coordinate gets closer and closer to that point.

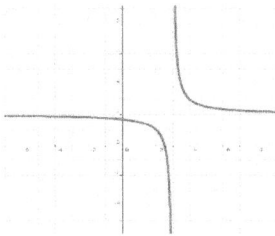

➢ **Jump discontinuity:** When moving along the curve of the function, we come across a jump.

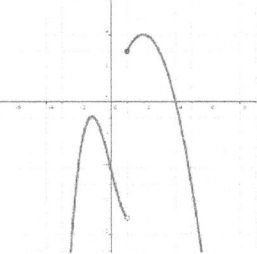

➢ **Point discontinuity:** At one point, the function is not defined or it is defined at a point outside the curve, but both lateral limits are equal.

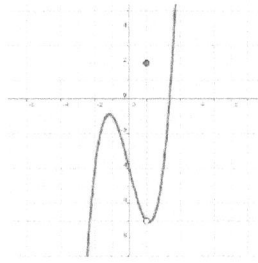

Notice a point discontinuity can be easily "repaired" by only adding the correct point in its definition:

For example, observe the curve of the function defined as $f(x) = \begin{cases} x \ \textit{if} \ x < 1 \\ 3 \ \textit{if} \ x = 1 \\ x \ \textit{if} \ x > 1 \end{cases}$

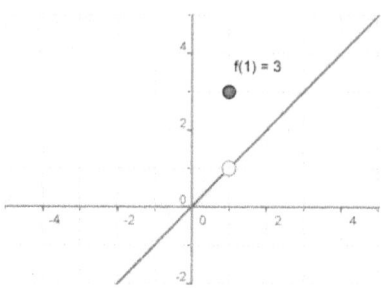

This function has a point discontinuity at x = 1, but it can be "repaired" if we change the 3 by a

1 in its definition: $f(x) = \begin{cases} x \ \textit{if} \ x < 1 \\ 1 \ \textit{if} \ x = 1 \\ x \ \textit{if} \ x > 1 \end{cases}$ Now, it is a continuous function.

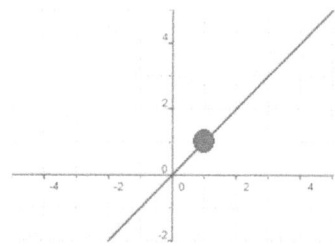

That is why point discontinuities are named ***avoidable discontinuities***.

$$\textit{Types of discontinuities} \begin{cases} \textit{Unavoidable discontinuities} \begin{cases} \textit{First kind} \rightarrow \textit{Asymptotic discontinuities} \\ \textit{Second kind} \rightarrow \textit{Jump discontinuities} \end{cases} \\ \textit{Avoidable discontinities} \rightarrow \textit{Point discontinuities} \end{cases}$$

6. Asymptotes of a function

For some functions, although we only know their curve in an interval, we can foresee their behaviour in far no studied zones, as they have continuous lines with a very clear trend. These lines are named asymptotes.

> **Asymptote:** A **line** that a curve approaches as it heads towards infinity.

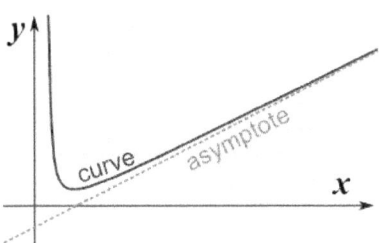

There are three types of asymptotes:

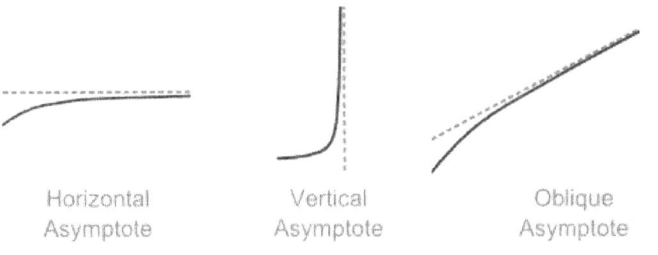

| Horizontal | Vertical | Oblique |
| Asymptote | Asymptote | Asymptote |

> ➢ Horizontal asymptotes: $y = b$
> ➢ Vertical asymptotes: $x = c$
> ➢ Oblique asymptotes: $y = mx + b$

Horizontal asymptote   **Vertical asymptote**   **Oblique asymptote**

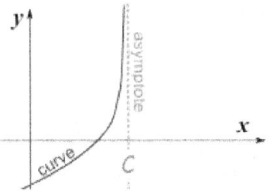

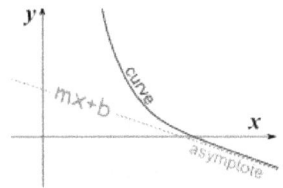

Location of asymptotes

> ➤ Horizontal asymptotes: $y = b$, where $b = \lim\limits_{x \to \pm\infty} f(x)$
>
> ➤ Vertical asymptotes: $x = c$. c is easily found, because $Domf = R - \{c_i\}$
>
> ➤ Oblique asymptotes: $y = mx + b$, where $\begin{cases} m = \lim\limits_{x \to \infty} \left(\dfrac{f(x)}{x} \right) \\ b = \lim\limits_{x \to \infty} \left(f(x) - m \cdot x \right) \end{cases}$

Example: Find out the asymptotes of the following functions:

a) $f(x) = \dfrac{x+5}{x-2}$　　　b) $f(x) = \dfrac{x^2+1}{x-2}$　　　c) $f(x) = \dfrac{x-4}{x^2+1}$

Solution:

a) - **Horizontal asymptote:** $\lim\limits_{x \to \pm\infty} \left(\dfrac{x+5}{x-2} \right) = 1$　So, there a horizontal asymptote: $y = 1$

- **Vertical asymptote:** As $Domf = R - \{2\}$, there is a vertical asymptote: $x = 2$.

$\begin{cases} \lim\limits_{x \to 2^-} \left(\dfrac{x+5}{x-2} \right) = \dfrac{6.9999}{-0.0001} \approx -\infty \\ \lim\limits_{x \to 2^+} \left(\dfrac{x+5}{x-2} \right) = \dfrac{7.0001}{0.0001} \approx +\infty \end{cases}$

- **Oblique asymptote:** As there is a horizontal asymptote, there is not an oblique one.

b) - **Horizontal asymptote:** $\lim\limits_{x \to \pm\infty} \left(\dfrac{x^2+1}{x-2} \right) = \infty$　So, there is not a horizontal asymptote.

- **Vertical asymptote:** As $Domf = R - \{2\}$, there is a vertical asymptote: $x = 2$.

$\begin{cases} \lim\limits_{x \to 2^-} \left(\dfrac{x^2+1}{x-2} \right) \approx \dfrac{5}{-0.0001} \approx -\infty \\ \lim\limits_{x \to 2^+} \left(\dfrac{x^2+1}{x-2} \right) \approx \dfrac{5}{0.0001} \approx +\infty \end{cases}$

- Oblique asymptote:

$$\begin{cases} m = \lim_{x \to \infty} \left(\dfrac{f(x)}{x} \right) = \lim_{x \to \infty} \left(\dfrac{x^2 + 1}{x^2 - 2x} \right) = 1 \\ b = \lim_{x \to \infty} (f(x) - m \cdot x) = \lim_{x \to \infty} \left(\dfrac{x^2 + 1}{x - 2} - 1 \cdot x \right) = \lim_{x \to \infty} \left(\dfrac{x^2 + 1 - x^2 + 2x}{x - 2} \right) = \lim_{x \to \infty} \left(\dfrac{2x + 1}{x - 2} \right) = 2 \end{cases}$$

So, there is an oblique asymptote: $y = x + 2$.

c) **- Horizontal asymptote:** $\lim_{x \to \pm\infty} \left(\dfrac{x - 4}{x^2 + 1} \right) = 0$ So, there a horizontal asymptote: $y = 0$

- **Vertical asymptote:** As *Domf* = R, there is not a vertical asymptote.

- **Oblique asymptote:** As there is a horizontal asymptote, there is not an oblique one.

---

9. Find out the asymptotes of these functions:

a) $y = \dfrac{x}{1 + x^2}$ b) $y = \dfrac{x^3}{1 + x^2}$ c) $y = \dfrac{4x^2 - 3}{x}$ d) $y = x^2 - x$

(Sol: a) y = 0; b) y = x; c) x = 0; d) No asymptotes)

Exercises

Review exercises

1. Given the following graph, calculate the limits below:

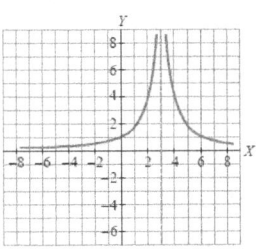

a) $\lim_{x \to +\infty} f(x)$ b) $\lim_{x \to -\infty} f(x)$ c) $\lim_{x \to 3^-} f(x)$ d) $\lim_{x \to 3^+} f(x)$ e) $\lim_{x \to 0} f(x)$

2. Given the following graph, calculate the limits below:

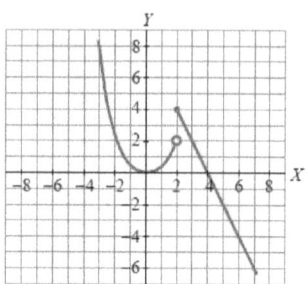

a) $\lim\limits_{x \to +\infty} f(x)$ b) $\lim\limits_{x \to -\infty} f(x)$ c) $\lim\limits_{x \to 2} f(x)$ d) $\lim\limits_{x \to 2^+} f(x)$ e) $\lim\limits_{x \to 0} f(x)$

3. The following graph corresponds to function f(x). Calculate the limits below:

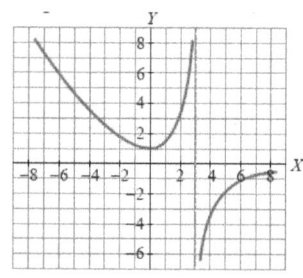

a) $\lim\limits_{x \to +\infty} f(x)$ b) $\lim\limits_{x \to -\infty} f(x)$ c) $\lim\limits_{x \to 3^-} f(x)$ d) $\lim\limits_{x \to 3^+} f(x)$ e) $\lim\limits_{x \to 0} f(x)$

4. Calculate the following limits:

a) $\lim\limits_{x \to -1} \dfrac{4}{x^2 + 2x + 3}$ b) $\lim\limits_{x \to 2} \sqrt{x^2 - 9}$ c) $\lim\limits_{x \to \pi/2} \cos x$ d) $\lim\limits_{x \to 1} \dfrac{x - 3}{x^2 + x + 1}$

e) $\lim\limits_{x \to 2} \sqrt{6 - 3x}$ f) $\lim\limits_{x \to 1} \log x$ g) $\lim\limits_{x \to 2} \left(-\dfrac{x^2}{2} + \dfrac{x^3}{4} \right)$ h) $\lim\limits_{x \to -2} 3^{x+1}$

i) $\lim\limits_{x \to \frac{\pi}{4}} tg\, x$ j) $\lim\limits_{x \to -2} (3 - x)^2$ k) $\lim\limits_{x \to -8} \left(1 + \sqrt{-2x} \right)$ l) $\lim\limits_{x \to \frac{\pi}{2}} sen\, x$

5. Calculate the following limits and draw a scheme:

1) $\lim\limits_{x\to+\infty}\dfrac{3x^2+1}{(2-x)^3}$

2) $\lim\limits_{x\to-\infty}\dfrac{2-x^3}{x^2-1}$

3) $\lim\limits_{x\to+\infty}\dfrac{3x}{5+3x}$

4) $\lim\limits_{x\to-\infty}\dfrac{3x}{5+3x}$

5) $\lim\limits_{x\to+\infty}\dfrac{1}{(1-x)^3}$

6) $\lim\limits_{x\to-\infty}\dfrac{3-x^3}{x^2}$

7) $\lim\limits_{x\to0}\dfrac{1}{x^2-x}$

8) $\lim\limits_{x\to1}\dfrac{x^2+x-2}{x^2-2x+1}$

9) $\lim\limits_{x\to0}\sqrt{4-x^2}$

10) $\lim\limits_{x\to3}\dfrac{-1}{2x-6}$

11) $\lim\limits_{x\to1}\dfrac{1}{x^2+1}$

12) $\lim\limits_{x\to-3}\dfrac{x+5}{x+3}$

13) $\lim\limits_{x\to1}\dfrac{x^2+2x-3}{x^2-1}$

14) $\lim\limits_{x\to+\infty}\left(-2x+3x^3\right)$

15) $\lim\limits_{x\to+\infty}\dfrac{3x^2+3x}{x^2-1}$

16) $\lim\limits_{x\to+\infty}\left(\dfrac{x}{2}-x^2\right)$

17) $\lim\limits_{x\to+\infty}\dfrac{x^4}{1+x^2}$

18) $\lim\limits_{x\to2}\dfrac{4-x^2}{3-\sqrt{x^2+5}}$

19) $\lim\limits_{x\to-\infty}\dfrac{x+1}{x^2-4}$

20) $\lim\limits_{x\to+\infty}\dfrac{2x^4-3x}{x^4+1}$

6. Calculate the following limits:

a) $\lim\limits_{x\to7}\dfrac{2-\sqrt{x-3}}{x^2-49}$

b) $\lim\limits_{x\to\infty}\left(\sqrt{x^2-3x}-x\right)$

c) $\lim\limits_{x\to\infty}\dfrac{\sqrt{3x^2+2x+1}}{2x+7}$

d) $\lim\limits_{x\to4}\left(\dfrac{x}{x-1}\right)^{\frac{2}{x-4}}$

e) $\lim\limits_{x\to1}\dfrac{\sqrt{2x-1}-1}{x^2-1}$

f) $\lim\limits_{x\to\infty}\left(\dfrac{2x+1}{2x-1}\right)^{x+1}$

g) $\lim\limits_{x\to2}\left(\dfrac{x+2}{2x}\right)^{\frac{1}{x-2}}$

h) $\lim\limits_{x\to2}\dfrac{x^3+2x^2-4x-8}{x^3+x^2-4x-4}$

i) $\lim\limits_{x\to1}\dfrac{x-1}{x^3-x^2-x+1}$

j) $\lim\limits_{x\to2}\dfrac{x+1}{x-2}$

k) $\lim\limits_{x\to-1}\dfrac{x+4}{(x+1)^2}$

l) $\lim\limits_{x\to+\infty}\sqrt{4x^2-3x+7}-2x$

7. Calculate the following limits:

a) $\lim\limits_{x\to4}\dfrac{3x^2-24x+48}{x-4}$

b) $\lim\limits_{x\to1}\dfrac{2x^3-14x^2+12x}{x^3-10x^2+27x-18}$

c) $\lim\limits_{x\to\infty}\left(\dfrac{x+a}{x+b}\right)^{x+c}$

d) $\lim\limits_{x\to4}\dfrac{x-4}{x^2-x-12}$

e) $\lim\limits_{x\to2}\dfrac{4-x^2}{3-\sqrt{x^2+5}}$

f) $\lim\limits_{x\to2}\dfrac{x^3-4x}{x^2-3x+2}$

g) $\lim\limits_{x\to\infty}\dfrac{2x^2+5x-1}{x^3+x}$

h) $\lim\limits_{x\to\infty}\dfrac{3x^2+1}{x+3}$

i) $\lim\limits_{x\to\infty}\dfrac{2x-3}{x^3-1}$

j) $\lim\limits_{x\to\infty}\dfrac{x-5}{\sqrt{x+4}-3}$

k) $\lim\limits_{x\to\infty}\left(\dfrac{x+2}{2x+3}\right)^{2x}$

l) $\lim\limits_{x\to\infty}\left(\dfrac{x^2+1}{x^2-1}\right)^{\frac{x}{2}}$

m) $\lim\limits_{x\to-\infty}\sqrt{x^2+x}-x$

n) $\lim\limits_{x\to3}\dfrac{x^3-3x^2+9x-27}{x^2-9}$

8. Calculate the limits of the following functions when x approaches 3 and represent your results:

a) $f(x) = \dfrac{x^3}{3} - 2x$ b) $f(x) = \dfrac{x^2}{x-3}$ c) $f(x) = \dfrac{x^2 - 6x + 9}{x^2 - 9}$

9. Given the following functions:

I) 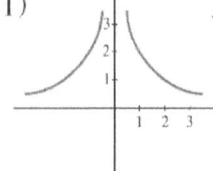   II) 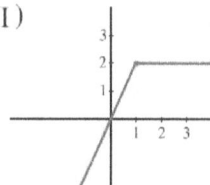   III) 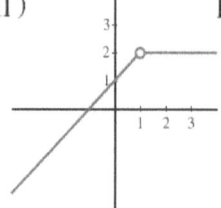   IV)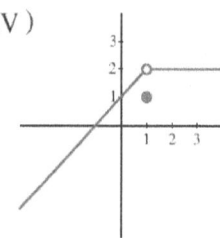

a) Indicate if they are continuous   b) Calculate f(1) for each one.

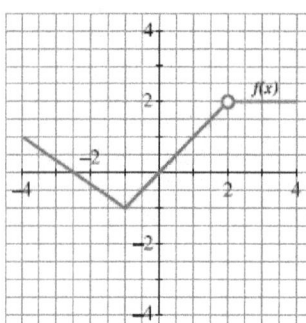

9. Given the following graph:

a) Indicate if f(x) is continuous or not. Justify your answer.

b) Find out f(-1), f(0), f(2) and f (3).

10. Are the following functions continuous at x = 2? If not, indicate the reason for the discontinuity.

a)

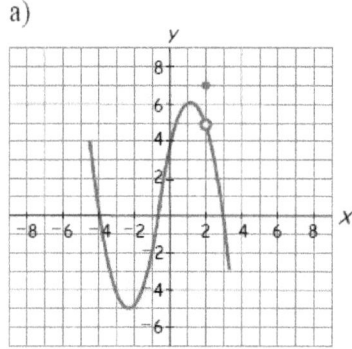

b)

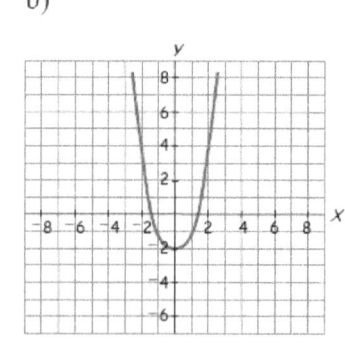

11. This is the graph for f(x). a) Is it continuous at x = -2?   b) and at x = 0?   If it is not continuous at any of the points, indicate the reason for the discontinuity.

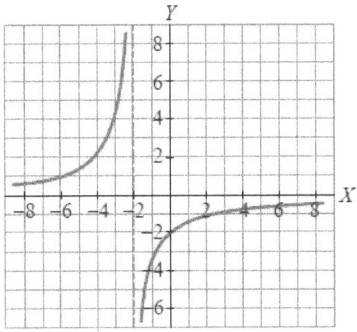

12. Find out the discontinuity points and intervals:

a) $y = \sqrt{\dfrac{x+5}{x^2 - 5x + 6}}$ b) $y = \dfrac{x+5}{x^2 - 5x + 6}$ c) $y = \sqrt{x^2 - 5x + 6}$

13. Study continuity of the following functions and represent them:

a) $f(x) = \begin{cases} \dfrac{x-1}{3} & si \ x \le 4 \\ x^2 - 15 & si \ x > 4 \end{cases}$ b) $f(x) = \begin{cases} x^2 - 2x & si \ 0 < x \le 1 \\ 3x - 1 & si \ x > 1 \end{cases}$ c) $f(x) = \begin{cases} x^2 & si \ x < 1 \\ \dfrac{3x - 1}{2} & si \ x > 1 \end{cases}$

d) $f(x) = \begin{cases} x^2 - 3 & si \ x \le 2 \\ 1 & si \ x > 2 \end{cases}$ e) $f(x) = \begin{cases} 2 - x^2 & si \ x \ne 0 \\ 1 & si \ x = 0 \end{cases}$ f) $f(x) = \begin{cases} 1 & si \ x = 0 \\ 1 - x^2 & si \ x \ne 0 \end{cases}$

14. Study continuity of the following functions and represent them:

a) $f(x) = \begin{cases} x + 3 & si \ -6 \le x < -2 \\ 1 & si \ -2 < x \le 1 \\ 2x + 1 & si \ 1 < x < 3 \\ -2x + 13 & si \ 3 \le x < 5 \\ 3 & si \ x > 5 \end{cases}$ b) $f(x) = \begin{cases} \dfrac{1}{x} & si \ x < 0 \\ x^2 + x & si \ 0 \le x < 1 \\ 2 & si \ x > 1 \end{cases}$ c) $f(x) = \begin{cases} \dfrac{2}{x+2} & si \ x < 0 \\ \dfrac{3}{x+3} & si \ x > 0 \end{cases}$

15. Calculate $\lim\limits_{x \to 2^-} f(x)$ and $\lim\limits_{x \to 2^+} f(x)$ if $f(x) = \begin{cases} 3 - x & if \ x \ge 2 \\ 0 & if \ x < 2 \end{cases}$

a) Does $\lim\limits_{x \to 2} f(x)$ exist?

b) Is f(x) continuous at x = 2?

16. Find out the value of *k* so that f(x) is continuous at x = 1.

$$f(x) = \begin{cases} 2x+1 & si \quad x \neq 1 \\ k & si \quad x = 1 \end{cases}$$

17. Find out the value of *m* so that $f(x) = \begin{cases} 3x^2 + mx - 1 & if \ x \leq 1 \\ 2x+3 & if \ x > 1 \end{cases}$ is continuous at R.

18. Find out the asymptotes of the following functions and draw a scheme of the functions:

a) $f(x) = \dfrac{1}{4-x^2}$

b) $f(x) = \dfrac{x+3}{x^2-x-2}$

c) $f(x) = \dfrac{2x^2}{(x+2)^2}$

d) $f(x) = \dfrac{-x^3+x}{2}$

e) $f(x) = \dfrac{x^3-1}{x+3}$

f) $f(x) = \dfrac{x}{x^2-9}$

g) $f(x) = \dfrac{x^3-2x^2}{2x+1}$

h) $f(x) = \dfrac{x^2+2}{x+1}$

i) $f(x) = \dfrac{x^4+2x}{x^2+1}$

j) $f(x) = \dfrac{1-3x}{2-x}$

k) $f(x) = \dfrac{1+x^2}{x^3}$

l) $f(x) = \dfrac{x}{x+2}$

Unit 11.- Derivatives

1. Introduction to derivatives. The point is slope

As you know, *slope*, *m*, of a straight line is defined as:

$$m = \frac{change\ of\ y}{change\ of\ x}$$

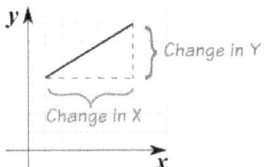

But, what about slope when graph is not a straight line but a curve? I that case, we can calculate an *"average slope"*.

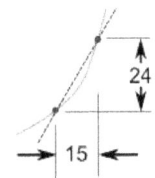

In this case, $average\ slope = \frac{24}{15} = 1.6$

But, what about slope at a point?

There is nothing to measure!!! $m = \frac{0}{0} = ???$

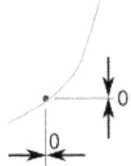

Well, a new operation, called **derivative, f '(x)**, solves this situation. And it does it by calculating an *average slope*, but considering a very thin interval, around your point.

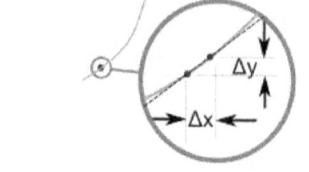

In fact, this interval is so thin that it is said to **shrink towards zero**.

We will use the expression $m = \dfrac{change\ of\ y}{change\ of\ x} = \dfrac{\Delta y}{\Delta x}$

Look at the graph to the right:

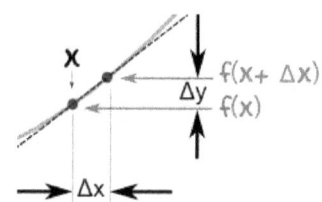

  x changes from x to $x+\Delta x$

  y changes from $f(x)$ to $f(x+\Delta x)$

$$\frac{\Delta y}{\Delta x} = \frac{f(x+h)-f(x)}{x+h-x} = \frac{f(x+h)-f(x)}{h}$$

So, slope of the function *f(x)* at the point *x*, or derivative of *f(x)*, *f '(x)*, is calculated as:

Definition of derivative: $\quad m = f'(x) = \lim_{h\to 0} \dfrac{f(x+h)-f(x)}{h}$

We will practice it with some examples:

Example: Calculate the derivative of function $f(x) = x^2$.

<u>Solution:</u>

$$f(x) = x^2$$
$$f(x+h) = (x+h)^2 = x^2 + 2xh + h^2$$
$$f'(x) = \lim_{h\to 0} \frac{x^2+2xh+h^2-x^2}{h} = \lim_{h\to 0} \frac{2xh+h^2}{h} = \lim_{h\to 0} \frac{h\cdot(2x+h)}{h} = \lim_{h\to 0} (2x+h) = \boxed{2x}$$

Example: Calculate the derivative of function $f(x) = x^3$.

Solution:

$$f(x) = x^3$$

$$f(x+h) = (x+h)^3 = x^3 + 3x^2h + 3xh^2 + h^3$$

$$f'(x) = \lim_{h \to 0} \frac{x^3 + 3x^2h + 3xh^2 + h^3 - x^3}{h} = \lim_{h \to 0} \frac{h \cdot (3x^2 + 3xh + h^2)}{h} = \lim_{h \to 0} (3x^2 + 3xh + h^2) = \boxed{3x^2}$$

What does $f'(x) = 2x$ mean?

It means that, for the function x^2, the slope or *"rate of change"* at any point is $2x$.

So, when x = 2 the slope is 2x = 4, as shown here:

Or when x = 5 the slope is 2x = 10, and so on.

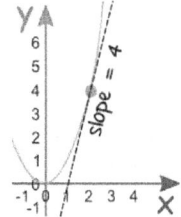

Exercises

1. a) Find out the derivative of the function $f(x) = x^2 - 5x$ and the slope of the tangent straight line to it in the points with abscise $x = 2$ and $x = 3$.

   b) This curve has a maximum or a minimum in the interval (2, 3). What do you think it is, a maximum or a minimum?

2. a) Find out the derivative of the function $f(x) = 3x^2 - 12x$ and the slope of the tangent straight line to it in the points with abscise $x = 1$, $x = 2$ and $x = 3$.

   b) Draw a brief scheme of this curve and the three tangent straight lines and justify if it has a maximum or a minimum at $x = 2$.

3. a) According to results of previous exercise, how can you find out the maximums and minimums *(extremes)* of a function?

   b) Benefits, in thousands euros, depend on the quantity of units that are sold, x *(x = thousands of units)*, according to function $B(x) = x^2 - 10x$. Find out the derivative of B(x), B'(x) and values of B'(1), B'(5) and B'(10). Draw a brief scheme of this curve and the three tangents and decide when the minimum benefits are obtained.

2. Derivative rules

We can use the same method to work out derivatives of other functions (like sine, cosine, logarithms, etc). But in practice the usual way to find derivatives is to use **derivative rules.**

We are studying these rules, one by one.

2.1. Constant function

$$y = a \quad \longrightarrow \quad y' = 0$$

Example: Find out the derivative of the following functions: $f(x)=8$, $f(x)=e^{-7}$, $f(x)=\dfrac{\sqrt[3]{32}}{\ln 7}$.

Solution: In all these cases, $f'(x) = 0$, as all of them are *constant* functions.

2.2. Potential function

$$y = x^n \quad \longrightarrow \quad y' = n \cdot x^{n-1}$$

Example: Find out the derivative of the following functions:

a) $f(x) = x^6$ $\qquad$ $f'(x) = 6x^{6-1} = \boxed{6x^5}$

b) $f(x) = x^3$ $\qquad$ $f'(x) = 3x^{3-1} = \boxed{3x^2}$

c) $f(x) = x^{\frac{5}{2}}$ $\qquad$ $f'(x) = \dfrac{5}{2}x^{\frac{5}{2}-1} = \dfrac{5}{2}x^{\frac{5}{2}-\frac{2}{2}} = \dfrac{5}{2}x^{\frac{3}{2}} = \dfrac{5\sqrt{x^3}}{2} = \boxed{\dfrac{5x\sqrt{x}}{2}}$

d) $f(x) = x^{-7}$ $\qquad$ $f'(x) = -7x^{-7-1} = -7x^{-8} = \boxed{\dfrac{-7}{x^8}}$

e) $f(x) = x^{\frac{-4}{7}}$ $\qquad$ $f'(x) = \dfrac{-4}{7}x^{\frac{-4}{7}-1} = \dfrac{-4}{7}x^{\frac{-4}{7}-\frac{7}{7}} = \dfrac{-4}{7}x^{\frac{-11}{7}} = \dfrac{-4}{7x^{\frac{11}{7}}} = \dfrac{-4}{7\sqrt[7]{x^{11}}} = \boxed{\dfrac{-4}{7x\sqrt[7]{x^4}}}$

f) $f(x) = x$ $\qquad$ $f'(x) = 1x^{1-1} = 1x^0 = \boxed{1}$

g) $f(x) = \dfrac{1}{x^3}$ $\qquad\qquad f'(x) = -3x^{-3-1} = -3x^{-4} = \boxed{\dfrac{-3}{x^4}}$

h) $f(x) = \sqrt[5]{x} = x^{\frac{1}{5}}$ $\qquad f'(x) = \dfrac{1}{5}x^{\frac{1}{5}-1} = \dfrac{1}{5}x^{\frac{1}{5}-\frac{5}{5}} = \dfrac{1}{5}x^{-\frac{4}{5}} = \dfrac{1}{5x^{\frac{4}{5}}} = \boxed{\dfrac{1}{5\sqrt[5]{x^4}}}$

2.3. $y = k \cdot f(x)$

$$y = k \cdot f(x) \quad \longrightarrow \quad y' = k \cdot f'(x)$$

Example: Find out the derivative of the following functions:

a) $f(x) = 4x$ $\qquad\qquad\qquad f'(x) = \boxed{4}$

b) $f(x) = 8x^3$ $\qquad\qquad\quad f'(x) = 8\cdot 3x^2 = \boxed{24x^2}$

c) $f(x) = 5x^{\frac{9}{2}}$ $\qquad\qquad f'(x) = 5\cdot\dfrac{9}{2}x^{\frac{9}{2}-1} = \dfrac{45}{2}x^{\frac{7}{2}} = \boxed{\dfrac{45}{2}\sqrt{x^7}}$

d) $f(x) = 3x^{-6}$ $\qquad\qquad f'(x) = 3(-6)x^{-7} = -18x^{-7} = \boxed{\dfrac{-18}{x^7}}$

e) $f(x) = 3\sqrt{x}$ $\qquad\qquad f'(x) = 3\cdot\dfrac{1}{2}\cdot x^{-1/2} = \dfrac{3}{2}\cdot\dfrac{1}{x^{1/2}} = \boxed{\dfrac{3}{2\sqrt{x}}}$

f) $f(x) = 2\sqrt[5]{x^3} = 2\cdot x^{3/5}$ $\quad f'(x) = 2\cdot\dfrac{3}{5}\cdot x^{-2/5} = \dfrac{6}{5}\cdot\dfrac{1}{x^{2/5}} = \boxed{\dfrac{6}{5\sqrt[5]{x^2}}}$

2.4. Sum of functions $\quad y = a{\cdot}f(x) + b{\cdot}g(x)$

$$y = a{\cdot}f(x) + b{\cdot}g(x) \quad \longrightarrow \quad y' = a{\cdot}f'(x) + b{\cdot}g'(x)$$

Example: Find out the derivative of the following functions:

a) $f(x) = x^3 + x^2 + x + 5$ $\qquad f'(x) = \boxed{3x^2 + 2x + 1}$

b) $f(x) = 5x^3 + 3x^2 + 6x + 5$ $\qquad f'(x) = \boxed{15x^2 + 6x + 6}$

c) $f(x) = x^3 + x^2 + x^{-1} + 7$ $\qquad f'(x) = -3x^{-4} + 2x - x^{-2}$

4. Find out the derivative of the following functions:

a) $y = 2x$ b) $f(x) = -5x$ c) $y = 0.001x$ d) $y = \dfrac{x}{2}$

e) $y = x$ f) $f(x) = \dfrac{2}{3}x$ g) $y = -x$ h) $y = -\dfrac{5x}{3}$

5. Find out the derivative of the following functions:

a) $y = x^2$ b) $f(x) = x^3$ c) $y = x^4$ d) $f(t) = t^5$

e) $y = x^{100}$ f) $y = \dfrac{1}{x}$ g) $y = \sqrt{x}$ h) $y = \sqrt[3]{x}$

i) $y = \sqrt[4]{x^3}$ j) $y = \sqrt[5]{x^2}$ k) $y = x^2\sqrt{x}$ l) $y = \dfrac{1}{\sqrt[3]{x}}$

l) $y = \dfrac{\sqrt{x}}{x^3}$

6. Find out the derivative of the following functions:

a) $y = 3x^2$ b) $y = 4x^3$ c) $f(x) = -2x^4$ d) $y = \dfrac{x^2}{2}$ e) $y = -x^5$

f) $y = \dfrac{2}{3}t^6$ g) $y = -x$ h) $y = 3\sqrt[3]{x^4}$ i) $y = \dfrac{\sqrt[4]{x}}{2}$ j) $y = -\dfrac{3x^4}{2}$

k) $f(t) = -2t^7$ l) $f(x) = \dfrac{x^3}{3}$ m) $y = 2x\sqrt[5]{x}$

7. Find out the derivative of the following functions:

a) $f(x) = x^2 + x^3$ b) $y = x^4 + 5$ c) $y = x^2 - 2$ d) $y = x - 2$

e) $f(t) = 3t - 5$ f) $y = 3x^2 - x^4$ g) $y = 2x^3 - 3x^4$ h) $y = 2t^4 - t^2 + 3$

i) $y = x^4 + x^3 + x^2 + x + 1$ j) $y = x^3 - 3x^2 + 5x - 8$

k) $f(x) = -3x^5 + 4x^3 - x + 2$ l) $y = \dfrac{x^4}{2} + 5x$ m) $y = \dfrac{x^3}{3} - \dfrac{x^2}{2} + \dfrac{x}{5} - \dfrac{1}{2}$

n) $f(x) = x^5 - \dfrac{x^4}{3} + \dfrac{x^3}{6} - 3x^2 + \dfrac{x}{3}$ q) $y = \dfrac{3x^6 - x^3 + 6x - 5}{3}$

o) $y = \dfrac{x^4 + x^2}{2}$ p) $f(x) = 0,05x^3 - 0,001x^2 + 0,1x - 0,02$

2.5. Product of functions $y = f(x) \bullet g(x)$

$$y = f(x) \bullet b \cdot g(x) \;\rightarrow\; y' = f'(x) \bullet g(x) + f(x) \bullet g'(x)$$

Example: Find out the derivative of the following products of functions:

a) $f(x) = (4x^3 - 6)(4x^2 + 4)$ $f'(x) = 12x^2(4x^2 + 4) + (4x^3 - 6)8x = 48x^4 + 48x^2 + 32x^4 - 48x =$
$$= 80x^4 + 48x^2 - 48x = \boxed{16x(5x^3 + 3x - 3)}$$

b) $f(x) = (3x^2 + 3)(2x^2 + 1)$

$f'(x) = 6x(2x^2 + 1) + (3x^2 + 3)4x = 12x^3 + 6x + 12x^3 + 12x = 24x^3 + 18x = \boxed{6x(4x^2 + 3)}$

Exercises

8. Find out the derivative of the following products of functions, in two ways:

  a) $y = (2x+3)(3x-2)$ b) $y = (x-2)(x+3)$ c) $f(x) = (2x+3)(x-5)$

9. Find out the derivative of the following products of functions:

  a) $f(x) = (x^2+2)(3x-1)$ b) $y = (x^2-5)(3x-1) + 7$ c) $y = (2x-3)^2$

  d) $f(x) = (x+2)^3$ e) $y = (1,2 - 0,001x^2)x$ f) $y = (2x^2-3)^2$

  g) $f(t) = 300t(1-t)$ h) $f(x) = (-4x^3-2x)^2$ i) $y = (t^2+t+1)^3$

  j) $y = (3x-2)(2x-3)(x+5)$

2.6. Quotient of functions $y = \dfrac{f(x)}{g(x)}$

$$y = \frac{f(x)}{g(x)} \;\rightarrow\; y = \frac{f'(x) \cdot g(x) - f(x) \cdot g'(x)}{g^2(x)}$$

Example: Find out the derivative of the following quotients of functions:

a) $f(x) = \dfrac{2x^3 + 5}{4x^2 + 7}$

b) $f(x) = \dfrac{4x^3 - 5x^2}{3x^2 - 4}$

a) $f'(x) = \dfrac{6x^2(4x^2+7) - (2x^3+5)8x}{(4x^2+7)^2} = \dfrac{24x^4 + 42x^2 - 16x^4 - 40x}{(4x^2+7)^2} = \dfrac{8x^4 + 42x^2 - 40x}{(4x^2+7)^2} = \boxed{\dfrac{2x(4x^3 + 21x - 20)}{(4x^2+7)^2}}$

b) $f'(x) = f(x) = \dfrac{(12x^2 - 10x)(3x^2 - 4) - (4x^3 - 5x^2)6x}{(3x^2 - 4)^2} = \dfrac{36x^4 - 48x^2 - 30x^3 + 40x - 24x^4 + 30x^3}{(3x^2 - 4)^2} =$

$= \dfrac{12x^4 - 48x^2 + 40x}{(3x^2 - 4)^2} = \boxed{\dfrac{4x(3x^3 - 12x + 10)}{(3x^2 - 4)^2}}$

---

10. Find out the derivative of the following quotients of functions:

a) $y = \dfrac{2x - 3}{3x + 2}$

b) $y = \dfrac{x^2 + 1}{x^2 - 4}$

c) $f(x) = \dfrac{x + 3}{x - 3}$

d) $y = \dfrac{x^2}{x^2 + 1}$

e) $y = \dfrac{x^2 + x + 1}{x}$

f) $f(x) = \dfrac{x^2 - 1}{x + 1}$

g) $f(x) = 3\dfrac{x^2 - 1}{x - 2}$

2.7. $y = lnf(x)$

$$y = lnf(x) \;\longrightarrow\; y' = \dfrac{f'(x)}{f(x)}$$

Example: Find out the derivative of the following functions:

a) $f(x) = Ln(x^2 + 3x - 10)$ $\qquad f'(x) = \dfrac{2x + 3}{x^2 + 3x - 10}$

b) $f(x) = lnx$ $\qquad f'(x) = \dfrac{1}{x}$

c) $f(x) = -5x^3 + 2Lnx - 3\,Ln(x^5 - 10x + 1)$ $\quad f'(x) = -15x^2 + \dfrac{2}{x} - \dfrac{15x^4 - 30}{x^5 - 10x + 1}$

Exercises	**11.** Find out the derivative of the following functions: a) $y = Ln(3x)$ b) $y = Ln(x^3)$ c) $y = Ln(\sqrt{x})$ d) $y = L(x^2 + 1)$ e) $y = Ln(2x^3 + 3x + 1)$ f) $y = Ln(x^2 - x + 5)^3$ g) $y = Ln(x^2 + x + 1)$ h) $y = Ln(x^2 - \sqrt{x})$ i) $y = Ln\left(5 - \dfrac{1}{x}\right)$ k) $y = Ln(\sqrt[3]{x} + \sqrt[4]{x})$

2.8. Exponential function

$$y = e^{f(x)} \longrightarrow y' = e^{f(x)} \cdot f'(x)$$

$$y = a^{f(x)} \longrightarrow y' = a^{f(x)} \cdot f'(x) \cdot \ln a$$

Example: Find out the derivative of the following functions:

a) $f(x) = e^{2x}$ $f'(x) = \boxed{2e^{2x}}$

b) $f(x) = e^{-4x}$ $f'(x) = (-4)e^{-4x} = \boxed{-4e^{-4x}}$

c) $f(x) = e^{\frac{2x}{3}}$ $f'(x) = \boxed{\dfrac{2}{3}e^{\frac{2x}{3}}}$

d) $f(x) = 7^{5x}$ $f'(x) = \boxed{5 \cdot 7^{5x} \cdot \ln 7}$

Exercises	**12.** Find out the derivative of the following functions: a) $y = 3^x$ b) $y = e^{3x+1}$ c) $y = e^{x^2+x+1}$ d) $y = e^{2x-3}$ e) $y = 7^{3x+1}$ f) $y = e^{4x}$ g) $y = e^{x^3-3x^2+2}$ h) $y = e^{3-x^2}$

2.9. Trigonometric functions

$$y = sinx \rightarrow y' = cosx \qquad y = sin(f(x)) \rightarrow y' = cos(f(x)) \cdot f'(x)$$

$$y = cosx \rightarrow y' = -sinx \qquad y = cos(f(x)) \rightarrow y' = -sin(f(x)) \cdot f'(x)$$

Example: Find out the derivative of the following functions:

a) $y = Sin(2x+1)$ $\qquad\qquad$ $y' = 2Cos(2x+1)$

b) $y = Cos(x^2 + e^x)$ $\qquad\qquad$ $y' = -(2x + e^x) \cdot Sin(x^2 + e^x)$

c) $y = tgx = \dfrac{Sinx}{Cosx}$

$$y' = \frac{Cosx \cdot Cosx - Sinx \cdot (-Cosx)}{Cos^2 x} = \frac{Cos^2 x + Sin^2 x}{Cos^2 x} = \frac{1}{Cos^2 x} = \boxed{Sec^2 x} = \boxed{1 + tg^2 x}$$

Exercises

13. Find out the derivative of the following functions:

a) $y = sen(3x + 1)$ $\qquad$ b) $y = sen(x^2 + 1)$ $\qquad$ c) $y = sen(5x^2 + 7x + 1)$

d) $y = sen(Lx)$ $\qquad$ e) $y = sen(senx)$ $\qquad$ f) $y = cos\left(\dfrac{x}{5}\right)$ $\qquad$ g) $y = cos(e^x)$

h) $y = cos(3x + 1)$ $\qquad$ i) $y = cos(x^2 - 1)$ $\qquad$ j) $y = cos(5x^2 - 7x + 1)$

k) $y = cos(x^3)$ $\qquad$ l) $y = cos(Lx)$ $\qquad$ m) $y = cos(cos x)$ $\quad$ n) $y = tg(2x + 1)$

ñ) $y = tg(x^2)$ $\qquad$ o) $y = tg(Lx)$ $\qquad$ p) $y = sen2x$

q) $y = cos(x^2 - 3)$ $\qquad$ r) $y = sen(-2x)$ $\qquad$ s) $y = sen(2x + 7)$

t) $y = sen(-3x + 6)$ $\qquad$ u) $y = sen(x^2 + 1)$ $\qquad$ v) $y = sen(x^{-2})$ $\qquad$ w) $y = sen(e^x)$

x) $y = sen(cos x)$ $\qquad$ y) $y = sen(tgx)$ $\qquad$ z) $y = cos(5x)$ $\qquad$ A) $y = cos(-3x)$

B) $y = cos(4x - 3)$ $\qquad$ C) $y = cos(-5x + 7)$ $\qquad$ D) $y = cos(senx)$

Arcsin, arccos and arctg functions

$$y = arcsinx \rightarrow y' = \frac{1}{\sqrt{1-x^2}} \qquad\qquad y = arcsin\{f(x)\} \rightarrow y' = \frac{f'(x)}{\sqrt{1-f^2(x)}}$$

$$y = arccosx \rightarrow y' = \frac{-1}{\sqrt{1-x^2}} \qquad\qquad y = arcsin\{f(x)\} \rightarrow y' = \frac{-f'(x)}{\sqrt{1-f^2(x)}}$$

$$y = arctgx \rightarrow y' = \frac{1}{1+x^2} \qquad\qquad y = arctg\{f(x)\} \rightarrow y' = \frac{f'(x)}{1+f^2(x)}$$

Example: Find out the derivative of the following functions:

a) $y = Arcsen(2x+1)$ $y' = \dfrac{2}{\sqrt{1-(2x+1)^2}}$

b) $y = Arc\cos\sqrt{x}$ $y' = \dfrac{1/2\sqrt{x}}{\sqrt{1-x}}$

c) $y = Arctg(Lnx)$ $y' = \dfrac{1/x}{1+Ln^2x}$

---

14. Find out the derivative of the following functions:

a) $y = arcsen(x^2+2)$

b) $y = (x^2 + 4x + 2) \cdot arcsen(x^4 + 3x^2 + 1)$

c) $y = (x^3 + e^{2x} + 3) \cdot arcsen(3x^2 - 2x + 2)$

d) $y = (x^2 + 1) \cdot arctg(x^3 + 5)$

e) $y = (x^4 + e^x + 1) \cdot arctg(3x^2 + x + 5)$

f) $y = (x^3 + 4x + 3) \cdot arctg(x^2 + e^{2x} + 1)$

Exercises

2.10. Power of a function $y = [f(x)]^n$

$$y = [f(x)]^n \qquad \rightarrow y' = n \cdot [f(x)]^n \cdot f'(x)$$

Example: Find out the derivative of the following functions:

a) $y = (x^3 + x^2)^4$ $\qquad$ $y' = 4(x^3 + x^2)^3 \cdot (3x^2 + 2x)$

b) $y = Sin^3(x^2 + e^x)$ $\qquad$ $y' = 3Sin^2(x^2 + e^x) \cdot (2x + e^x)$

c) $y = Ln^2(5x + 1)$ $\qquad$ $y' = 2Ln(5x + 1) \cdot \dfrac{5}{5x + 1}$

---

15. Find out the derivative of the following functions:

a) $f(x) = (x^2 + 1)^7$ $\qquad$ b) $f(x) = \sqrt{\sqrt{\sqrt{x}}}$ $\qquad$ c) $f(x) = sen^2(2x + 1)$

d) $f(x) = sen^{-2}(x^2 + x)$ $\quad$ e) $f(x) = \cos^3(5x + 1)$ $\quad$ f) $f(x) = \dfrac{1}{\sqrt[3]{(x^2 + 1)}}$

g) $f(x) = \dfrac{1}{\sqrt{tg(2x + 1)}}$ $\quad$ h) $f(x) = \sqrt{\cot g(3x + 1)}$ $\quad$ i) $f(x) = \dfrac{1}{\sqrt{\cot gx}}$

j) $f(x) = Ln^4(4x^3 + x^2 + 3x + 5)$ $\qquad$ k) $f(x) = Ln^5(3x^4 - 6x^2 + \dfrac{7}{2}x^{-4} + 6)$

Exercises

Resume of derivatives

Simple functions		Chain rule	
$y = a$	$\rightarrow$ $y' = 0$		
$y = x$	$\rightarrow$ $y' = 1$		
$y = x^n$	$\rightarrow$ $y' = n \cdot x^{n-1}$	$y = [f(x)]^n$	$\rightarrow$ $y' = n \cdot [f(x)]^{n-1} \cdot f'(x)$
$y = e^x$	$\rightarrow$ $y' = e^x$	$y = e^{f(x)}$	$\rightarrow$ $y' = e^{f(x)} \cdot f'(x)$
$y = senx$	$\rightarrow$ $y' = cosx$	$y = sen(f(x))$	$\rightarrow$ $y' = cos(f(x)) \cdot f'(x)$
$y = cosx$	$\rightarrow$ $y' = -senx$	$y = cos(f(x))$	$\rightarrow$ $y' = -sen(f(x)) \cdot f'(x)$
$y = lnx$	$\rightarrow$ $y' = \dfrac{1}{x}$	$y = ln(f(x))$	$\rightarrow$ $y' = \dfrac{f'(x)}{f(x)}$
$y = \sqrt{x}$	$\rightarrow$ $y' = \dfrac{1}{2 \cdot \sqrt{x}}$	$y = \sqrt{f(x)}$	$\rightarrow$ $y' = \dfrac{f'(x)}{2 \cdot \sqrt{f(x)}}$

Mixed exercises

16. Find out the derivative of the following functions:

a) $f(x) = 3x^4 - 2x + 5$ b) $f(x) = e^x$ c) $f(x) = 2x^3 - x^2 + 1$ d) $f(x) = Lnx$

e) $f(x) = 2x^5 + \dfrac{x}{3}$ f) $f(x) = senx$ g) $f(x) = x^3 - 3x^2 + \dfrac{1}{5}$ h) $f(x) = \cos x$

i) $f(x) = 4x^3 - 3x^2 + 2$ j) $f(x) = tgx$ k) $f(x) = \dfrac{x^2 + 2}{2x + 1}$ l) $f(x) = xe^x$

m) $f(x) = \dfrac{3x - 1}{x^2 - 2}$ n) $f(x) = x^2 senx$ ñ) $f(x) = \dfrac{1 - x^2}{x - 3}$ o) $f(x) = xLnx$

p) $f(x) = \sqrt{x} + \dfrac{2}{x}$ q) $f(x) = \dfrac{3x + 1}{e^x}$ r) $f(x) = \dfrac{3x^2}{2x + 3}$ s) $f(x) = \sqrt[3]{x} \cdot senx$

17. Find out the derivative of the following functions:

a) $f(x) = \left(3x^2 + x\right)^4$ b) $f(x) = \sqrt{4x^3 + 1}$ c) $f(x) = e^{4x^3 - 2x}$ d) $f(x) = Ln\left(3x^4 - 2x\right)$

e) $f(x) = sen\left(\dfrac{x + 1}{2x - 3}\right)$ f) $f(x) = 3x^4 - \dfrac{9x^2}{3}$ g) $f(x) = \dfrac{3x^2 - 2}{x^2 - 1}$ h) $f(x) = xe^x$

i) $f(x) = 8x^5 - 2x^3 + \dfrac{1}{3}$ j) $f(x) = \left(x^4 - 3x\right)e^x$ k) $f(x) = sen\left(\dfrac{x}{x^2 - 1}\right)$ l) $f(x) = \dfrac{3x^4}{2} - \dfrac{6x^3}{5}$

m) $f(x) = \dfrac{x^2 - 3}{2x^3 + 1}$ n) $f(x) = Ln\left(x^4 - 2x\right)$ ñ) $f(x) = \dfrac{-2x^4 + 3x^2}{5}$ o) $f(x) = \dfrac{3x - 4}{x^2 + 3x}$

p) $f(x) = \sqrt{2x^3 - 3}$ q) $f(x) = \dfrac{1}{2}x^4 - \dfrac{3}{5}x^7$ r) $f(x) = e^x \cdot senx$ s) $f(x) = \cos\left(\dfrac{x}{x^2 + 2}\right)$

t) $f(x) = 4x^5 - \dfrac{2x}{3}$ u) $f(x) = \left(x^2 - 3x\right)e^x$ v) $f(x) = sen\left(\dfrac{x - 1}{x^2 + 1}\right)$

w) $f(x) = -x^7 + \dfrac{3}{4}x - 1$ x) $f(x) = \dfrac{4x^3 - 3}{x^2 - 1}$ y) $f(x) = e^{7x^4 - 3}$

z) $f(x) = 9x^2 - 3x^4 + \dfrac{1}{3}$ 1) $f(x) = \dfrac{3x^3}{4 - x^2}$ 2) $f(x) = Ln\left(2x^5 + 3x\right)$

3) $f(x) = \dfrac{-3x^5 + 2x}{7}$ 4) $f(x) = x^4 \cos x$ 5) $f(x) = e^{\frac{x^2 + 1}{x - 1}}$

6) $f(x) = \dfrac{4x^6}{3} - 2x + 5$ 7) $f(x) = \dfrac{2x}{x^2 + 1}$ 8) $f(x) = \sqrt{2x - 3x^4}$

3. Equation of the tangent straight line to a curve

If you remember the beginning of the unit, the geometrical interpretation of the derivative of a function at a point is the slope of the tangent straight line at that point.

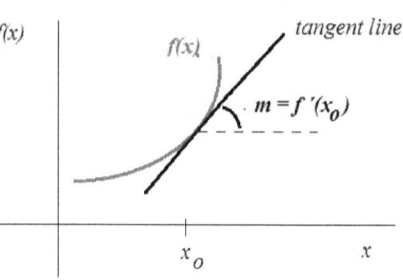

This allows us to determine the equation of the straight line that is tangent to the curve.

Remember how the equation of a straight line is obtained from its slope, *m*, and one of its points, $A(a_x, a_y)$:

Slope-point equation: $\quad y - a_y = m \cdot (x - a_x)$

In this new context,

> ➤ **Slope:** $m = f'(x_0)$
> ➤ **Point:** $(x_0, f(x_0))$
>
> $$y - f(x_0) = f'(x_0) \cdot (x - x_0)$$

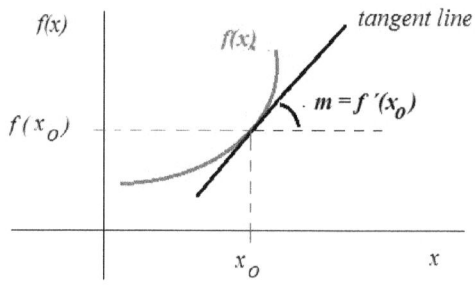

We are seeing it with an example:

Example: Write the equation of the straight line that is tangent to the curve of the function $f(x) = x^3 + 2x$ at the point with abscise $x_0 = 2$.

Solution: $f'(x) = 3x^2 + 2$

$$\left. \begin{array}{l} x_0 = 2 \\ m = f'(2) = 3 \cdot 2^2 + 2 = 14 \\ y_0 = f(2) = 2^3 + 2 \cdot 2 = 12 \end{array} \right\} \quad \rightarrow \quad y - 12 = 14(x - 2) \quad \rightarrow \quad \boxed{y = 14x - 16}$$

Equation of the "normal" straight line to a curve

Sometimes, you may be asked to obtain the *"normal"* straight line to a curve. In geometry, the word *"normal"* means perpendicular.

To obtain the normal straight line, do the same as in tangent line, but changing the slope:

Slope of a normal straight line: Given a straight line with slope *m*, slope of a straight line, perpendicular to it, *m′*, can be calculated as:

$$m' = -\frac{1}{m}$$

Example: Write the equations of the tangent and normal straight lines to the curve of the function $f(x) = x \cdot e^x$ at the point with abscise $x_0 = 1$.

Solution: $f'(x) = 1 \cdot e^x + x \cdot e^x = e^x \cdot (x+1)$

a) Tangent:
$\left. \begin{array}{l} x_0 = 1 \\ m = f'(1) = e^1 \cdot (1+1) = 2e \\ y_0 = f(1) = 1 \cdot e^1 = e \end{array} \right\}$ → $y - e = 2e(x-1)$ → $\boxed{y = 2ex - e}$

b) Normal:
$\left. \begin{array}{l} x_0 = 1 \\ m' = -1/2e \\ y_0 = f(1) = 1 \cdot e^1 = e \end{array} \right\}$ → $y - e = -\frac{1}{2e}(x-1)$ → $\boxed{y = -\frac{x}{2e} + \frac{2e^2+1}{2e}}$

Exercises

18. Find out the equations of the tangent and normal straight lines to these functions:

 a) $f(x) = x^3 - 5$ in x = 1   b) $f(x) = \frac{1}{x}$ in x = (-2)   c) $f(x) = \sqrt{x^2+1}$ in x = 0

 d) $f(x) = \frac{x}{x^2+1}$ in x = 1   e) $f(x) = \frac{-3}{x^2}$ in x = 0

19. Find out the equation of the tangent straight line to these functions:

 a) $f(x) = 3x^2+8$ in x=1   b) $y = 2x^5+4$ in x=-1   c) $f(x) = x^4-1$ en x=0

 (Sol: a) 6x-y+5=0; b) 10x-y+12=0; c) y=-1)

20. At what point of the graph of the parable $f(x) = x^2 - 6x + 8$ its tangent line is parallel to abscises axis? What is the name of this particular point? Draw a scheme.

(Sol: y = – 1; vertex (3,-1))

21. At what point of the graph of the previous parable its tangent line is parallel to bisectrix of the first quadrant? Draw a scheme. *(Sol: (7/2,-3/4))*

22. (PAU) Find out the points of the curve $y = x^3 + 9x^2 - 9x + 15$ at which its tangent line is parallel to the straight line $y = 12x + 5$. *(Sol: (1,16) and (-7,176))*

4. Maximums and minimums. Increasing and decreasing intervals

If we consider a point, x_0, at which $f(x)$ is increasing, its tangent straight line is also increasing, its slope is positive. As this slope is given by the derivative of the function at that point, $f'(x_0)$, has a positive value.

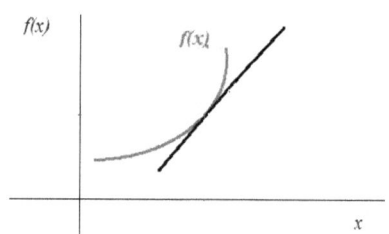

The same but with a negative sign will happen at a point at which the curve is decreasing:

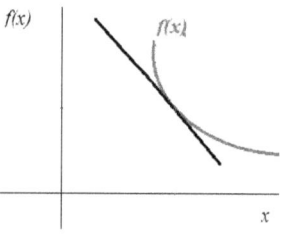

> ➤ If $f'(x_0) > 0 \;\rightarrow\; f(x)$ is **increasing** at $x = x_0$
> ➤ If $f'(x_0) < 0 \;\rightarrow\; f(x)$ is **decreasing** at $x = x_0$

But, perhaps even more important, *what is the slope of the tangent line exactly at a maximum or a minimum of the curve?*

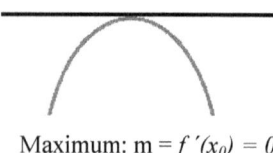

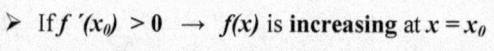

Maximum: $m = f'(x_0) = 0$ Minimum: $m = f'(x_0) = 0$

- 268 -

Exactly,

> If at $x = x_0$, the curve of $f(x)$ has a **maximum** or a **minimum** (**extremes**) then:
>
> $$f'(x_0) = 0$$

So, here you are the *tool* to find out the extremes of a function. You only need to know if the point you have found out is a maximum or a minimum. Pay attention to the following examples:

Example: Find extremes and increasing and decreasing intervals of function $f(x)=3x^2-2x+1$.

Solution: $Domf = \mathbf{R}$.

- **Extremes:** $f'(x) = 6x - 2$ $\rightarrow$ $f'(x_0) = 0$ $\rightarrow$ $6x - 2 = 0$ $\rightarrow$ $x = 1/3$.

- **Intervals:** Build a table below the Number Line, in which you must indicate *extremes* and, if there are, *vertical asymptotes*:

	$-\infty$	1/3	$+\infty$
Sign of $f'(x)$			

As you can see, there are 2 intervals; $(-\infty, 1/3)$ and $(1/3, +\infty)$. Consider one value at each interval and study the sign of f'(x) at each interval:

- Interval $(-\infty, 1/3)$: let's consider $x = 0$ $\rightarrow$ $f'(0) = 6\cdot0 - 2 = -2 < 0$
- Interval $(1/3, +\infty)$: let's consider $x = 1$ $\rightarrow$ $f'(1) = 6\cdot1 - 2 = 4 > 0$

	$-\infty$	1/3	$+\infty$
Sign of $f'(x)$	< 0		> 0
Monotony	Decreasing		Increasing
	↘		↗

MIN

So, $f(x)$ is decreasing at $(-\infty, 1/3)$, increasing at $(1/3, +\infty)$ and it has a minimum at $x = 1/3$, whose value is $f\left(\frac{1}{3}\right)=3\left(\frac{1}{3}\right)^2 -2\left(\frac{1}{3}\right)+1 = \frac{1}{3}-\frac{2}{3}+1 = \frac{2}{3}$.

> Study of extremes and increasing intervals is called **monotony** of the function.

Example: Study monotony of the function $f(x) = \dfrac{x^2+3}{x-1}$.

Solution: *Domf* = $\mathbf{R}$ − {1}.   Notice there is a vertical asymptote at x = 1.

- Extremes: $f'(x) = \dfrac{2x(x-1)-(x^2+3)\cdot 1}{(x-1)^2} = \dfrac{2x^2-2x-x^2-3}{(x-1)^2} = \dfrac{x^2-2x-3}{(x-1)^2} \rightarrow f'(x_0) = 0 \quad \rightarrow$

$\rightarrow \dfrac{x^2-2x-3}{(x-1)^2} = 0 \rightarrow x^2-2x-3=0 \rightarrow \begin{cases} x_2 = 3 \\ x_1 = -1 \end{cases}$

- **Intervals:** Build a table below the Number Line, in which you must indicate *extremes* and, if there are, *vertical asymptotes*, this is, x = (-1), x = 1 and x = 3.

	−∞	(-1)	1	3	+∞
Sign of $f'(x)$					

- Interval $(-\infty, -1)$: consider $x = -2 \rightarrow f'(-2) = \dfrac{(-2)^2-2(-2)-3}{(-2-1)^2} = \dfrac{4+4-3}{(-3)^2} = \dfrac{5}{9} > 0$

- Interval $(-1, 1)$: consider $x = 0 \rightarrow f'(0) = \dfrac{0^2-2\cdot 0-3}{(0-1)^2} = \dfrac{-3}{1} = -3 < 0$

- Interval $(1, 3)$: consider $x = 2 \rightarrow f'(2) = \dfrac{2^2-2\cdot 2-3}{(2-1)^2} = \dfrac{4-4-3}{(1)^2} = \dfrac{-3}{1} = -3 < 0$

- Interval $(3, +\infty)$: consider $x = 5 \rightarrow f'(5) = \dfrac{5^2-2\cdot 5-3}{(5-1)^2} = \dfrac{25-10-3}{(4)^2} = \dfrac{12}{16} > 0$

	−∞	(-1)	1	3	+∞
Sign of $f'(x)$		> 0	< 0	< 0	> 0
Monotony		Increasing	Decreasing	Decreasing	Increasing
		↗	↘	↘	↗

MAX   Vert. asym.   MIN

So, *f(x)* is increasing at $(-\infty, (-1)) \cup (3, +\infty)$, decreasing at $((-1), 1) \cup (1, 3)$. It has a maximum at x = (-1) whose value is $f(-1) = \dfrac{(-1)^2+3}{-1-1} = -2$ and a minimum x = 3 whose value is

$f(3) = \dfrac{3^2+3}{3-1} = 6$.

23. Determine increasing and decreasing intervals of the following functions:

a) $f(x) = x^2$

b) $f(x) = x^4 - 2x^2$

c) $y = x^3 - 3x^2 + 1$

d) $f(x) = x^3 - 6x^2 + 9x - 8$

e) $f(x) = x^3 - 4x^2 + 7x - 6$

k) $y = 2x^3 - 9x^2$

l) $f(x) = x^3 - 6x^2 + 9x$

f) $f(x) = x^3$

g) $f(x) = x^4 + 8x^3 + 18x^2 - 10$

h) $y = x^3 - 3x^2 - 9x + 1$

i) $f(x) = x^4 - 4x^3 + 1$

j) $y = \dfrac{x^3}{3} - \dfrac{x^2}{2} - 6x + 3$

m) $y = x^3 - 12x$

(Sol: a) decr at $(-\infty,0)$ and inc at $(0,+\infty)$;

b) inc at $(-1,0)U(1,+\infty)$ and decr at $(-\infty,-1)U(0, 1)$;

c) incr at $(-\infty, 0)U(2,+\infty)$ and decr at $(0,2)$;

d) incr at $(-\infty,0)U(3,+\infty)$ and decr at $(1,3)$;

e) incr at R;   f) incr at R;   g) decr at $(-\infty, 0)$ and inc at $(0, +\infty)$;

h) incr at $(-\infty,-1)U(3,+\infty)$ and decr at $(-1,3)$;

i) decr at $(-\infty,3)$ and inc at $(3, +\infty)$)

24. Find and classify the extremes of the following functions:

a) $f(x) = \dfrac{2x^2}{x^2 - 1}$ **b)** $f(x) = \dfrac{9x}{x^2 + 9}$ **c)** $f(x) = (x-1)^2(x+5)$

d) $f(x) = x^3 + 6x^2 - 15x$

(Sol: a) max (0, 0);   b) min (-3, -3/2), max (3, 3/2);   c) max (-3, 32), min (1, 0);

d) max (-5, 100), min (1, -8))

25. Find out extremes of the following functions:

a) $f(x) = \dfrac{3 - x^2}{x + 2}$ **b)** $f(x) = \dfrac{x^3}{x + 2}$

(Sol: a) (-1, 2), (-3, 6);   b) (0, 0), (-3, 27))

5. Optimization problems

Example: Cost of fuel spent by a boat depends on its speed according to the expression:

$$C(x) = \frac{x^2}{60} + \frac{450}{x},$$

where x is speed (miles/h). Cost is given in €. Calculate the most economical speed and its equivalent cost .

Solution: We must obtain the derivative of the function and make it 0:

$$C'(x) = \frac{x}{30} - \frac{450}{x^2} = 0 \quad \rightarrow \quad \frac{x}{30} = \frac{450}{x^2} \quad \rightarrow \quad x^3 = 13500 \quad \rightarrow \quad x = \sqrt[3]{13500} = \boxed{23.81 \text{ mile/h}}$$

Minimum cost is: $C(23.81) = \dfrac{(23.81)^2}{60} + \dfrac{450}{23.81} = \boxed{28.35 \text{ €}}$.

Exercises

26. It is estimated that the monthly benefits of a candy factory, in thousands of euros, are given by the function $B(x) = -0,1x^2 + 2,5x - 10$, when x tons of product are sold.

    a) Calculate the number of tons that must be sold to obtain the maximum benefit.

    b) What is the maximum benefit?   *(Sol: 12.5 ton , 5625 €)*

27. Water consumption, in cubic meters per month, for a company varies during the first half (January-June) according to the function

    $$C(t) = 8t^3 - 84t^2 + 240t; \quad \text{with } 0 \le t \le 6$$

    a) At what months the maximum and minimum consumptions are produced?

    b) Determine the value of the maximum and minimum consumptions.

    (Sol: maximum 2 months and 5 months minimum)

28. The number of people using the facilities of a summer pool is expressed by the function: $f(t) = 10t^3 - 120t^2 + 450t$

    where t expresses the time from opening, at 12 am (t = 0), until closing of the pool, which occurs 19 h. At what periods the number of people is increasing or decreasing ?

    (Sol: Grows from 12 to 15 and from 17 to 19 h, decreases from 15 to 17 h)

6. Study of continuity and derivability of a function

> ➤ Study of derivability of $f(x)$ is equivalent to study of continuity of its derivative, $f'(x)$
>
> ➤ When a function is not continuous at $x = x_0$, it cannot be derivable at that point. Continuity is a previous condition for derivability.

Example: Study continuity and derivability of the function: $f(x) = \begin{cases} x^2 + 4x + 3 & if \ x \leq 0 \\ 4e^x - 1 & if \ x > 0 \end{cases}$

Solution: a) Continuity:

$$\begin{cases} \lim_{x \to 0^-} (x^2 + 4x + 3) = 3 \\ \lim_{x \to 0^+} \left(4e^x - 1 \right) = 4 \cdot 1 - 1 = 3 \end{cases}$$ So, $f(x)$ is continuous at R.

b) Derivability: Derive: $f'(x) = \begin{cases} 2x + 4 & if \ x \leq 0 \\ 4e^x & if \ x > 0 \end{cases}$

$$\begin{cases} f'(0^-) = 2 \cdot 0 + 4 = 4 \\ f'(0^+) = 4 \cdot e^0 = 4 \end{cases}$$ So, $f(x)$ is derivable at R.

Example: Study continuity and derivability of the function: $f(x) = \begin{cases} x^2 & if \ x \leq 1 \\ e^x - 1 & if \ x > 1 \end{cases}$

Solution: a) Continuity:

$$\begin{cases} \lim_{x \to 1^-} (x^2) = 1 \\ \lim_{x \to 1^+} \left(e^x - 1 \right) = e - 1 \end{cases}$$

- As $e - 1 \neq 1$, $f(x)$ is NOT continuous at $x = 1$.
- As $f(x)$ is not continuous at $x = 1$, it cannot be derivable at $x = 1$.

Example: Find out a and b so that f(x) is continuous and derivable at R: $f(x) = \begin{cases} x + a & if \ x \leq 1 \\ ax - b & if \ x > 1 \end{cases}$

Solution: $\lim_{x \to 1^-} f(x) = 1 + a;$ $\lim_{x \to 1^+} f(x) = a - b;$ → $1 + a = a - b$ → $b = -1$

$f'(x) = \begin{cases} 1 & if \ x \leq 1 \\ a & if \ x > 1 \end{cases}$ $f'(1^-) = 1;$ $f'(1^+) = a$ → $a = 1$

So, $a = 1$ and $b = -1$.

29. Given function $f(x) = \begin{cases} 3 - ax^2, & \text{if } x \leq 1 \\ 2/(ax), & \text{if } x > 1 \end{cases}$

a) Find out values of a that makes it to be a continuous function.

b) For what values of a is $f(x)$ derivable? *(Sol: a) 1 and 2; b) 1)*

30. Given function $f(x) = \begin{cases} 5 + 2senx & \text{if } x \leq 0 \\ -x^2 + ax + b & \text{if } x > 0 \end{cases}$

a) Find out values of a and b that makes it to be a continuous function.

b) For what values of a and b is $f(x)$ derivable at $x = 0$? *(Sol: b = 5, a = 2)*

31. Study continuity and derivability of the function $f(x) = \begin{cases} e^x, & \text{when } x \leq 0 \\ 2x + 1, & \text{when } x > 0 \end{cases}$

7. Representation of functions

In previous units you have already studied some aspects related to the graph of a function, such as domain, symmetries, asymptotes… In this unit you have learnt about their monotony. That is why now you are ready for a complete study and representation of functions. Pay attention to the following example, made for you.

Complete study of a function

1.- Domain	**4.-** Symmetries.
2.- Asymptotes	**5.-** Monotony
3.- Intersection points with axes	**(*)** *Draw at the same time you study each item*.

Example: Study and represent the following function: $y = \dfrac{3x}{x-2}$

Solution:

1.- Domain: $Domf = \mathbf{R} - \{2\}$ ← *(Vertical asymptote)*

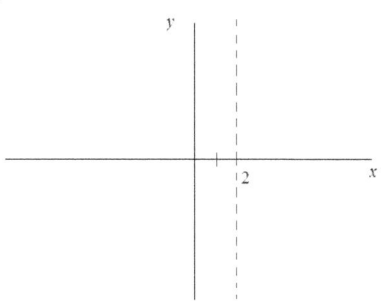

2.- Asymptotes:

- **Horizontal asymptote:** $\displaystyle\lim_{x\to\pm\infty}\left(\dfrac{3x}{x-2}\right) = 3$ So, there is a horizontal asymptote, *y = 2.*

- **Vertical asymptote:** As $Domf = \mathbf{R} - \{2\}$, there is a vertical asymptote: *x = 2.*

$$\left|\begin{array}{l}\displaystyle\lim_{x\to 2^-}\left(\dfrac{3x}{x-2}\right) \approx \dfrac{6}{-0.0001} \approx -\infty \\[4mm] \displaystyle\lim_{x\to 2^+}\left(\dfrac{3x}{x-2}\right) \approx \dfrac{6}{0.0001} \approx +\infty\end{array}\right.$$

- **Oblique asymptote:** As there is a horizontal asymptote, there is **NOT** an oblique one.

3. Intersection with axes:

- For x = 0 → $y = \dfrac{3 \cdot 0}{0-2} = 0$ →

  P(0, 0)

- For y = 0 → $0 = \dfrac{3x}{x-2}$ → $3x = 0$ → $x = 0$

  Q(0, 0)

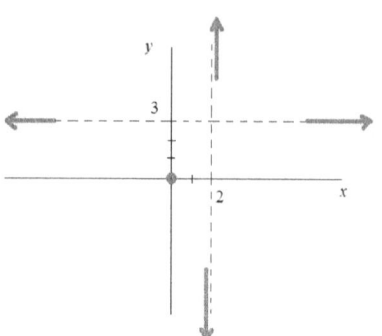

4. Symmetries:

Substitute x by -x: $f(-x) = \dfrac{3 \cdot (-x)}{-x-2} = \dfrac{-3x}{-x-2}$

Notice *f(-x)* is not *f(x)* or *–f(x)*. So, this function has not any symmetry.

5.- Monotony:

$y' = \dfrac{3 \cdot (x-2) - 3x \cdot 1}{(x-2)^2} = \dfrac{-6}{(x-2)^2}$ → $f'(x_0) = 0$ → $\dfrac{-6}{(x-2)^2} = 0$ → $-6 = 0$

This equation has no solution. This means *f(x)* <u>has not maximum or minimum</u>.

	$-\infty$	2	$+\infty$
Sign of $f'(x)$			

- Interval $(-\infty, 2)$: let's consider $x = 0$ → $f'(0) = \dfrac{-6}{(0-2)^2} = \dfrac{-6}{4} < 0$

- Interval $(2, +\infty)$: let's consider $x = 3$ → $f'(3) = \dfrac{-6}{(3-2)^2} = \dfrac{-6}{1} < 0$

	$-\infty$	2	$+\infty$
Sign of $f'(x)$		< 0	< 0
Monotony		Decreasing	Decreasing
		↘	↘

Vert. asymp.

So, *f(x)* is always decreasing. It is said to be ***decreasing at all its domain***.

Now, you only have to join lines, observing
you draw according to obtained monotony.

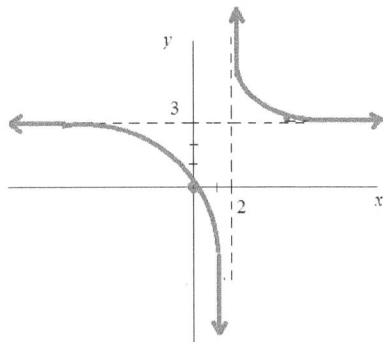

Exercises

32. Study and represent the following functions:

a) $y = x^3 + 3x^2$

b) $y = x^4 - 2x^2$

c) $y = \dfrac{x^2}{x+1}$

d) $y = \dfrac{x^3}{x-2}$

e) $y = \dfrac{x^3 + 2}{x}$

f) $y = \dfrac{x^3}{x+2}$

g) $y = \dfrac{x^4 - 1}{x}$

h) $y = \dfrac{x^2 - 4}{x^2 - 1}$

i) $y = \dfrac{x^2}{x^2 + 1}$

j) $y = \dfrac{2x^2 + 1}{x^2}$

k) $y = \dfrac{x^3}{x^2 + 1}$

l) $y = \dfrac{x^3 + 4}{x^2}$

m) $y = \dfrac{x^3}{x^2 - 1}$

n) $y = \dfrac{x^4}{x^2 + 1}$

Review exercises

Mean variation rate

1. Calculate the mean variation rate of the function $f(x) = \dfrac{x^2 - 1}{2}$ at the interval [0, 2]. Does it increase or decrease?

2. Calculate the mean variation rate of the function $f(x) = \dfrac{3}{x}$ at the interval [-3, -1] and decide if it increases or decreases at this interval.

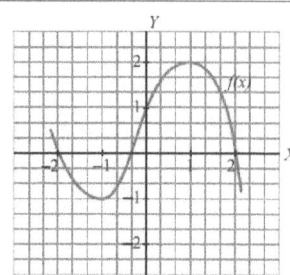

3. Calculate the mean variation rate of *f(x)* and decide if it increases or decreases at the intervals:

　　a) [-2, -1]　　　b) [0, 1].

Definition of derivative

4. Find out, by using the definition, the derivative of the following functions:

a) $f(x) = x^2 + 2x$　　　　b) $f(x) = x^2 + 1$　　c) $f(x) = \dfrac{2x+1}{4}$　　　d) $f(x) = \dfrac{3}{x}$

5. Find out, by using the definition, the derivative of the following functions, at the indicated

points:　　　　a) $f(x) = \dfrac{3x+1}{2}$　in $x = -1$　　　　　　b) $f(x) = \dfrac{1}{x}$ in $x = 2$

　　　　c) $f(x) = 3x^2 + 2x$　in $x = 1$　　　　　　d) $f(x) = \dfrac{x^2}{3}$ in $x = 1$

Derivatives

6. Calculate the derivative of the following functions:

a) $f(x) = \ln\dfrac{1 - \cos x}{1 + \cos x}$　　　b) $y = \sqrt{(x^2 + 5x)^3}$　　c) $y = \sqrt[3]{(5x^2 - 2x)^2}$　d) $f(x) = x^2 \ln(3x - 4)$

e)　$f(x) = (x-1)\ln(x^2 - 1)$　　　f) $f(x) = \dfrac{\ln(x^2 - 1)}{x^3}$　g) $y = 2^{3x^2 - 1}$　h) $y = e^{-x^2 + 3}$　i) $y = x^2 e^{2x+1}$

j) $y = \dfrac{e^x}{x+2}$　　　k) $y = 2^{x^2 - 3}$　　　l) $y = 3^{2x - x^2}$　m) $y = e^{-x+3}$　n) $y = 2e^{5x}$

ñ) $y = (2x+1)e^{2x+1}$　　　　　　o) $y = \dfrac{e^x}{x}$

(Sol: a) $\dfrac{2}{senx}$; b) $y' = \dfrac{3}{2}(2x+5)\sqrt{x^2 + 5x}$; c) $y' = \dfrac{2(10x - 2)}{3\sqrt[3]{5x^2 - 2x}}$; d) $2x\ln(3x - 4) + \dfrac{3x^1}{3x - 4}$;

e) $\ln(x^2 - 1) + \dfrac{2x}{x+1}$; f) $\dfrac{2x^2 - 3(x^2 - 1)\ln(x^2 - 1)}{x^4(x^2 - 1)}$; g) $y' = 6x\cdot 2^{3x^2 - 1}\ln 2$; h) $y' = -2xe^{-x^2 + 3}$;

i) $y' = 2x(1 + x^2)e^{2x+1}$; j) $y' = \dfrac{e^x(x+1)}{(x+2)^2}$; k) $2x\cdot 2^{x^2 - 3}\ln 2$; l) $(2 - 2x)\cdot 3^{2x - x^2}\ln 3$; m) $-e^{-x+3}$;

n) $10e^{5x}$ ñ) $(4x + 4)e^{2x+1}$; o) $\dfrac{e^x(x-1)}{x^2}$)

7. Calculate the derivative of the following functions:

a) $f(x) = (3x^5 - 4x^2 + 7)^4$

b) $f(x) = \dfrac{2x^3 - 1}{x^2 - 4x}$

c) $f(x) = \dfrac{2}{x^3} - \dfrac{3}{x^4} + \dfrac{4}{x^5}$

d) $y = \dfrac{2}{3x^2 - 5x}$

e) $y = \dfrac{3x}{(x^2 - 1)^5}$

(Sol: a) $f'(x) = 4(3x^5 - 4x^2 + 7)^3 \cdot (15x^4 - 8x)$; *b)* $f'(x) = \dfrac{2(x^4 - 8x^3 + x - 2)}{(x^2 - 4x)^2}$;

c) $f'(x) = -\dfrac{6}{x^4} + \dfrac{12}{x^5} - \dfrac{20}{x^6}$; *d)* $y' = \dfrac{10 - 12x}{(3x^2 - 5x)^2}$; *e)* $y' = \dfrac{-27x^2 - 3}{(x^2 - 1)^6}$ *)*

8. Calculate the derivative of the following functions:

a) $y = \log(4x^2 - x + 2)$ b) $y = \log(x^3 - 5x)^7$ c) $f(x) = \log\dfrac{1}{x^2}$ d) $f(x) = \log(x - 2\sqrt{x})$

(Sol: a) $\dfrac{8x - 1}{4x^2 - x + 2} \log e$; *b)* $\dfrac{21x^2 - 35}{x^3 - 5} $; *c)* $\dfrac{1}{x}\log e$; *d)* $\dfrac{1}{x - 2\sqrt{x}}\left(1 - \dfrac{1}{\sqrt{x}}\right)\log e$ *)*

9. Calculate the derivative of the following functions:

a) $f(x) = x^2 e^{\cos x}$ b) $f(x) = \cos^2 e^x$ c) $f(x) = \cos^2 x^3$ d) $f(x) = \cos\dfrac{1}{x^2}$ e) $f(x) = \dfrac{x}{\operatorname{sen} x}$

(Sol: a) $xe^{\cos x}(2 - x\operatorname{sen} x)$; *b)* $-2e^x \operatorname{sen} e^x \cdot \cos e^x$; *c)* $-6x^2 \operatorname{sen} x^3 \cdot \cos x^3$;

d) $2x^{-3}\operatorname{sen}(x^{-2})$; *e)* $\dfrac{\operatorname{sen} x - x\cos x}{\operatorname{sen}^2 x}$ *)*

10. Calculate the derivative of the following functions:

a) $y = \dfrac{1}{5x}$

b) $y = \dfrac{-3}{x^2}$

c) $y = \dfrac{2}{x^3}$

d) $y = \dfrac{-1}{x^2 - 2x}$

e) $y = \dfrac{5}{2x^2 - 7x}$

(Sol: a) $y' = \dfrac{-1}{5x^2}$; *b)* $y' = \dfrac{6}{x^3}$; *c)* $y' = \dfrac{-6}{x^4}$; *d)* $y' = \dfrac{2x - 2}{(x^2 - 2x)^2}$; *e)* $y' = \dfrac{-5(4x - 7)}{(2x^2 - 7x)^2}$ *)*

11. Calculate the derivative of the following functions:

a) $y = \sqrt{3x^2 + 4x - 5}$

b) $y = \sqrt{x^4 + 4x}$

c) $y = \sqrt{(1 + 5x)^3}$

d) $y = \dfrac{3}{7}\sqrt{x^2 - x}$

(Sol: a) $y' = \dfrac{3x + 2}{\sqrt{3x^2 + 4x - 5}}$; *b)* $y' = \dfrac{2x^3 + 2}{\sqrt{x^4 + 4x}}$; *c)* $y' = \dfrac{15}{2}\sqrt{1 + 5x}$; *d)* $y' = \dfrac{6x - 3}{14\sqrt{x^2 - x}}$ *)*

12. Calculate the derivative of the following functions:

a) $y = \dfrac{1}{\sqrt{x}}$ b) $y = \sqrt{\dfrac{x^2 + 3x}{2}}$ c) $y = \sqrt{\dfrac{2x - 3}{x^2}}$

(Sol: a) $y' = \dfrac{-1}{2\sqrt{x^3}}$; b) $y' = \dfrac{2x + 3}{2\sqrt{2}\sqrt{x^2 + 3x}}$; c) $y' = \dfrac{-x^2 + 3x}{x^3\sqrt{2x - 3}}$)

13. Calculate the derivative of the following functions:

a) $y = 2^{x^2 - 3}$ b) $y = 3^{2x - x^2}$ c) $y = e^{-x + 3}$ d) $y = 2e^{5x}$ e) $y = (2x + 1)e^{2x + 1}$

(Sol: a) $y' = 2x \cdot 2^{x^2 - 3} \ln 2$; b) $y' = (2 - 2x) \cdot 3^{2x - x^2} \ln 3$; c) $y' = -e^{-x + 3}$; d) $y' = 10e^{5x}$

e) $y' = (4x + 4)e^{2x + 1}$)

14. Calculate the derivative of the following functions:

a) $y = \dfrac{e^x}{x}$ b) $y = \dfrac{x}{e^x}$ c) $y = \dfrac{3e^x}{2x + 1}$ d) $y = \dfrac{xe^x}{1 - x}$ e) $y = e^{\sqrt{x}}$ f) $y = \sqrt{e^x}$

(Sol: a) $y' = \dfrac{e^x(x - 1)}{x^2}$; b) $y' = \dfrac{1 - x}{e^x}$; c) $y' = \dfrac{(6x - 3)e^x}{(2x + 1)^2}$;

d) $y' = \dfrac{(1 + x - x^2)e^x}{(1 - x)^2}$; e) $y' = \dfrac{1}{2\sqrt{x}}e^{\sqrt{x}}$; f) $y' = \dfrac{1}{2}e^{x/2}$)

15. Calculate the derivative of the following functions:

a) $y = \log(x^2 + 3x)$ b) $y = \log(3x + 4)^7$ c) $y = \log(5x)$ d) $y = \log(5x^2)$

e) $y = \log(5x)^2$ f) $y = (\log(5x))^2$ g) $y = \log\left(\dfrac{2x - 1}{x^2}\right)$ h) $y = \dfrac{\log(2x - 1)}{\log x^2}$

(Sol: a) $y' = \dfrac{2x + 3}{x^2 + 3x}\log e$; b) $y' = \dfrac{21}{3x + 4}$; c) $y' = \dfrac{1}{x}\log e$; d) $y' = \dfrac{2}{x}\log e$; e) $y' = \dfrac{2}{x}\log e$;

f) $y' = \dfrac{2\log(5x)\log e}{x}$; g) $y' = \left(\dfrac{2}{2x - 1} - \dfrac{2}{x}\right)\log e$; h) $y' = \dfrac{\dfrac{2}{2x - 1}\log(x^2) - \log(2x - 1)\cdot\dfrac{2}{x}\log e}{\left(\log x^2\right)^2}$)

16. Calculate the derivative of the following functions:

a) $y = \ln(2x^2 + 3)$ b) $y = 2\ln(x^2 + 3)$ c) $y = \ln(x^2 + 3)^2$ d) $y = \ln(2x^2 + 3)^2$

(Sol: a) $y' = \dfrac{4x}{2x^2 + 3}$; b) $y' = \dfrac{4x}{x^2 + 3}$; c) Idem; d) $y' = \dfrac{8x}{2x^2 + 3}$)

17. Calculate the derivative of the following functions:

a) $y = \ln \sqrt{3x}$ b) $y = \sqrt{\ln 3x}$ c) $y = \ln(3\sqrt{x})$ d) $y = \ln(3 - \sqrt{x})$

(Sol: a) $y' = \dfrac{1}{2x}$; b) $y' = \dfrac{1}{2x\sqrt{\ln 3x}}$; c) $y' = \dfrac{1}{2x}$; d) $y' = \dfrac{-1}{6\sqrt{x} - 2x}$)

18. Calculate the derivative of the following functions:

a) $y = \ln\left(\dfrac{x^2}{3}\right)$ b) $y = \dfrac{\ln x^2}{3}$ c) $y = \dfrac{\ln x^2}{\ln 3}$

(Sol: a) $y' = \dfrac{2}{x}$; b) $y' = \dfrac{2}{3x}$; c) $y' = \dfrac{2}{x\ln 3}$)

19. Calculate the derivative of the following functions:

a) $y = 3senx - 5\cos x$ b) $y = xsen\, 3x$ c) $y = \cos x\cdot sen\, x$ d) $y = \cos 3x\cdot sen\, x$

(Sol: a) $y' = 3\cos x + 5senx$; b) $y' = sen3x + 3x\cos 3x$; c) $y' = os2x$;

d) $y' = -3sen3x\cdot senx + \cos 3x\cdot \cos x$)

20. Calculate the derivative of the following functions:

a) $y = x^2 \cos 4x$ b) $y = 2x^3 - sen5x$ c) $y = sen^2(3x-1)$ d) $y = \dfrac{\cos 2x}{x}$

(Sol: a) $y' = 2x\cos 4x - 4x^2 sen4x$; b) $y' = 6x^2 - 5\cos 5x$;

c) $y' = 2sen(3x-1)\cdot 3\cos(3x-1)$; d) $y' = -\dfrac{2xsen2x + \cos 2x}{x^2}$)

21. Calculate the derivative of the following functions:

a) $y = \dfrac{1}{senx}$ b) $y = \dfrac{1}{\cos x}$ c) $y = \dfrac{\cos x}{senx}$

(Sol: a) $y' = \dfrac{-\cos x}{sen^2 x}$; b) $y' = \dfrac{senx}{\cos^2 x}$; c) $y' = -\dfrac{1}{sen^2 x}$)

22. Calculate the derivative of the following functions:

a) $y = e^{2x} sen3x$ b) $y = \cos e^x$ c) $y = e^{\cos x}$

(Sol: a) $y' = 2e^{2x} sen3x + e^{2x}\cdot 3\cos 3x$; b) $y' = -e^x sene^x$; c) $y' = -senxe^{\cos x})2$

23. Calculate the derivative of the following functions:

a) $y = sen(\ln x)$ b) $y = \cos(\ln x)$ c) $y = \cos\dfrac{1}{x}$ d) $y = \sqrt{senx}$

(Sol: a) $y' = \dfrac{1}{x}\cos(\ln x)$; b) $y' = -\dfrac{1}{x}sen(\ln x)$; c) $y' = \dfrac{1}{x^2}sen\dfrac{1}{x}$; d) $y' = \dfrac{\cos x}{2\sqrt{senx}}$)

24. Calculate the derivative of the following functions:

a) $y = tag(x^2 - 1)$ b) $y = tag(x-1)^2$ c) $y = 2tag(x-1)$ d) $y = tag^2(x-1)$

$$(Sol: a) \quad y' = \frac{2x}{cos^2(x^2-1)} \; ; b) \quad y' = \frac{2(x-1)}{cos^2(x-1)^2} \; ; c) \quad y' = \frac{2}{cos^2(x-1)}$$

$$d) \quad y' = 2tag(x-1)(1 + tag^2(x-1)))$$

25. Calculate the derivative of the following functions:

a) $y = arcsen\, 2x$ b) $y = arcsen(2+x)$ c) $y = arccos\, x^2$

d) $y = arccos\, e^x$ e) $y = arctag(3x+2)$ f) $y = arctag(x^2)$

$$(Sol: a) \quad y' = \frac{2}{\sqrt{1-4x^2}} \; ; b) \quad y' = \frac{1}{\sqrt{1-(2+x)^2}} \; ; c) \quad y' = -\frac{2x}{\sqrt{1-(x^2)^2}}$$

$$d) \quad y' = -\frac{e^x}{\sqrt{1-(e^x)^2}} \; ; e) \quad y' = \frac{3}{1+(3x+2)^2} \; ; f) \quad y' = \frac{2x}{1+x^4})$$

26. Calculate the derivative of the following functions:

a) $y = 3x^5 - 4x^3 + 3x + 7$

b) $y = \dfrac{3x^4}{4} - \dfrac{5x^3}{3} + \dfrac{9x^2}{2} + 5x - 15$

c) $y = \dfrac{x^2 - 3x + 7}{5}$

d) $y = (3x^3 - 5x + 1).(x + x^2)$

e) $y = \dfrac{2}{x^2 + 2x}$

f) $y = \dfrac{x^3}{3x+2}$

g) $y = \left(\dfrac{3x-2}{7-9x}\right)^2$

h) $y = \dfrac{(5-x)^2}{3x-1}$

i) $y = \dfrac{1}{x} + \dfrac{x}{2}$

j) $y = \sqrt{x^9} \cdot 4x^5$

k) $y = \dfrac{2}{x^5} + \sqrt{3}$

l) $y = \sqrt{12x} + e^{2x+1} + \log_2 3x$

m) $y = (3x - 1)^2.(1 - 4x)$

n) $y = \dfrac{x^5\sqrt{x}}{x^{-3}(x^2)^5}$

ñ) $y = (3x^3 - 5x + 2)^4$

o) $y = (3x^2 - x)^{-4}$

p) $y = \sqrt{3x^2} - \sqrt{5x}$

q) $y = \sqrt{1-x^2}$

r) $y = \left(\dfrac{x+3}{x-1}\right)^3$

s) $y = (2x - 4)^4 + 2.\sqrt{x^2 - 1}$

t) $y = \sqrt{\dfrac{x+1}{x^2}}$

u) $y = \dfrac{\sqrt{x^2 - 3}}{x}$

v) $y = Ln(x^2 + 2x) + e^{-x}$

w) $y = \log_3 x + 3^x$

x) $y = 2.sen(3x+4)$

y) $y = 3\cos^3(3x)$

z) $y = tag(x^2+1)$

1) $y = \sqrt[5]{x^3 - x}$

2) $y = x.e^x$

3) $y = \dfrac{Lnx}{senx}$

4) $y = 4.(2x^3-1)^5$

5) $y = e^{\sqrt{x+3}}$

6) $y = \sqrt[3]{Ln(3x+5)}$

7) $y = \dfrac{e^x - e^{-x}}{2}$

8) $y = tag\sqrt{3x+2}$

Tangent straight line

27. Write the equation of the straight line that is tangent to the curve $y = x^3 - 2x$ at the point with abscise $x = 2$.

28. Write the equation of the straight line with slope 7 that is tangent to the curve

$y = 3x^2 + x - 1$.

29. Find out the points at which tangent straight line to the function $f(x) = x^3 + 6x^2 - 15x$ is horizontal. Decide if they are maximums or minimums.

30. Find out the points at which tangent straight line to the function $f(x) = \dfrac{3 - x^2}{x + 2}$ is horizontal.

31. Write the equation of the straight line that is tangent to the curve $y = 2x^2 + 3x - 1$ at the point with abscise $x = 1$.

32. Write the equation of the straight line that is tangent to the curve $y = x - 4x^2$ and parallel to straight line $y = -7x + 3$.

33. Find out the equation of the straight line with slope -4 that is tangent to the curve

$y = x^4 + 2$.

34. Write the equation of the straight line that is tangent to the curve $y = 2x^3 + x$ at $x = -1$.

35. a) Write the equation of the straight line that is tangent to the curve $f(x) = x^2 - 3x$ at the point with abscise $x = -1$.

b) Is it increasing or decreasing at $x = 2$?

Monotony

36. Study the monotony and locate the relative extremes of the function: $f(x) = x^4 - 2x^2$.

37. Given function $f(x) = 4x^2 - 2x + 1$:

a) Is it increasing or decreasing at $x = 0$? And at $x = 1$?

b) Determine the intervals at which $f(x)$ increases and decreases.

38. a) Write the equation of the straight line that is tangent to the curve $f(x) = 2x - 3x^2 + x$ at the point with abscise $x = 2$.

b) Determine the intervals at which $f(x)$ increases and decreases.

39. Given function $f(x) = 5x^2 - 3x$:

a) Is it increasing or decreasing at $x = -1$? And at $x = 1$?

b) Determine the intervals at which $f(x)$ increases and decreases.

40. Find out the intervals at which the following functions are increasing and decreasing:

a) $f((x) = 8x - x^2$ b) $f((x) = \dfrac{x^2 - 3x}{4}$

41. Given function $f(x) = 14x - 7x^2$:

a) Is it increasing or decreasing at $x = 0$? And at $x = 4$?

b) Determine the intervals at which $f(x)$ increases and decreases.

Optimization problems

42. The benefits, in hundreds of thousands of euros, estimated for a company for the next 5 years, are given by the function: $b(t) = \dfrac{t^2 - 6}{t + 4}$, with $0 \leq t \leq 5$, where t is time in years.

Study how benefits will increase or decrease for the next 5 years.

(Sol: they increase for the 5 years)

43. An investment fund generates a return that depends on the amount of money invested by the formula: $R(x) = -0,002x^2 + 0,8x - 5$; where $R(x)$ represents the yield obtained when investing a quantity *"x"* of money (in hundreds of euros). Determine, justifying the answers:

a) How much money (in euros) we must invest to get the maximum return ?

b) What is the value of that maximum profit? *(Sol: a) € 20,000 ; b) € 7,500)*

44. The value of a product t months after having been set at the market, during the first year, is given by: $v(t) = t^2 - 6t + 10$. In what month should you buy the product to acquire it at the most advantageous price? *(Sol: in March)*

45. Number of people attending an exposition in a day is given by: $p(t) = 12t - 2t^2$, where t is the elapsed time from opening. If it is open from 15 to 21 h, at what time are attending more people? How many people? *(Sol: 18 people, at 18h)*

46. Fuel consumption of a car (in L/km) is given by the expression: $f(v) = \dfrac{3e^{v/90}}{v}$ (where v is the speed, in km/h). Determine the minimum consumption and the speed to get it.

(Sol: 0,09 L/km; 90 km/h)

47. Production of a certain vegetable in a greenhouse depends on the temperature by the expression: $Q(t) = (t + 1)^2 \cdot (32 - t)$, where t is the temperature in °C and $Q(t)$ is the production in kg. Find out the optimus temperature and the production at that temperature. *(Sol: 5.324 kg, 21°C)*

48. After t hours studying, yield of a student (at ascale from 0 to 100) is given by:

$y(t) = \dfrac{380 \cdot t}{t^2 + 4}$ At what intervals does yield increase and at what does it decreases, during

the first 7 hours studying? When is maximum benefit obtained?

(Sol: a) Increases from 0 to 2 and decreases from 2 to 7; maximum at 2 hours)

Continuity and derivability

49. Given function $f(x) = \begin{cases} sen(x) & \text{if } x \leq 0 \\ x - ax^2 & \text{if } x > 0 \end{cases}$ find out values of a so that f(x) is continuous

and derivable at all R.

50. Given function $f(x) = \begin{cases} x^2 + ax + b & \text{if } x < 2 \\ 2x & \text{if } x \geq 2 \end{cases}$ find out values of a so that f(x) is

continuous and derivable at all R.

Representation of functions

51. Draw the graph of the function *f(x)*, knowing that:

- Its derivative has value 0 at (0, 0).

- It intersects axes only at (0, 0).

- Its asymptotes are $x = -2$, $x = 2$ and $y = 0$.

$- \lim_{x \to -2^-} f(x) = -\infty; \quad \lim_{x \to -2^+} f(x) = +\infty; \quad \lim_{x \to 2^-} f(x) = +\infty; \quad \lim_{x \to 2^-} f(x) = -\infty$

52. Draw the graph of the function *f(x)*, knowing that:

- It is continuous.

- Its derivative has value 0 at (-3, -2), (0, 2) and at (2, -3).

- It intersects axes at points (-4, 0), (-2, 0), (1, 0), (3, 0) and (0, 2).

$- \lim_{x \to +\infty} f(x) = +\infty; \quad \lim_{x \to -\infty} f(x) = +\infty$

53. Study and represent the following functions:

a) $f(x)= x^3 -12x$

b) $f(x)= x^3 - 4x^2 + 4x$

c) $f(x)= x^4 + 2x^2 +1$

d) $f(x)= \dfrac{x+3}{x-1}$

e) $f(x)= \dfrac{3x}{x-3}$

f) $f(x)= \dfrac{x^2}{x-2}$

g) $f(x)= \dfrac{x^3-2}{x}$

h) $f(x)= \dfrac{x^2}{x^2-1}$

i) $f(x)= \dfrac{2x^2}{x^2-4}$

j) $f(x)= \dfrac{2x^3}{x^2+2}$

k) $f(x)= \dfrac{x^4-4}{x^2-1}$

l) $f(x)= \dfrac{x^4-2x^2+1}{x^2}$

m) $f(x)= \dfrac{2x^5}{x^2+1}$

54. Study and represent the following functions:

a) $f(x)= (x-1)^2(x+8)$

b) $f(x)= 2x^4 - 4x^2 +1$

c) $f(x)= x^3 + 3x^2 - 9x$

d) $f(x)= 4x^2 - 2x^4 + 2$

e) $f(x)= x^3 + 2x^2 + x$

f) $f(x)= \dfrac{x^2+x+1}{x+1}$

g) $f(x)= \dfrac{x^2-2x+1}{x-3}$

h) $f(x)= \dfrac{x^2-2x-3}{x^2-2x}$

i) $f(x)= \dfrac{2x^2+4x+2}{x^2+2x-3}$

j) $f(x)= \dfrac{x^3-4}{x^2}$

k) $f(x)= \dfrac{x^3}{3} - x^2 - 3x$

l) $f(x)= x^3 + 3x^2 + 3x$

m) $f(x)= x^4 - 2x^2 + \dfrac{1}{2}$

n) $f(x)= x(x-3)^2$

ñ) $f(x)= x^4 - 8x^2$

o) $f(x)= \dfrac{x^2+6x+12}{x+4}$

p) $f(x)= \dfrac{x^2}{1-x^2}$

q) $f(x)= \dfrac{x+1}{x^2}$

r) $f(x)= \dfrac{3}{x^2-4x}$

s) $f(x)= \dfrac{x^2}{x+2}$

Unit 12.- Two-dimensional statistics

1. Introduction to statistics. Review of first concepts

Statistics have become an important part of everyday life. We are confronted by them in newspapers and magazines, on television and in general conversations. We encounter them when we discuss the cost of

living, unemployment, medical breakthroughs, weather predictions, sports, politics and the state lottery. Although we are not always aware of it, each of us is an informal statistician. We are constantly gathering, organizing and analysing information and using this data to make judgments and decisions that will affect our

actions. This unit will start with a brief review of main concepts you already studied previous years, as measures of central tendency (mean, median and mode) and measures of dispersion (standard deviation), and we will focus on *two-dimensional distributions*.

Variables

Data are numbers or measurements that are collected. Data may include numbers of individuals that make up the census of a city, ages of pupils in a certain class, temperatures in a town during a given period of time, sales made by a company, or test scores made by ninth graders on a standardized test.

> *Variables* are characteristics or attributes that enable us to distinguish one individual from another. They take on different values when different individuals are observed.

Some variables are height, weight, age and price. Variables are the opposite of *constants,* whose values never change.

Statistical variables can be classified in several ways: quantitative and qualitative variables.

Quantitative and qualitative variables

> ➢ A *quantitative or numerical variable* is that one whose data are numbers. For example, heights, number of books, incomes in €, etc.
>
> ➢ A *qualitative or categorical variable* is that one whose data have labels (i.e. words). For example, your favourite musical group, a list of the products bought by different families at a grocery store, eyes colour, etc.

Discrete and continuous variables

Numerical variables can be classified into discrete and continuous variables.

> ➤ *Discrete variables* are those whose data are whole numbers, and are usually a count of objects. For instance, number of pets in a house, number of children in a family, ... It does not make sense to have 3.5 children.

> ➤ *Continuous variables* are those that may take any real value. For example, the amount of time a group of children spent watching TV, heights, weights, etc. Your weight may be 74 kg, 74.5 kg or 74.568 kg.

1.1. Measures of central tendency

In the following sections, *xi* will denote an isolated value as well as the mid-point of the interval class for grouped data.

a) Mean

The *average* or *mean* of a list of numbers is the total of all values divided by the number of values.

To calculate the mean, we can use the absolute frequencies of values, multiplying every value by its absolute frequency, then adding all these products and finally dividing by the number of values. It is shown in the formula:

$$\bar{x} = \frac{x_1 \cdot f_1 + x_2 \cdot f_2 + ... + x_n \cdot f_n}{N} = \frac{\sum x_i \cdot f_i}{N}$$

Example: Find the mean of 10, 11, 7 and 8: Solution: $\bar{x} = \frac{10 + 11 + 7 + 8}{4} = \frac{36}{4} = \boxed{9}$.

Example: Find the mean of the following values:

4 4 4 4 5   5 6 6 6 6   6 7 8 8 8   9 9 9 9 10

Solution: We can use a frequency table:

x_i	f_i
4	4
5	2
6	5
7	1
8	3
9	4
10	1

$$\overline{x} = \frac{4 \cdot 4 + 5 \cdot 2 + 6 \cdot 5 + 7 \cdot 1 + 8 \cdot 3 + 9 \cdot 4 + 10 \cdot 1}{20} = \frac{16 + 10 + 30 + 7 + 24 + 36 + 10}{20} = \frac{133}{20} = \boxed{6.65}.$$

Exercises

1. Calculate the mean of: 5, 3, 54, 93, 83, 22, 17 and 19.

2. Calculate the mean of values given as a table:

x_i	f_i
2	5
3	5
4	7
5	8

Median, quartiles and percentiles

- The median, ***Me***, of a list of values is found by ordering them from lower to higher. Once they are arranged, median is the central value. Median is the value that has as 50% of the values lower than it.

- Quartiles, ***Q₁*** and ***Q₃*** are the values that have 25% and 75% of the values lower than them.

- A percentile, ***Pi***, is the value that has a *i*% of the values lower than it.

Example: Money weekly given by parents to a group of students is:

$$10 - 15 - 12 - 20 - 25 - 18 - 12 - 30 - 22 - 19 - 18 - 15 - 13 - 20 - 24$$

Calculate the median, the quartiles and the percentile 40.

Solution: First of all, we have to arrange the data:

$$10 - 12 - 12 - 13 - 15 - 15 - 18 - 18 - 19 - 20 - 20 - 22 - 24 - 25 - 30$$

There are 15 students:

- Me: $15 \cdot \dfrac{1}{2} = 7.5 \;\rightarrow\; Me$ is between 7th and 8th data. As both of them are 18, $\boxed{Me = 18}$.

- Q₁: $15 \cdot \dfrac{1}{4} = 3.75 \;\rightarrow\; Q_1$ is between 3rd and 4th data. They are different: $Q_1 = \dfrac{12+13}{2} = \boxed{12.5}$.

- Q₃: $15 \cdot \dfrac{3}{4} = 11.25 \;\rightarrow\; Q_3$ is between 11th and 12th data. As they are different, $Q_3 = \dfrac{20+22}{2} = \boxed{21}$.

- P₄₀: $15 \cdot \dfrac{40}{100} = 6 \;\rightarrow\; P_{40}$ is the 6th value, $\boxed{P_{40} = 15}$.

Mode

> The **mode** in a list of numbers is the most repeated number (or numbers). It is the value with the highest absolute frequency.

Example: Find the mode of following sets:

    a) 2, 3, 3, 6, 4, 3, 2, 5, 6, 3;   b) 2, 2, 2, 3, 4, 4, 4, 5, 6, 6.

Solution:

a) The mode is 3, because it is the number that appears more often in the list. $\boxed{Mo = 3}$.

b) We have two modes in this case, 2 and 4, because they appear three times.

Both numbers have the highest absolute frequency. $\boxed{Mo = 2, 4}$.

3. Find the mean, median and mode for the following data:

    10, 12, 13, 12, 13, 10, 14 and 13.

4. Twenty families are asked about how many children they have. These are the answers:

    3 3 4 1 2 3 2 5 1 0

    2 2 3 2 4 2 5 3 4 3

    Find the mean, median and mode.

5. Eight people work in an office. They are paid hourly rates of

    $12, $15, $15, $14, $13, $14, $13, $13

    Find: a) the mean, b) the median, c) the mode.

6. The following table shows the number of children that 100 families in a town have.

Children	1	2	3	4	5	6
Families	48	25	16	4	5	2

    Find the mean, mode, median, Q_1, Q_3 and P_{30}.

7. A gardener buys 10 packets of seeds from two different companies. Each pack contains 20 seeds and he records the number of plants which grow from each pack.

    Company A: 20 5 20 20 20 6 20 20 20 8

    Company B: 17 18 15 16 18 18 17 15 17 18

    a) Find the mean, median and mode for each company's seeds.

    b) Which company does the mode suggest is best?

    c) Which company does the mean suggest is best?

8. In a beauty contest, the scores awarded by eight judges were:

    5.9   6.7   6.8   6.5   6.7   8.2   6.1   6.3

    Determine: a) the mean, b) the median, c) the mode.

    d) Only six scores are to be used. Which two scores may be omitted to leave the value of the median the same?

9. A manager keeps a record of the number of calls she makes each day on her cell phone.

Number of calls per day   0   1   2   3   4   5   6   7   8

Frequency   3   4   7   8   12   10   14   3   1

Calculate the mean number of calls per day.

10. A class conduct an experiment in biology. They place a number of 1 m by 1 m square grids on the playing field and count the number of plants in each grid. The results obtained are given below.

6	3	2	1	3	2	1	3	0	1
0	3	2	1	1	4	0	1	2	0
1	1	2	2	2	4	3	1	1	1
2	3	3	1	2	2	2	1	7	1

    a) Calculate the mean number of plants.

    b) How many times was the number of plants seen greater than the mean?

11. Hannah drew this bar chart to show the number of repeated cards she got when she opened packets of football stickers. Calculate the mean number of repeats per packet.

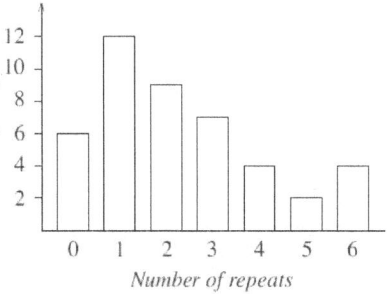

Number of repeats

12. The number of days that students were missing from school due to sickness in one year was recorded. Estimate the mean.

Number of days off sick   (1 – 5)   (6 – 10)   (11 – 15)   (16 – 20)   (21 – 25)

Frequency   12   11   10   4   3

Exercises

13. The table shows the distribution of scores of 40 students on a Mathematics test. Estimate the mean score obtained on the test.

Score	10 - 12	13 - 15	16 - 18	19 - 21	22 - 24
Frequency	4	6	13	9	8

Median, quartiles and percentiles for grouped data

Example: In a petrol station, they are recording the number of cars depending on the time. This is the result:

Hours:	[0, 4)	[4, 8)	[8, 12)	[12, 16)	[16, 20)	[20, 24)
Number of cars:	6	14	110	120	150	25

Calculate the median and Q_3.

Solution: Build the cumulative frequency table:

Interval higher extreme	Fi	Hi	Hi (%)
0	0	0	0
4	6	0,0141	1,41
8	20	0,0471	4,71
12	130	0,3059	30,59
16	250	0,5882	58.82
20	400	0,9412	94,12
24	425	1	100

Now, we are working on the frequency polygon, building the accurate triangles, and applying what we already know about similar triangles:

Me:

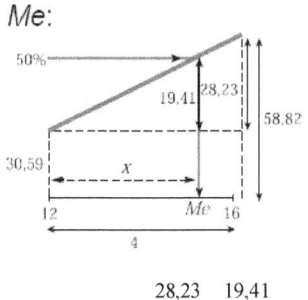

$$\frac{28,23}{4} = \frac{19,41}{x}$$

$$x = 2,75$$

$$Me = 12 + 2,75 = \boxed{14,75}$$

Q₃:

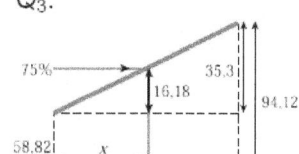

$$\frac{35,3}{4} = \frac{16,18}{x}$$

$$x = 1,83$$

$$Q_3 = 16 + 1,83 = \boxed{17,83}$$

<table>
<tr><td rowspan="...">Exercises</td><td>

14. Time taken by employees of a firm to go from home to the office is shown in the next table:

Time (min)	[0, 15)	[15, 30)	[30, 45)	[45, 60)	[60, 75)	[75, 90)
Workers	10	23	32	5	6	4

Calculate Me, Q_1 and Q_3.

15. The table below gives data on the heights, in cm, of 51 children.

Class Interval	$140 \leq h < 150$	$150 \leq h < 160$	$160 \leq h < 170$	$170 \leq h < 180$
Frequency	6	16	21	8

Estimate the mean height and the median.

16. Incomes, in M€, of 500 firms, are shown in the following table:

Incomes	[1, 2)	[2, 3)	[3, 4)	[4, 5)	[5, 6)	[6, 7)
Firms	50	80	170	90	56	54

Calculate the mean, Me, Q_1 and Q_3.

17. The weights of a number of students were recorded in kg.

Weight	$30 \leq w < 35$	$35 \leq w < 40$	$40 \leq w < 45$	$45 \leq w < 50$	$50 \leq w < 55$
Frequency	10	11	15	7	4

Estimate the mean weight and the median.

</td></tr>
</table>

1.2. Measures of dispersion

Variance

The Variance (σ^2) is a measure of how spread our numbers are. Its symbol is the squared of Greek letter sigma.

The **variance** is calculated as the average of the **square** of the differences from the mean. Its formula is:

$$\sigma^2 = \frac{(x_1 - \bar{x})^2 + (x_2 - \bar{x})^2 + \dots + (x_n - \bar{x})^2}{n} = \frac{\Sigma (x_i - \bar{x})^2}{n}$$

There is an equivalent formula to calculate variance. Sometimes it is easier to use:

$$\sigma^2 = \frac{x_1^2 + x_2^2 + \dots + x_n^2}{n} - (\bar{x})^2 = \frac{\Sigma x_i^2}{n} - (\bar{x})^2$$

Standard deviation

The **standard deviation** is defined as the square root of variance. So, it is denoted by σ.

$$\sigma = \sqrt{\sigma^2}$$

Example: Calculate the variance and standard deviation of this weights distribution:

$$83, 65, 75, 72, 70, 80, 75, 90, 68, 72$$

Solution: First of all, we calculate the mean, because we will need it later:

$$\bar{x} = \frac{65 + 68 + 70 + 72 + 72 + 75 + 75 + 80 + 83 + 90}{10} = \frac{750}{10} = 75$$

- Variance: $\sigma^2 = \frac{(65 - 75)^2 + (68 - 75)^2 + (70 - 75)^2 + \dots + (90 - 75)^2}{10} = \frac{506}{10} = \boxed{50.6}$.

- Standard deviation: $\sigma = \sqrt{\sigma^2} = \sqrt{50.6} = \boxed{7.11}$.

Variation coefficient

Standard deviation gives us information about dispersion of data. But, sometimes, we want to compare dispersions of different sets of data. If magnitudes of data in different sets of data are very distinct, standard deviation might not be useful.

For example, imagine we have two sets of weights: set A are weights of cars and set B are weights of oranges. Obviously, standard deviation of set B is going to be less than in set A, but that does not mean that in this set there is a lower dispersion of data, it is just because its data are quite lower.

In these cases, we need a new parameter, the *variation coefficient*.

The **variation coefficient** is defined as $$VC = \frac{\sigma}{\bar{x}}$$

As you can imagine, the higher VC is, the higher dispersion is.

Example: In three restaurants, A, B and C, we are investigating salaries (€ per weekend).

	A	B	C
Mean	943	132	37
Standard deviation	148	22	12

In what restaurant do they have the most homogenous salaries?

Solution:

$$VC_A = \frac{943}{148} = 6.37 \qquad VC_B = \frac{132}{22} = 6 \qquad VC_C = \frac{37}{12} = 3.08$$

So, as the lowest VC has been obtained for restaurant C, this is the one having the most homogenous salaries.

18. Find the mean, range, variance, standard deviation and coefficient of variation of the numbers: 6, 7, 8, 5, 9

19. The table below gives the number of road traffic accidents per day in a small town.

Accidents per day	0	1	2	3	4	5	6
Frequency		5	8	6	3	2	1	1

Find the mean and standard deviation of this data.

20. Find the mean and standard deviation of each set of data given below.

A 51 56 51 49 53 62

B 71 76 71 69 73 82

C 102 112 102 98 106 124

Calculate the coefficient of variation of the previous sets.

21. Two machines, A and B, fill empty packets with soap powder. A sample of packets was taken from each machine and the weight of powder (in kg) was recorded.

A 2.27 2.31 2.18 2.2 2.26 2.24

B 2.78 2.62 2.61 2.51 2.59 2.67 2.62 2.68 2.70

   a) Find the mean and standard deviation for each machine.

   b) Which machine is most consistent?

22. Two groups of students were trying to find the acceleration due to gravity. Each group conducted 5 experiments.

Group A	9.4	9.6	10.2	10.8	10.1
Group B	9.5	9.7	9.6	9.4	9.8

Find the mean and standard deviation for each group, and comment them.

23. In two groups, 3rd A and 3rd B, we have done the same exam. These are the marks:

	Mean	Std. Dev.
3rd A	5.8	2.9
3rd B	6.3	1.2

   a) Calculate their variation coefficients and indicate in which group results have more dispersion.

   b) In a group, there were 6 nines and 5 ones, while in the other one, there were only 1 nine and 2 ones. Could you indicate these groups?

2. Two-dimensional distributions

So far we have studied a single characteristic of a population *(size, weight, ...)*, but we could study several of them simultaneously. For example:

1. Weight and height of a sample of 100 people.

2. Number of hours students spend watching television and academic results.

4. Production and sales of a factory.

This type of statistical variable is called ***two-dimensional statistics***.

Often, we want to study the relationship between two characteristics of a population. This is the object of the ***"linear regression"***.

Two-dimensional variables

They are those resulting from the observation of a phenomenon on two ***modalities***, the pair *(X, Y)*, where X is a one-dimensional variable taking values $x_1, x_2, ... x_n$, and Y is another one-dimensional variable which takes the values $y_1, y_2, ... y_n$. So, the two-dimensional statistical variable (X, Y) takes values: $(x_1, y_1), (x_2, y_2), ...(x_n, y_n)$.

If we represent these pairs in a system of Cartesian axes a set of points on the plane called ***scatter plot*** or ***point cloud*** is obtained.

Two-dimensional frequency tables

Example: We have measured ages and weighs of 5 children and these are the results:

Edad (años)	2	4'5	6	7'2	8
Peso (Kg)	15	19	25	33	34

Its scatter plot is:

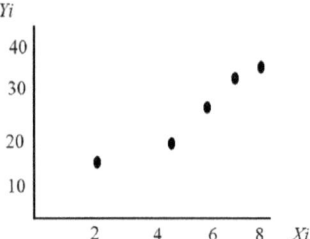

Example: 50 families have been classified according to the sex of their children. Let X the number of boys and Y the number of girls. The results are:

X \ Y	0	1	2	3	4	5	6	
0	2	-	4	3	1	-	-	10
1	3	-	9	-	-	3	-	15
2	-	6	-	6	-	-	1	13
3	1	4	-	-	2	1	-	8
4	-	-	2	-	1	-	-	3
5	-	-	-	1	-	-	-	1
	6	10	15	10	4	4	1	50

This kind of table are called **2-input data table**.

3. Calculation of statistical parameters

2-dimensional data are organized in tables as following one:

VARIABLE X x_i	VARIABLE Y y_i	ABSOLUTE FREQUENCY f_i
x_1	y_1	f_1
x_2	y_2	f_2
x_3	y_3	f_3
....	....	....
x_n	y_n	f_n
		$N = \Sigma f_i$ = Total quantity of data

Remember expressions to calculate mean, variance and standard deviation for a one-dimensional distribution:

	Variable X	Variable Y
Mean	$\bar{x} = \dfrac{\sum\limits_{i=1}^{n} x_i \cdot f_i}{N}$	$\bar{y} = \dfrac{\sum\limits_{i=1}^{n} y_i \cdot f_i}{N}$
Variance	$\sigma_x^2 = S_x^2 = \dfrac{\sum\limits_{i=1}^{n} f_i \cdot (x_i - \bar{x})^2}{N}$	$\sigma_y^2 = S_y^2 = \dfrac{\sum\limits_{i=1}^{n} f_i \cdot (y_i - \bar{y})^2}{N}$
	$\sigma_x^2 = S_x^2 = \dfrac{\sum\limits_{i=1}^{n} f_i \cdot x_i^2}{N} - \bar{x}^2$	$\sigma_y^2 = S_y^2 = \dfrac{\sum\limits_{i=1}^{n} f_i \cdot y_i^2}{N} - \bar{y}^2$

Remember standard deviation, σ, equals squared root of variance, $\sigma = \sqrt{\sigma^2}$

In this case, we have $\sigma_x = \sqrt{\sigma_x^2}$ and $\sigma_y = \sqrt{\sigma_y^2}$.

Covariance

Covariance is calculated as: $\sigma\, xy = S\, xy = \dfrac{\sum\limits_{i=1}^{n} f_i \cdot (x_i - \bar{x})(y_i - \bar{y})}{N} = \dfrac{\sum\limits_{i=1}^{n} f_i \cdot x_i y_i}{N} - \bar{x} \cdot \bar{y}$

Example: We have measured ages and weighs of 5 children and these are the results:

Age (years)	2	4'5	6	7'2	8
Weigh (kg)	15	19	25	33	34

Calculate means, variances, standard deviations and covariance.

Solution: First, build the following table:

x_i	y_i	f_i	$x_i \cdot f_i$	$y_i \cdot f_i$	$x_i^2 \cdot f_i$	$y_i^2 \cdot f_i$	$x_i \cdot y_i \cdot f_i$
2	15	1	2	15	4	225	30
4'5	19	1	4'5	19	20'25	361	85'5
6	25	1	6	25	36	625	150
7'2	33	1	7'2	33	51'84	1089	237'6
8	34	1	8	34	64	1156	272
		$N = 5$	$\Sigma x_i = 27'7$	$\Sigma y_i = 126$	176'09	3456	775'1

a) Mean of X: $\bar{x} = \dfrac{26'5}{5} = \boxed{5.54}$ 　　　　　　Mean of Y: $\bar{y} = \dfrac{27}{5} = \boxed{25.2}$

c) Variance of X: $\sigma_x^2 = \dfrac{176'09}{5} - 5.54^2 = 4.526 \Rightarrow \sigma_x = +\sqrt{4'526} = \boxed{2.1275}$

d) Variance of Y: $\sigma_y^2 = \dfrac{3456}{5} - 25.2^2 = 56.16 \Rightarrow \sigma_y = +\sqrt{56'16} = \boxed{7.4939}$

e) Covariance: $\sigma_{XY} = \dfrac{775'1}{5} - 5.54 \cdot 25.2 = 155.02 - 139.608 = \boxed{15.412}$

Example: Marks in Maths and Physics obtained by 40 students are:

X = Maths marks	3　4　5　6　6　7　7　8　10
Y = Physics marks	2　5　5　6　7　6　7　9　10
Number of students	4　6　12　4　5　4　2　1　2

Calculate means, variances, standard deviations and covariance.

Solution: Do it yourself!!!

x_i	y_i	f_i	$x_i \cdot f_i$	$x_i^2 \cdot f_i$	$y_i \cdot f_i$	$y_i^2 \cdot f_i$	$x_i.y_i.f_i$
3	2	4					
4	5	6					
5	5	12					
6	6	4					
6	7	5					
7	6	4					
7	7	2					
8	9	1					
10	10	2					
		N=40					

4. Pearson linear correlation coefficient

When scatterplots were presented, we observed there is a linear correlation between the variables. Now, it is interesting to quantify this correlation. This is the objective of **Pearson linear correlation coefficient**, which is defined by the following equation:

Pearson linear correlation coefficient: $r = \dfrac{S_{xy}}{S_x S_y}$ $\qquad -1 \leq r \leq 1$

Sign of **r** indicate the type of correlation between the variables, X and Y:

> If r > 0, the correlation is **direct**
> If r < 0, the correlation is **inverse**
> If r = 0, there is **no correlation**

So:

- If $r = 1$ → there is a perfect direct linear correlation between X and Y (Figure 1).

- If $r = (-1)$ → there is a perfect inverse linear correlation between X and Y (Figure 2).

- If $r = 0$ → there is not any correlation between X and Y (Figure 3).

Obviously, *r* will have values that will be closer to 1 or (-1) as correlation between variables is bigger.

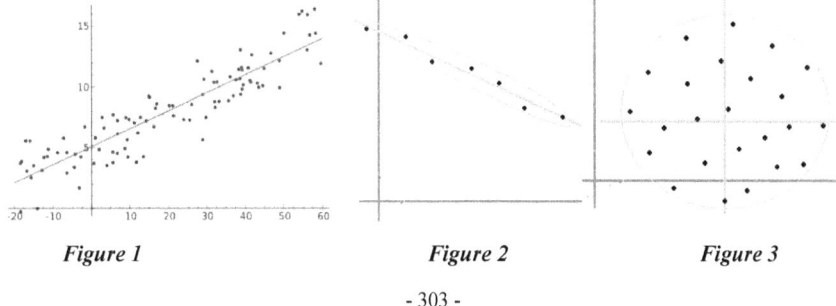

Figure 1   Figure 2   Figure 3

If two variables are strongly correlated, the scatterplot condenses around a line. x is the independent variable and y the dependent variable. Now, the problem is finding the equation of the line that best fits to the scatterplot.

To obtain the equation of the straight line that best fits there are several methods, but the **least squares** method is the most widely used.

Regression straight line of "y on x"

$$y - \bar{y} = \frac{S_{xy}}{S_x^2}(x - \bar{x})$$

By substituting *"x"* by values, we can obtain, in an approximated way, the **estimated** or *foreseen* values of *"y"*.

Regression straight line of "x on y"

$$x - \bar{x} = \frac{S_{xy}}{S_y^2}(y - \bar{y})$$

With this equation, we can **estimate** values of *"x"* from known values of *"x"*.

How good is the correlation?

Remember the closer is |r| to 1, the closer are the experimental points to the regression line.

24. A group of 10 friends was brought to a test. They noted the number of hours they spent studying the week before the exam and the mark obtained in the test. The information is contained in the following table:

Hours	21	15	10	15	20	30	18	20	25	16
Mark	9	7	5	2	7	8	8	6	5	4

Represent data as a scatterplot and indicate which of these values you think is the most likely to be the correlation coefficient: 0.92; –0.44; –0.92; 0.44.

25. A survey has been conducted by asking about the number of people living in the family home and the number of rooms of the house. The following table lists the information obtained:

People	3	5	4	6	5	4
Rooms	2	3	4	4	3	3

Calculate covariance and correlation coefficient. How is correlation between both variables?

26. It has been studied, in several printer models, what the costs per page (in cents) in black and white (B) and in colour (C) are. The following table gives the first six pairs of data:

X: B	8	11	17	21	14	10
Y: C	33	49	95	106	58	53

a) Find out the regression straight line of Y on X.

b) What is the cost of printing one colour page in a printer model in which a black and white page had a price of 12 cents? Is estimation trustable?

(It is known that r = 0,97).

Exercises

Review exercises

One-dimensional statistics *(section 1)*

1. A test with 25 questions has been answered by 120 students. Following table shows the results:

Number of correct answers	Percentage
5	10%
15	45%
20	25%
25	

   a) Calculate the number of students that answered correctly all the questions.

   b) Calculate the mean of correct answers of population.

   c) Calculate the standard deviation.

2. What will happen to mean and standard deviation if we add a same number to all data? And if we multiply them by a same number? Check with these data: 3, 5, 6, 3, 4, 2, 3, 6.

3. If two distributions have the same mean and first one has a higher standard deviation, which of them will have a higher variation coefficient?

4. If two distributions have the same standard deviation and first one has a higher mean, which of them will have a higher variation coefficient?

5. Mean of marks in 3rd A was 6.2, and in 3rd B, 4. If there are 15 students in 3rd A and 35 in 3rd B, what would the mean be if calculated for both groups together?

6. My mark is calculated as the mean of four exams. If in the three first I have a mean of 4.2, what should be my mark at the fourth exam to pass the subject?

7. Complete the following table, knowing it mean is 2.7.

x_i	f_i
1	3
2	...
3	7
4	5

8. My final mark in a subject is calculated from marks of three partial exams: second exam's mark has double value, and third one's value is three times the first one.

   a) If my marks have been 5, 6 and 4, What will be my final mark?

   b) What would my final mark be if each exam had an importance of 10%, 40% and 50%, respectively?

9. In a company there are 34 employees and 6 directives. Mean salary of all them is 909 €. What is the mean salary of directives if we know that mean of salaries of the rest of workers is 780 €?

10. A stopwatch was used to find the time that it took a group of students to run 100 m.

Time (seconds)	$10 \le t < 15$	$15 \le t < 20$	$20 \le t < 25$	$25 \le t < 30$
Frequency	6	16	21	8

Estimate the mean and the median.

11. The distances that students in a year group travelled to school is recorded.

Distance (km)	$0 \le d < 0.5$	$0.5 \le d < 1.0$	$1.0 \le d < 1.5$	$1.5 \le d < 2.0$
Frequency	30	22	19	8

Estimate the median and the mean.

12. The ages of the people at a youth camp are summarised in the table below.

Age (years)	$6 - 8$	$9 - 11$	$12 - 14$	$15 - 17$
Frequency	8	22	29	5

Estimate the mean age, the median, Q_1 and Q_3.

13. The length of telephone calls from an office was recorded. The results are given in the table below.

Length of call (min)	$0 < t \le 0.5$	$0.5 < t \le 1.0$	$1.0 < t \le 2.0$	$2.0 < t \le 5.0$
Frequency	8	10	12	4

Estimate the mean and standard deviation using this table.

14. The charges (to the nearest $) made by a jeweller for repair work on jewellery in one week are given in the table below.

Charge ($)	20 – 29	30 – 49	50 – 99	100 – 149	150 – 199	200 –300
Frequency	10	22	6	2	4	1

Use this table to estimate the mean and standard deviation.

15. a) Calculate the standard deviation of the numbers 3, 4, 5, 6, 7.

b) Show that the standard deviation of every set of five consecutive integers is the same as the answer to part (a).

16. Ten boys had a test which was marked out of 50. Their marks were

$$28, 42, 35, 17, 49, 12, 48, 38, 24 \text{ and } 27$$

a) Calculate: the mean and the standard deviation of the marks.

b) Ten girls had the same test. Their marks had a mean of 30 and a standard deviation of 6.5. Compare the performances of the boys and girls.

17. There are twenty students in class A and twenty students in class B. All the students in class A were given an I.Q. test. Their scores on the test are given below.

$$100, 104, 106, 107, 109, 110, 113, 114, 116, 117,$$
$$118, 119, 119, 121, 124, 125, 127, 127, 130, 134.$$

a) The mean of their scores is 117. Calculate the standard deviation.

b) Class B takes the same I.Q. test. They obtain a mean of 110 and a standard deviation of 21. Compare the data for class A and class B.

d) Class C has only 5 students. When they take the I.Q. test they all score 105. What is the value of the standard deviation for class C?

Two-dimensional statistics *(sections 2 to 4)*

18. In six secondary schools they have studied marks of their students in Maths and in English and they have obtained the following results:

x: Maths	6,5	5,2	6	6,5	7	6
y: English	7	5	5	6	7,5	5

a) Find out the regression straight line of Y on X.

b) Calculate y(5.5). Is estimation trustable? (It is known that r = 0,87).

19. In a driving school they have studied the number of weeks (X) that students had attended and the number of weeks (Y) it takes them to pass the theory test (since it was aimed at the driving school). Data for six students are:

X:	6	1	4	3	5	8
Y:	6	5	5	6	5	10

Find out the regression straight line of Y on X.

20. Marks of ten students in Maths and Physics are the following:

Maths	7	6	4	5	9	10	3	1	10	6
Physics	8	6	3	6	10	9	1	2	10	5

Represent data as a scatterplot and indicate which of these values you think is the most likely to be the correlation coefficient: 0.23; 0.94; –0.37; –0.94.

21. We have measured power (in kW) and the expense (litres/100 km) of six different car models, obtaining the following results:

Power	81	85	66	85	104	83
Expense	7,5	10,6	8,2	9,2	10,7	8,7

Calculate the covariance and the correlation coefficient.

22. A group of six athletes have had a set of tests for length (L) and height (H) jumps. They have had scores between 0 and 5. The results are the following:

Find out the regression straight line of Y on X.

23. We have studied, for different types of yogurts, the percentage of fat they contain and kilocalories per 100 g. These are the results:

X: fat %	2,2	2	1,9	3,1	3	2
Y: kcal	64	55	58	79	65	52

a) Find out the regression straight line of Y on X.

b) How many kilocalories would you expect a yogurt to have if it has 2.5% of fat?

c) And if it has 10% of fat?